The Agricultural Groundwater Revolution
Opportunities and Threats to Development

Contents

Contributors

J.A. Tony Allan, *King's College London & SOAS University of London, SOAS, Thornhaugh Street, London, WC1H 0XG, UK, Fax/Tel: +44 20 78984058, E-mail: tony.allan@kcac.uk*

Yamileth Astorga, *Coordinator of Global Water Partnership of Central America (GWP-CA), Oficial Tēcnico, GWP-Centroamérica, Correo electrónico, Fax: +240-9934, Tel: +506-241 0101, E-mail: tempis@racsa.co.cr*

Maureen Ballestero, *Coordinator of Global Water Partnership of Central America (GWP-CA), Oficial Técnico, GWP-Centroamérica, Correo electrónico, Fax: +240-9934, Tel: +506-241 0101, E-mail: tempis@racsa.co.cr*

Amelia Blanke, *Department of Agricultural and Resource Economics, University of California, Davis, CA 95616, USA, Fax: +530 752 6770, Tel: +530 752 6770, E-mail: blanke@primal.ucdavis.edu*

Imogen Fullagar, *Charles Sturt University, c/o CSIRO Land and Water, GPO Box 1666, Canberra, ACT 2601, Australia, Tel: +2 6246 5858, Mobile: +423 149 497, E-mail: Imogen.Fullagar@csiro.au*

Alberto Garrido, *Department of Agricultural Economics, Technical University of Madrid, 28040 Madrid, Spain, Fax: +34 91 3944-845, E-mail: alberto.garrido@upm.es*

Mark Giordano, *International Water Management Institute, 127, Sunil Mawatha, Pelawatte, Battaramulla, Sri Lanka, Fax: 94112786854, E-mail: mark.giordano@cgiar.org*

Jikun Huang, *Center for Chinese Agricultural Policy, Institute of Geographical Sciences and Natural Resources Research, Chinese Academy of Sciences, Jia 11 Datun Road, Anwai, Beijing 100101, PR China, Fax: +861064856533, E-mail: ccap@igsnrr.ac.cn*

Qiuqiong Huang, *Department of Applied Economics, University of Minnesota, 249e Classroom Office Building, Saint Paul, MN 55108, USA, E-mail: qhuan@um.edu.*

Karin Erika Kemper, *World Bank (South Asia Sustainable Development Department), 1818H Street NW, Washington, DC 20433, USA, Fax: +1 202 614 1074, E-mail: kkemper@worldbank.org*

Avinash Kishore, *Woodrow Wilson School of Public and International Affairs, Princeton University, Robertson Hall, Princeton, NJ 08544-1013, USA, E-mail: akishore@Princeton.EDU*

M. Ramón Llamas, *Department of Geodynamics, Complutense University of Madrid, 28040 Madrid, Spain, Fax: +3491 3944-845, E-mail: mrllamas@geo.ucm.es*

Mutsa Masiyandima, *International Water Management Institute, 141, Cresswell Street, Weavind Park 0184, Pretoria, South Africa, Fax: +27 12 8459110, E-mail: M.masiyandima@cgiar.org*

Marcus Moench, *Institute for Social and Environmental Transition (ISEI), 948 North Street, Suite 7, Boulder, CO 80304, USA, Fax +720 564 0653, E-mail: moenchm@I-s-e-t.org*

Srinivas Mudrakartha, *Vikram Sarabhai Centre for Development Interaction (VIKSAT) Nehru Foundation for Development, Ahmedabad 380054, India, Fax: +91-79-26856220, E-mail: mudrakarthas@yahoo.com*

John C. Peck, *University of Kansas School of Law, Lawrence, KS 66045; Special Counsel, Foulston Siefkin LLP, Overland Park, KS 66210, USA, Tel: +1 785 864 9228, E-mail: jpeck@ku.edu*

Virginia Reyes, *Coordinator of Global Water Partnership of Central America (GWP-CA), Oficial Técnico, GWP-Centroamérica, Correo electrónico, Fax: +240 9934, Tel: +506-241 0101, E-mail: vreyes@gwpcentroamerica.org*

Scott Rozelle, *Department of Agricultural and Resource Economics, University of California, Davis, CA 95616, USA, Fax: +530 752 6770, Tel:+530 752 6770, E-mail: rozelle@primal.ucdavis.edu*

Ramaswamy Sakthivadivel, *33 First East Street, Kamrajnagar, Thiruvanmiyur, Chennai 600 041, Tamil Nadu, India, E-mail: sakthivadivelr@yahoo.com*

Edella Schlager, *University of Arizona, School of Public Administration and Policy, McClelland Hall, Room 405, Tucson, AZ 85721, USA, E-mail: bluff2u@aol.com*

Christopher Scott, *2601 Spencer Road, Chevy Chase, MD 20815, USA, Tel: +301 587 5485, E-mail: a.chris.scott@gmail.com*

Tushaar Shah, *IWMI-TATA Water Policy Program, Anand Field Office, Elecon Premises, Anand-Sojitra Road, Vallabh Vidyanagar 388 129, Anand, Gujarat, India, Fax: +91 26923 229310, E-mail: T.Shah@cgiar.org*

Abhishek Sharma, *PriceWaterhouseCoopers, PricewaterhouseCoopers Pvt. Ltd., PwC Centre Saidulajab, Opposite D-Block Saket, Mehrauli Badarpur Road, New Delhi–110 030, Telephone: [91](11) 41250000, Fax: [91](11)4125 0250 Email: shrma_abhishek_22@yahoo.co.in, locus1@rediffmail.co*

Hugh Turral, *International Water Management Institute, 127, Sunil Mawatha, Pelawatte, Battaramulla, Sri Lanka, Fax: 94112786854, E-mail: H.Turral@cgiar.org*

Jac A.M. van der Gun, *International Groundwater Resources Assessment Centre (IGRAC), PO Box 80015, 3508 TA Utrecht, The Netherlands, Fax: +31 30 256 4755, E-mail: j.vandergun@nitg.tno.nl*

Karen G. Villholth, *International Water Management Institute, 127, Sunil Mawatha, Pelawatte, Battaramulla, Sri Lanka, Fax: 94112786854, E-mail: k.villholth@cgiar.org*

Jinxia Wang, *Center for Chinese Agricultural Policy, Institute of Geographical Sciences and Natural Resources Research, Chinese Academy of Sciences, Jia 11 Datun Road, Anwai, Beijing 100101, PR China, Fax: +861064856533, E-mail: jxwang.ccap@igsnrr.ac.cn*

Series Foreword
Comprehensive Assessment of Water Management in Agriculture

There is broad consensus on the need to improve water management and to invest in water for food, as these are critical to meeting the millennium development goals (MDGs). The role of water in food and livelihood security is a major issue of concern in the context of persistent poverty and continued environmental degradation. Although there is considerable knowledge on the issue of water management, an overarching picture on the water–food–livelihood–environment nexus is missing, leaving uncertainties about management and investment decisions that will meet both food and environmental security objectives.

The Comprehensive Assessment of Water Management in Agriculture (CA) is an innovative multi-institute process aimed at identifying existing knowledge and stimulating thought on ways to manage water resources to continue meeting the needs of both humans and ecosystems. The CA critically evaluates the benefits, costs and impacts of the last 50 years of water development and the challenges currently facing communities. It assesses innovative solutions and explores consequences of potential investment and management decisions. The CA is designed as a learning process, engaging networks of stakeholders to produce knowledge synthesis and methodologies. The main output of the CA is an assessment report that aims to guide investment and management decisions in the near future considering their impact over the next 50 years in order to enhance food and environmental security to support the achievement of the MDGs. This assessment report is backed by CA research and knowledge-sharing activities.

The primary assessment research findings are presented in a series of books that will form the scientific basis for the CA. The books will cover a range of vital topics in the areas of water, agriculture, food security and ecosystems – the entire spectrum of developing and managing water in agriculture, from fully irrigated to fully rain-fed lands. They are about people and society, why they decide to adopt certain practices and not others and, in particular, how water management can help poor people. They are about ecosystems – how agriculture affects ecosystems, the goods and services ecosystems provide for food security, and how water can be managed to meet both food and environmental security objectives. This is the second book in the series.

Effectively managing water to meet food and environmental objectives will require the concerted action of individuals from across several professions and disciplines – farmers, fisherfolk, water managers, economists, hydrologists, irrigation specialists, agronomists and social scientists. This book represents an effort to bring a diverse group of people together to present a truly cross-disciplinary perspective on water, food and environmental issues within the coastal zone. The complete set of books would be invaluable for resource managers, researchers and field implementers. These books will provide source material from which policy statements, practical manuals and educational and training material can be prepared.

The CA is done by a coalition of partners that includes 11 Future Harvest agricultural research centres supported by the Consultative Group on International Agricultural Research (CGIAR), the Food and Agriculture Organization of the United Nations (FAO) and partners from some 80 research and development institutes globally. Co-sponsors of the assessment, institutes that are interested in the results and help frame the assessment, are the Ramsar Convention, the Convention on Biological Diversity, FAO and the CGIAR.

We appreciate the financial support in the production of this book from the governments of the Netherlands and Switzerland for the Comprehensive Assessment.

David Molden
Series Editor
International Water Management Institute
Sri Lanka

1 The Agricultural Groundwater Revolution: Setting the Stage

MARK GIORDANO AND KAREN G. VILLHOLTH

International Water Management Institute, 127, Sunil Mawatha, Pelawatte, Battaramulla, Sri Lanka

Over the last 50 years groundwater development has played a fundamental role in agricultural production in many parts of the developing world. For example, groundwater now accounts for approximately 50% of all irrigation supply in South Asia and perhaps two-thirds of supply in the grain belts of North China. The rapid growth in use in these and other regions has played a vital role in maintaining the rise in grain output associated with the Green Revolution, transforming production and livelihood strategies for millions of small farmers and providing food to growing urban centres.

Groundwater use of course has not come without problems, both in terms of sustainability and quality. In India and the areas of China just mentioned, plummeting water tables are notorious and bring into question the future use of the resource. In other regions, particularly in the Middle East and North Africa, groundwater is taken from fossil sources with no chance of recharge and so any utilization has to be considered part of a longer-term development strategy rather than 'sustainable use'. In addition to use within the agricultural sector, competition for groundwater from cities makes it harder for farmers to maintain supplies, and drawdown by farmers themselves can make critical rural domestic supplies more costly to obtain. Pollution from the very farming systems supported by groundwater, as well as from industry, increasingly degrade the utility of groundwater resources. High rates of use in many areas have severed the links between surface and groundwater resources, damaging ecosystems in their own right and reducing the livelihood generation options that rural residents rely on.

While it is the problem of overuse that garners the most attention, other parts of the developing world have yet to take full advantage of the livelihood-generating and poverty-reducing potential of groundwater. Some countries in sub-Saharan Africa and Central America provide such examples. However, even within regions commonly associated with overuse, there are areas with little utilization. While fears of a groundwater boom turning to bust are perhaps highest in India, the eastern regions of the country have hardly tapped the development potential. As shown in this book, similar areas of limited utilization exist even in the North China Plain, another region where concern over overexploitation is the norm.

The problems for groundwater management in the developing world are not unique to either groundwater or the region. However, groundwater management, especially in a developing country context, does pose unique sets of management challenges. Unlike surface water, groundwater is 'invisible' and difficult to conceptualize and to measure. Further, it must be understood in terms of timescales much longer than those used in surface water management, with lags between rainfall, storage and use best thought of in terms of years, decades or even centuries. Groundwater is in many respects an open-access resource, with any particular aquifer typically underlying multiple farmers, any of whom can extract the resource with relatively cheap and simple pump technology. In many densely populated developing countries, this means that management would involve the coordination of hundreds or even thousands of users, a challenge made even more difficult for user groups or governments by insufficient resources for basic measurement and coordination. Understanding the physics of groundwater movement and measurement, the sociology of groundwater users, the political economy of the water and agricultural sectors and the laws and institutions that have been, or might be, brought to bear is necessary if we are to come to terms with the challenges of groundwater use and management.

Recognizing the value that groundwater can have for poverty alleviation, livelihood generation and global food supply, the threats to, and continued opportunities for, groundwater use and the multi-disciplinary nature of groundwater management problems, the Comprehensive Assessment of Water Management in Agriculture funded a research programme on the state of groundwater use and governance in the developing world and suggested options for the future. The goal of this book is to synthesize that work and provide an overview of the issues and options in agricultural groundwater use across the developing world. To accomplish this goal, the book calls on the work of regional and subject matter experts from Asia, Africa, Australia, Europe, Latin America and North America with expertise in anthropology, biology, economics, geography, hydrology, law and political science. Together this group of authors must be one of the most geographically and disciplined diverse groups ever to come together to address agricultural groundwater challenges.

The book is divided into three parts plus two final chapters. Part I provides regional overviews of agricultural groundwater systems covering much of the developing world. The regions include 'traditional' use areas such as South Asia, China and the Middle East/North Africa where groundwater already contributes substantially to existing agricultural economies. It also includes two areas, sub-Saharan Africa and Central America, where groundwater contributions to agriculture have been less well recognized or developed. Together these chapters tell a story of the similarity and difference in both the present contribution of groundwater to agriculture and livelihood across the developing world and the challenges for the future.

Part II presents an overview of three major governance paradigms that have been, or might be, employed to address the challenges facing groundwater management. The paradigms include what might at first be considered two opposing schools of thought, one focusing on community approaches or self-management by users and the other on instrumental approaches, i.e. formal laws, regulations and pricing requiring higher government involvement. Chapter 9, the final one in Part II, highlights the possibilities for 'adaptive approaches' that acknowledge the possibilities and limitations of community and instrumental solutions, and pushes both users and researchers to focus not just on groundwater resources but also on the livelihood of groundwater users and their options, including options outside of agriculture.

Part III provides a series of case studies that examine how these governance paradigms have actually functioned on the ground. Chapters include community management of groundwater resources through water harvesting, the role of indirect instruments such as electricity policy and farmer adaptation to changing resource conditions. The chapters also highlight the experience of the varying approaches in the three main agricultural groundwater economies of the developed world: Spain, the USA and Australia. They provide an opportunity to reflect on why approaches to groundwater management that might work under one set of socio-ecological conditions might not work in another.

One of the key points woven through all of the chapters is that information is a critical element for groundwater management, no matter what region. Chapter 16 further highlights this issue by examining global information sharing in groundwater management.

Chapter 17 concludes the book with an overview of the state of affairs in agricultural groundwater use in the developing world, the opportunities and threats for the resource to continue contributing to global food supply and to lifting farmers out of poverty and into the future, agricultural or not, that they most desire. The conclusion is based on the contents of the book, but also draws in additional resources of the comprehensive assessment so as to cover topics not directly addressed, or not addressed in great detail, in this book.

While it is impossible to fully summarize the findings of so many authors with so many different perspectives studying such a wide geographic area, a generalized set of conclusions does emerge from the chapters as a whole:

1. **Information gaps but underestimated use** – there are substantial gaps in basic information on groundwater availability and agricultural use. In general, however, agricultural groundwater use appears to be substantially underestimated in most published figures.
2. **Major impacts across settings** – groundwater use in agriculture has increased substantially over the last 50 years and played a significant role in increased food production and livelihood security across broad areas of the world with wide-ranging climatic and socio-economic settings.
3. **Threats to the future but continued expansion** – while there are well-grounded fears that overdraft may threaten the future of the resource and the activities it supports in many regions (e.g. western and southern India, parts of northern China), overall agricultural groundwater use continues to grow, sometimes at

an increasing rate. Still, there are regions where possibilities for use have not been fully exploited (e.g. parts of sub-Saharan Africa, eastern India).

4. **Boom and bust trajectories** – there is a tendency for agricultural groundwater economies to move along a trajectory of initial utilization, agrarian boom, growing scarcity and eventually, in some cases, bust as groundwater tables fall.

5. **Special management challenges** – developing institutions to better manage the boom and bust trajectory is complicated by the physical properties of the resource and their interaction with social systems. Groundwater is often 'invisible' and thus difficult to measure. It must be understood in terms of years, decades or centuries. The vast areal extent and number of users of many aquifers, and low-cost means of abstraction, make groundwater subject to 'open access' problems. The political economy of agriculture and the sometimes stark social trade-offs inherent in decisions to control use make institutional action difficult.

6. **Emerging paradigms** – paradigms for tackling the challenges of groundwater management are emerging and being used with limited success. These paradigms can be thought of as mitigation – e.g. community-based agreements to regulate use as well as more formal economic and legal measures–and adaptation measures to facilitate farmers' ability to adjust to changing groundwater conditions.

7. **Influence and options from other sectors** – irrespective of the management approach taken, it must be remembered that groundwater outcomes are often driven in part by forces outside the water sector including agricultural and energy policy as well as broader political aims and processes. This increases the complexity of the problem but also the range of possible solutions.

8. **No single 'best practice'** – groundwater management approaches that work in one country or region may not do so in another. This is because of variation in the physical properties of groundwater as well as in the socio-economic and political strengths, weaknesses and desires of the societies involved. In this sense, groundwater management solutions will be local solutions.

9. **Costs of failure and value of success** – the establishment and maintenance of groundwater management regimes clearly has costs and can pit current use against future options. However, without active management, the total number of 'losers' including resource users, the environment and society as a whole is likely to be higher and more skewed towards the poorest user groups, than if action were taken.

We hope you will enjoy the book and that it will inspire new thinking, criticism, debate and, hopefully, positive action.

I

The Situation: Overview of Regional and Topical Issues

2 The Groundwater Economy of South Asia: An Assessment of Size, Significance and Socio-ecological Impacts

Tushaar Shah

IWMI-TATA Water Policy Program, Anand Field Office, Elecon Premises, Anand-Sojitra Road, Vallabh Vidyanagar 388 120, Anand, Gujarat, India

Introduction

Groundwater has come to be the mainstay of irrigated agriculture in many parts of Asia, especially in populous South Asia and the North China Plain. Between them, India, Pakistan, Bangladesh and North China use over 380–400 km^3 of groundwater annually, over half of the world's total annual use. However, there are large variations in the patterns of Asian groundwater use. Groundwater irrigation is of little importance in South-east Asia and southern China, which have abundant surface water. On the other hand, nearly all of India, northern Sri Lanka, Pakistan Punjab and Sind, and the North China Plain represent regions where groundwater has come to play a unique and increasingly critical role in supporting a dynamic smallholder peasant agriculture. In fact, while the bulk of the rest of the world's groundwater use is urban and industrial, most South Asian groundwater use is in agriculture. The importance of groundwater to the agricultural economies of South Asia can easily be seen in figures from the region's two most populous countries. In India, some 60% of the irrigated areas are served by groundwater wells.[1] In Pakistan – which inherited the world's oldest and largest continuous system of canal irrigation 57 years ago and today serves some 16 million hectares in the Indus basin – it has been commonly thought so far that groundwater provides over 40% of the total crop water requirements in the highly populous province of Punjab, which produces 90% of the country's food (Qureshi and Barrett-Lennard, 1998). A 2001 International Water Management Institute (IWMI) survey of 180 farmers in Rechna Doab, however, showed that more than 70% of the farmers received 80–100% of their irrigation water from wells and tube wells (Shah *et al.*, 2003).

Throughout South Asia, the history of protective well irrigation goes back to the millennia. However, intensive groundwater use on the scale we find today is a story of the last 50–nay, 30–years. In India, the total number of mechanized wells and tube wells rose from less than a million in 1960 to an estimated 19 million in

 The Agricultural Groundwater Revolution: Opportunities and Threats to Development (M. Giordano and K.G. Villholth)

2000. In Pakistan Punjab, it increased from barely a few thousands in 1960 to 0.5 million in 2000. In Bangladesh, which hardly had any groundwater irrigation until 1960, the area irrigated by groundwater wells shot up from 4% in 1972 to 70% in 1999 (Mainuddin, 2002).

Hydrogeology and Resource Availability

This explosive growth in groundwater irrigation has had little relationship with the pattern of occurrence of the groundwater resource. Figure 2.1 presents the first ever groundwater recharge map of the world prepared by researchers at the University of Kassel (Germany). It shows that in terms of long-term groundwater recharge, South Asia and the North China Plain are less well endowed compared to South America, pockets of sub-Saharan Africa and South-east Asia.

Many scientists argue that in the long run, groundwater development is self-regulating; people cannot pump more water than there is in the aquifers. According to them, long before the hydrogeology of aquifers imposes a check on further development, the economics of pumping water from deep aquifers would do so.[2] It is therefore ironic that global pockets of intensive groundwater use have emerged in regions that are not amongst the best endowed for it. Many of these regions have alluvial aquifers of high quality. The entire Indo-Gangetic plain that encompasses Pakistan Punjab and Sind, all of Northern India, Nepal Terai and Bangladesh are examples; so are areas of the North China Plain. However, all these are arid or semiarid, receiving little rainfall to provide natural recharge. Two-thirds of India (nearly half of the Indian subcontinent), in contrast, is doubly disadvantaged: it has semiarid climate with limited rainfall for recharging the aquifers; and hard-rock, basaltic aquifers with low storativity values. Peninsular India therefore is amongst the worst candidates for intensive groundwater irrigation; and yet, this is the region that has followed the Indo-Gangetic plain in ushering in a tube well revolution.

This paradox is global. High levels of sunlight combined with frequently lower levels of pest and disease problems can create optimal conditions for intensive agriculture – as in California, Spain and Israel. In contrast, many humid regions do not have as intensive agriculture despite – or perhaps because of – abundant water from groundwater or other sources (M. Moench, 2005, e-mail communication). In arid areas without resources for recharge, however, stringent limits to intensive groundwater irrigation are accessed early, leading to severe depletion, and at times, corrective measures as in Israel (which achieved high agricultural water productivity) and Saudi Arabia (which for some time had a vibrant wheat economy based on irrigation with fossil groundwater that has been progressively shrunk (Abderrahman, 2003)).

With this backdrop in mind, Fig. 2.2 attempts to highlight the irony of Asia's groundwater boom in the last 50 years. It is a common knowledge that hydrogeologic features of a terrain vary greatly even within a square mile, especially in hard-rock aquifers. So the classificatory approach we have used in Fig. 2.2 oversimplifies the great hydrogeologic diversity found in Asia, and can be justified only from the viewpoint of understanding aggregate patterns at a sub-

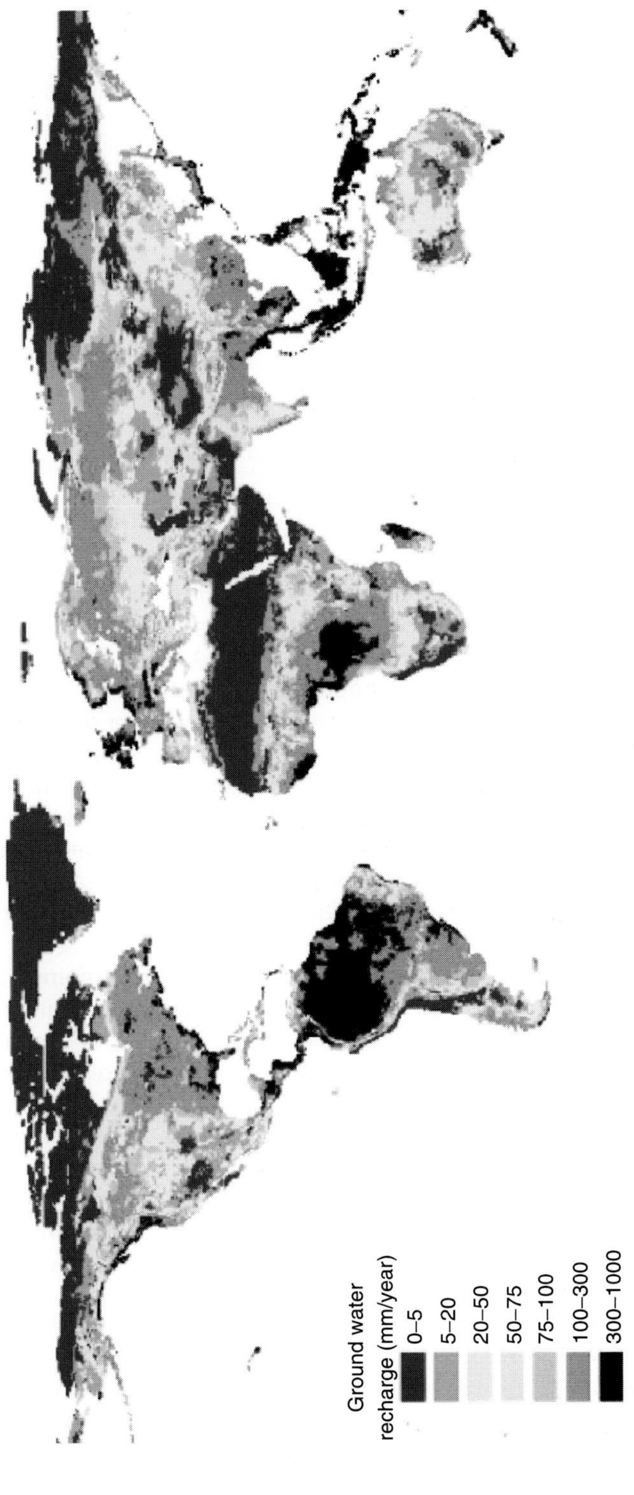

Fig. 2.1. Long-term average groundwater recharge. (Döll *et al.* 2002.)

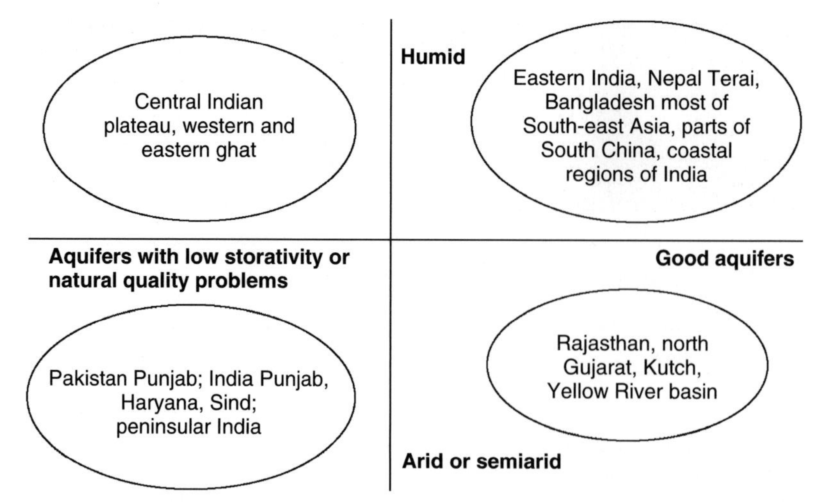

Fig. 2.2. Hydrogeologic patterns in Asia.

continental level. Regions best suited for this boom are those with high rainfall and good aquifers (North-West quadrant); however, except for Bangladesh and parts of eastern India, the groundwater boom has left these regions untouched. The groundwater irrigation economy is insignificant in South China and much of South-east Asia, which can sustain much more intensive groundwater irrigation than they currently practise. In contrast, it has assumed boom proportions in all the other three quadrants, none of which has 'appropriate' hydrogeologic and climatic conditions for intensive groundwater irrigation.

Around the world, intensive groundwater development without appropriate resource management regimes has resulted in resource degradation. In South Asia, this threat is growing. Besides non-point pollution of groundwater through chemical fertilizers and pesticides, intensive use of groundwater in agriculture gives rise to four resource management challenges: (i) controlling resource depletion; (ii) optimal management of conjunctive use of surface and groundwaters; (iii) managing the productivity impacts of secondary salinization; and (iv) managing natural groundwater quality concerns. The seriousness of each of these varies across regions depending upon their hydrogeology and the degree of groundwater development as set out in Fig. 2.3. It is clear that even in upper-right quadrant regions, which provide robust hydrogeologic platforms for intensive groundwater irrigation, socio-ecological and public health problems need to be managed as groundwater irrigation expands. In the eastern Gangetic basin, for instance, groundwater development is associated with mobilization of (geogenic) arsenic. Coastal areas are typically humid and have good alluvial aquifers; but salinity ingress or sea-water intrusion into coastal aquifers is a common problem, sometimes even at early stages of groundwater development. Likewise, in all humid areas (or arid areas with large volumes of surface water movement) with intensive groundwater irrigation, conjunctive management of surface and groundwaters remains a major challenge as well as an opportunity.

Hydro-geological Settings		**Socio-economic and Management Challenges**			
		Resource Depletion[a]	**Optimizing conjunctive use**[b]	**Secondary salinization**[c]	**Natural Groundwater Quality Concerns**
Major alluvial plants	**Arid**	●●	●○	●●●	●
	Humid	●	●●●		●●
Coastal plains		●●	●	●●○	●
Inter-Montane valleys		●	●●	○	●
Hard-rock areas		●●	○	●	●●●

[a]Related to aquifer recharge rates and storate availability.
[b]Implies both abundant surface and groundwater availability.
[c]Implies limited fresh groundwater availability and presence of saline groundwater and/or land drainage problems.

Fig. 2.3. Resource management challenges of intensive groundwater use in Asian agriculture.

As mentioned earlier, the geology of central and peninsular India is different and far more complex compared with that of the Indo-Gangetic basin, which consists of extensive alluvial aquifers throughout. Figure 2.4, showing a map of major aquifers of India by the Central Ground Water Board, suggests the dominance of basalt and crystalline rock formation in peninsular India. The water-bearing and -conveying properties of these aquifers vary greatly even over small distances, making scientific resource management critical and difficult at the same time (GoI, 1995). Overall, however, the yields of these aquifers are quite modest and, in fact, much smaller than much of sub-Saharan Africa; yet, there is a heavy and growing dependence on groundwater irrigation even in these regions.

Scale and Significance of South Asia's Groundwater Economy

Historical underpinnings

Rapid growth in groundwater use is a central aspect of the world's water story, especially since 1950. Shallow wells and muscle-driven lifting devices have been in vogue in many parts of the world for millennia. In British India (which includes India, Pakistan and Bangladesh), these accounted for over 30% of irrigated land even in 1903 (http://dsal.uchicago.edu/statistics/1894_excel) when only 14% of cropped area was irrigated. With the rise of the tube well technology and modern pumps, groundwater use soared to previously unthinkable levels after 1950; as a result, by the mid-1990s, groundwater-irrigated areas in

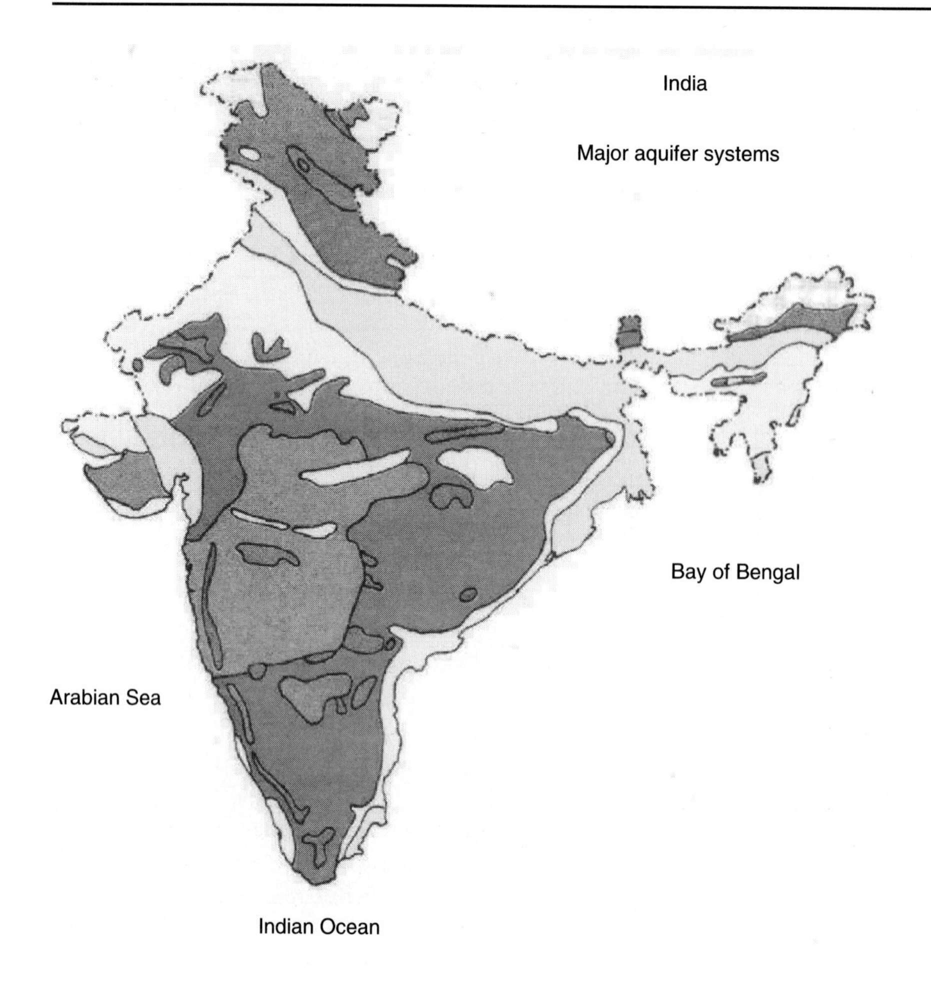

Explanations

Aquifer systems	Yield potential (LPS)
Alluvium, extensive	>40
Alluvium and sandstones, discont.	10–40
Limestones	5–25
Crystalline rocks	1–40
Basalts	1–25
Aquifers in hilly areas	<1

Fig. 2.4. Major aquifer systems of India. (From CGWB, 1995, p. 145.)

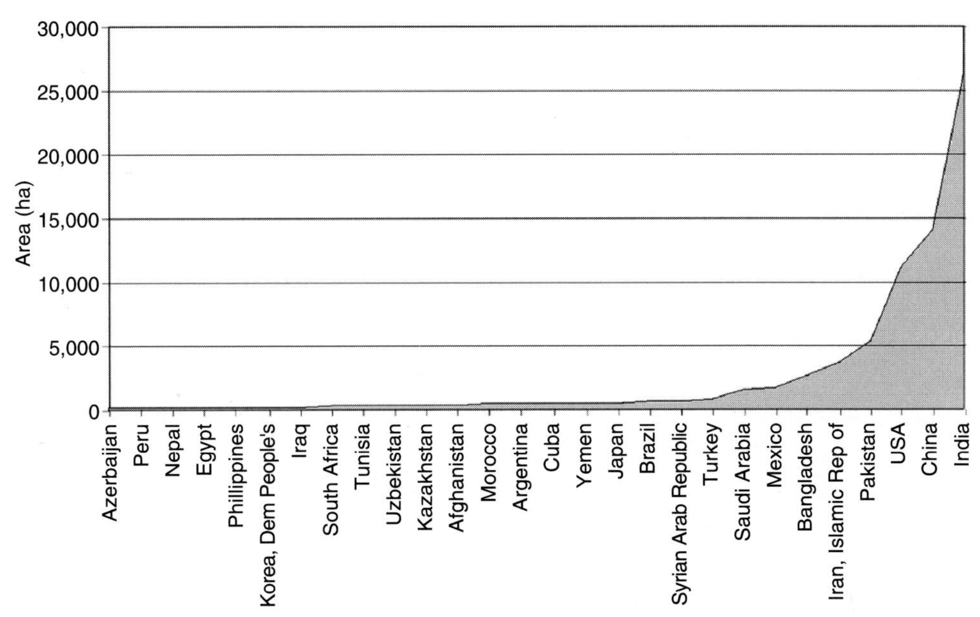

Fig. 2.5. Groundwater-irrigated area in countries with intensive groundwater use in agriculture. (From Food and Agricultural Organization, 2003.)

India, Pakistan and Bangladesh together were much larger than anywhere else in the world (Fig. 2.5). Indeed, one might surmise that of the 270–300 million hectares of global irrigation economy, more than one-third – around 110 million hectares – likely comprises groundwater-irrigated areas in the Indian subcontinent alone. Other groundwater economies of the world seem small by South Asian standards. In Spain, groundwater use increased from 2 km^3/year in 1960 to 6 km^3/year in 2000 before it stabilized (Martinez-Cortina and Hernandez-Mora, 2003). In western USA, which is larger in geographic area than the Indian subcontinent, although growth in total agricultural water use has tapered off, groundwater's share in irrigation has increased from 23% in 1950 to 42% in 2000, and has stabilized at around 107 km^3 (http://water.usgs.gov/pubs/circ/2004/circ1268/). In the Indian subcontinent, groundwater use soared from around 10–12 km^3 before 1950 to 240–260 km^3 in 2000 (Shah, 2005). Despite its growing pre-eminence, data on groundwater use are hard to find; however, Fig. 2.6 uses patchy data available from several countries to backcast the probable trajectories of growth in groundwater use in selected countries. While in the USA, Spain, Mexico, and African countries like Morocco and Tunisia total groundwater use peaked during the 1980s, in South Asia and the North China Plain, the upward trend began during the 1970s and is still growing (see Wang *et al.*, Chapter 3, this volume).

The striking aspect of South Asia's (and China's) groundwater boom is that it has acquired its present prominence only after 1970. Figure 2.7 shows the growth in the number of irrigation pumps in India during 1951–1993 and projects these to 2005. Figure 2.8 shows the corresponding change in the relative

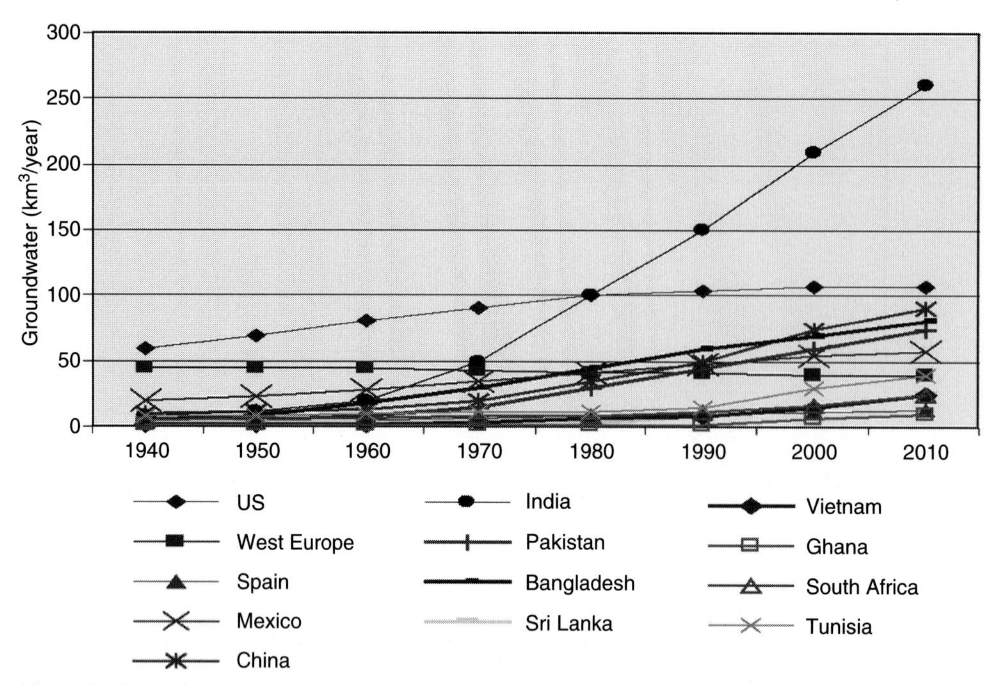

Fig. 2.6. Growth in groundwater use in selected countries. (From authors' estimates.)

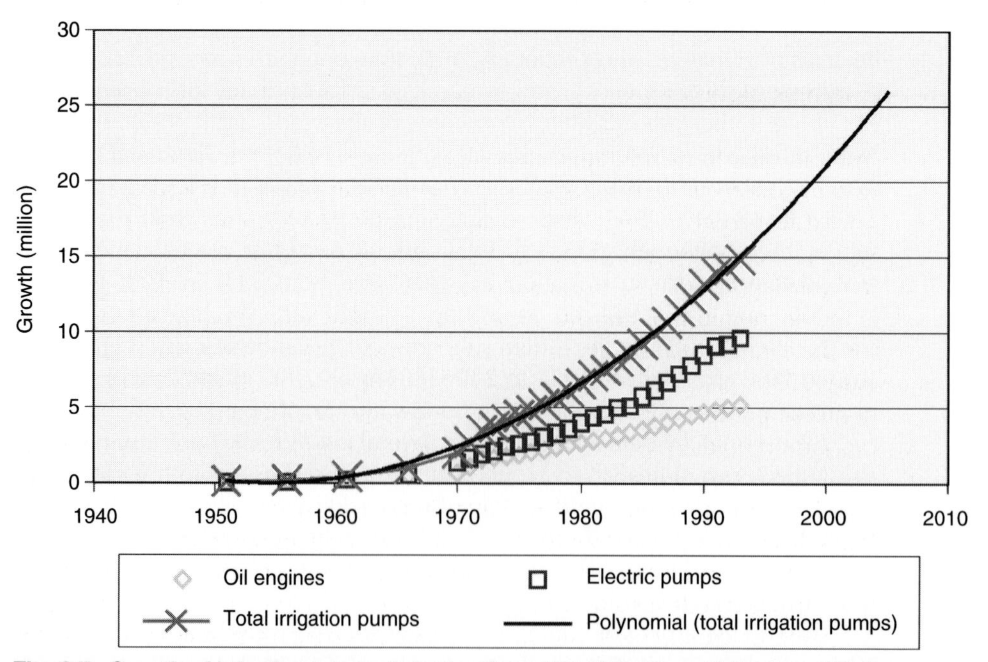

Fig. 2.7. Growth of irrigation pumps in India. (From World Bank and Ministry of Water Resources, 1998.)

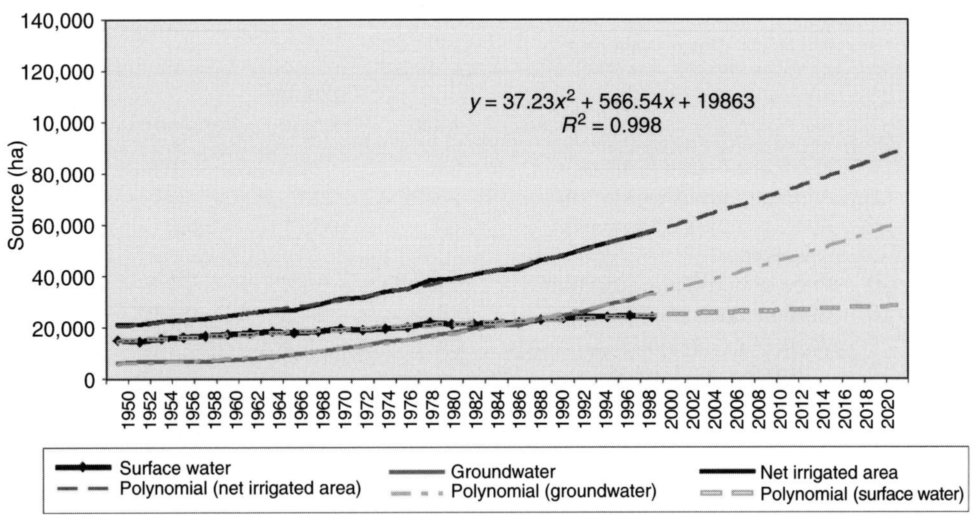

Fig. 2.8. Changing share of different sources in India's irrigated area: 1951–1998–2020.

share of different sources of irrigation in total irrigated area in India, indicating clearly that groundwater wells that irrigated just around 10 million hectares in 1970 are now serving over 35 million hectares of net irrigated area in India. Surface irrigation sources – tanks and canals – that had dominated irrigated agriculture in India for decades now gave way to groundwater irrigation. How did this role reversal affect the economics of South Asian agriculture?

Socio-economic significance

In these predominantly agrarian regions of South Asia, the booming groundwater economies have assumed growing significance from viewpoints of livelihood and food security; however, their significance as engines of rural and regional economic growth has remained understudied. There are several ways to consider the scale of the groundwater economy; but one practical measure is the economic value of the groundwater production. An unpublished report for the United States Agency for International Development (USAID) in the early 1990s placed the contribution of groundwater irrigation to India's gross domestic product (GDP) at around 10% (Daines and Pawar, 1987); if the same proportion holds now, the size of the groundwater irrigation economy of India would be approximately \$50–55 billion. In Table 2.1, we attempt a rough estimation of the market value of groundwater use in the Indian subcontinent. India, Pakistan and Bangladesh have active markets in pump irrigation service in which tube well owners sell groundwater irrigation to their neighbours at a price that exceeds their marginal cost of pumping. This price offers a market valuation of groundwater use in irrigation. We have used available estimates of the number of irrigation wells and estimates from sample surveys on average yield of wells and annual hours of operation of irrigation tube wells in the

Table 2.1. Proximate size of the agricultural groundwater economy of South Asia and the North China Plain (2002).

		India	Pakistan Punjab	Bangladesh	Nepal Terai
A	Number of wells (million)	21	0.5	0.8	0.06
B	Average output/well (m^3/h)	25–27	100	30	30
C	Average hours of operation/ well/year	360	1090	1300	205
D	Price of pump irrigation ($/h)	1–1.1	2	1.5	1.5
E	Groundwater used (km^3)	189–204	54.5	31.2	0.37
F	Value of groundwater used per year in billion dollars	7.6–8.3	1.1	1.6	0.02

countries covered. In India, for instance, a large number of farmers paid their neighbouring bore well owners $0.04/m³ for purchased groundwater irrigation in around 2000[3]; applying this price to the annual groundwater use of say 200 billion cubic metres gives us $8 billion as the economic value of groundwater used in Indian agriculture per year. For the Indian subcontinent as a whole, the corresponding estimate is around $10 billion. In many parts of water-scarce India, water buyers commonly enter into pump irrigation contracts offering as much as one-third of their crop share to the irrigation service provider; in water-abundant areas, in contrast, purchased pump irrigation cost amounts generally to 15–18% of the gross value of the output it supports. This can be used to draw the general inference that the agricultural output that groundwater irrigation supports is 4–5 times its market value.

Impact on agricultural growth: the case of India

Table 2.2 provides a synopsis of more detailed evidence of the size of India's groundwater economy, which is more explicitly described in DebRoy and Shah (2003).[4] In short, a regression equation was fit to cross-section data for 273 districts in which the dependent variable was the average value of gross farm output per hectare; and independent variables were average fertilizer use per hectare, percent of net sown area under surface irrigation and percent of net sown area under groundwater irrigation. Regressions were estimated for 1970–1973 and 1990–1993 data-sets. These showed that adding a hectare under groundwater irrigation made smaller contribution to increasing average value of output per hectare compared with adding a hectare under canal irrigation because farmers in South Asian canal commands are doubly blessed: they use cheap canal water to cut irrigation costs and costly groundwater to give their crops 'irrigation-on-demand'. However, the increase in groundwater irrigated area in an average Indian district after 1970 has been so large that groundwater irrigation contributed much more to increased value of agricultural output per hectare compared with surface irrigation. Table 2.3 summarizes the results; it shows that in the scenario of growing productivity of farmland, the contribution of surface

Table 2.2. Contribution of surface water irrigated and ground water irrigated area to total agricultural output, all India: 1970–1973 and 1990–1993. (From DebRoy and Shah, 2003.)

Year/indicators (at 1990 dollar/rupee exchange rate)	1970–1973	1990–1993	Change (%)
Average agricultural productivity ($/ha)	261.4	470.3	79.9
Contribution of SW ($/ha)	41.3	62.6	51.6
Contribution of GW ($/ha)	13.3	74.0	456.4
Contribution of SW (million $)	4,680	7,005	49.7
Contribution of GW (million $)	1,320	7,297	452.8
Contribution of SW as percent of total agricultural output	15.5	13.9	−1.6% points
Contribution of GW as percent of total agricultural output	4.4	14.5	+10.1% points
Total agricultural output/year (million $)	28,282	49,891	76.4

Table 2.3. Groundwater use per hectare in South Asia. (From IWMI survey of 2629 farmers in 2002.)

	Horse power hours	Total crop water requirements (m³)*	Estimated average application of irrigation water by sample farmers
Wheat	656	4,000	1,476.00 (36.9)
Kharif paddy	1,633	12,000	3,674.25 (30.6)
Boro paddy	3,266	18,000	7,348.50 (40.8)
Oilseeds	816	5,500	1,836.00 (33.4)
Coarse cereals	811	5,000	1,824.75 (36.5)

*Michael 2001

irrigation to aggregate farm output increased by 50% over 1973–1993, but that of groundwater irrigation soared by 450% over the same period. Interestingly, at $7.3 billion, groundwater contribution to agricultural output is close to $8 billion, which is our rough estimate of the economic value of groundwater irrigation in India in Table 2.2. To place this number in perspective, it is useful to note that this contribution of groundwater development to annual farm output in India is four times the annual public investment in irrigation projects, and more than all expenditures incurred by governments in India on poverty alleviation and rural development programmes.

Population pressure as the driver of tube well density

When the colonial government began building large run-of-the-river irrigation systems in northern and North-western India (which included the present Pakistan) in the early 19th century, these led to the decline in the tradition of well irrigation in Uttar Pradesh but stimulated it in North-western India.

During the latter half of the 20th century, these canal-irrigated areas led the charge in creating South Asia's groundwater boom, resulting in a widely held belief that large-scale tube well irrigation development occurs only in canal-irrigated areas. There was a time perhaps when this was largely true; however, the groundwater reality of South Asia has transcended this stage. In fact, as Figs 2.7 and 2.8 show, the density of tube wells – and groundwater irrigation in India and Pakistan Punjab – seems to have less to do with availability of surface water for recharge than with population pressure on agriculture. The figures show that tube well density is high throughout the Gangetic basin in India, which does have high groundwater availability but also very high population density. However, tube well density in Pakistan Punjab is highest in the most densely populated districts (Qureshi *et al.*, 2003). It is also high in many other parts of India such as Tamilnadu, Andhra Pradesh and Karnataka where water resources are limited but population density is high. On the other hand, in many parts of central India, little of the available resource is developed; yet tube well density is low because these regions are sparsely populated (DebRoy and Shah, 2003). China too has a similar pattern: groundwater development is low in South China, which has abundant surface water and low population density (except in the eastern coastal region); but tube well densities are high in the North China Plain, which has low surface water resource and high population density. Compared to large public irrigation projects that are driven by hydrologic opportunity, groundwater development is democratic, providing irrigation wherever people are.

Regional equity and drought-proofing

This pattern of groundwater development has brought much succour to the rural economy of the region. Without groundwater development, agriculture would have stagnated or declined in peninsular and eastern India and Bangladesh; food security would of course be endangered; but a more critical problem would be supporting rural livelihood during the decades these regions would take to transfer a sufficient proportion of their agrarian populations to off-farm livelihood systems. South Asia emerged out of British rule with a pattern of irrigation development that showed high regional inequality. The colonial government of India invested in large irrigation projects as a response to recurring famines that caused millions of starvation deaths; but these investments were concentrated in the North-western parts of British India and the Cauvery delta in the South while irrigation development in central and eastern regions was neglected (Whitcombe, 1984; Roy 2004). In the post-colonial era, too, public investments in canal irrigation projects were concentrated in pockets, leaving the rest of the region to rain-fed farming. In contrast, the development of groundwater irrigation had a significant 'equalizing effect'. It also emerged as the biggest drought-mitigator; during the 1960s, a major drought reduced India's food production by 30–40%, forcing India into embarrassing 'ship-to-mouth' dependence on US PL 480 wheat. Since the 1990s, food production has hardly been affected by a single drought (Sharma and Mehta, 2002), though a string of 2–3 drought years can still have an impact. Groundwater development has thus been a major restorer of India's national pride and confidence in feeding its people, and it has helped Bangladesh to transform from an endemic rice importer into a rice exporter (Palmer-Jones, 1999). Throughout the region, the easing of the obsessive sense of insecurity about national food self-sufficiency is explained in no small measure by the development of groundwater irrigation.

Supplemental nature of groundwater irrigation in South Asia

In order to better understand the nature of groundwater irrigation in South Asia, IWMI, in collaboration with several partners, undertook a large-scale survey of 2600 well owners from 300 villages selected to represent all regions of India, Pakistan, Bangladesh and 20 districts of Nepal Terai (see DebRoy and Shah, 2003, and Shah *et al.*, 2006, for details of the survey design and results). One of the aims was to find out if intensive groundwater irrigation occurs in regions with large-scale canal irrigation. Figure 2.9, which summarizes the results, shows that almost everywhere in the subcontinent, groundwater contribution to irrigated areas exceeds that of surface water; that outside of Pakistan Punjab and Sind, conjunctive use of surface and groundwaters at the farmer level is small; that in North-western India, despite massive investments in canal irrigation, the bulk of the irrigation is delivered by wells and tube wells. Figure 2.10, again based on the IWMI survey, suggests that thanks to the groundwater revolution, rain-fed regions, districts or even villages are rare in South Asia; there are just rain-fed and irrigated plots. Just around 5% of the 278 villages covered reported completely rain-fed agriculture; nearly half of the villages had groundwater-dominated irrigated agriculture; pure canal irrigation (i.e. with no wells or tube wells) accounted for just 10% of the villages and 20% of the irrigated area in the sample.

Another key feature of groundwater irrigation in South Asia is its predominantly supplemental nature. The IWMI survey of 2002 collected information from 2629 sample farmers about the depth of pumping water level, hours pumped for different crops and the capacity of pumps. Using these data, rough estimates were made of the actual average application of irrigation water for key crops. When these are compared with CROPWAT recommendations, we find that farmers provide around one-third of the crop-water requirements through groundwater.

Other studies show that such supplemental groundwater irrigation is also significantly more productive compared with surface irrigation, because it offers individual farmer irrigation 'on demand' which few surface systems can offer; and because its use entails significant incremental cost of lift, farmers tend to economize on its use and maximize application efficiency. Evidence in India suggests that crop yield per cubic metre of water applied on groundwater-irrigated farms tends to be 1.2–3 times higher than that applied on surface water–irrigated farms (Dhawan, 1989, p. 167).[5] In terms of return on investment, groundwater irrigation in South Asia has done very well. In Pakistan Punjab, capital investment in private tube wells is estimated to be of the order of Pak Rs. 25 billion[6] ($0.4 billion at 2001 prices), whereas, according to one

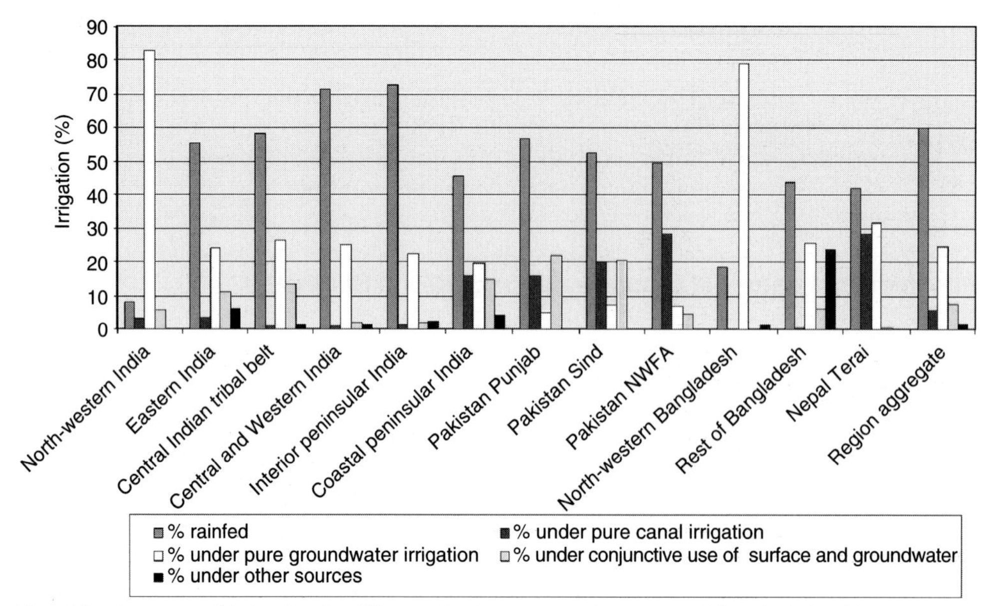

Fig. 2.9. Sources of irrigation in different hydro-economic zones of South Asia. (From IWMI–Tata Survey.)

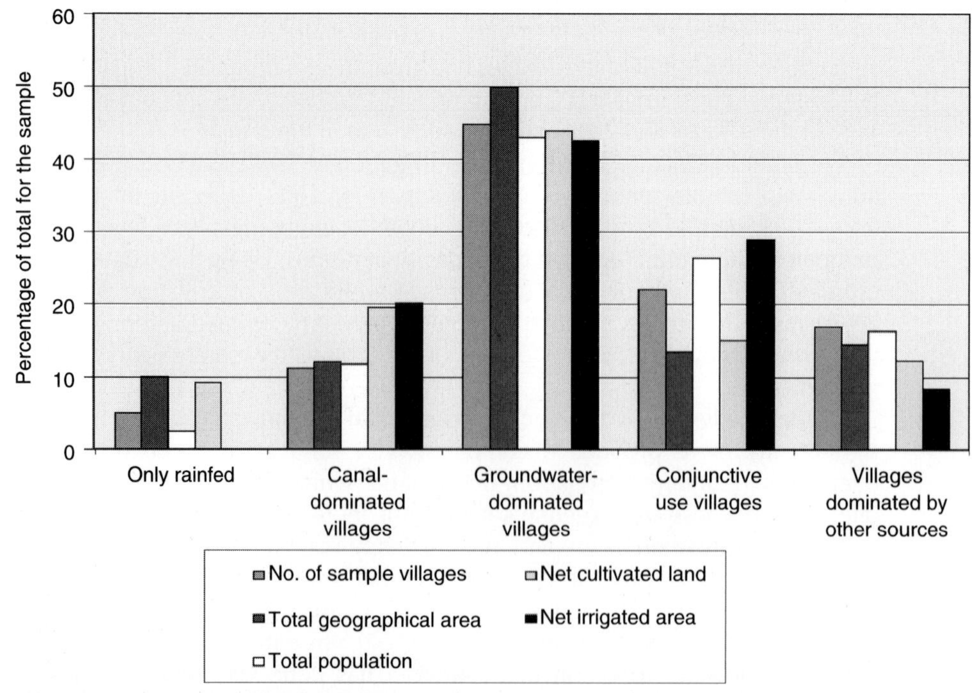

Fig. 2.10. Relative importance of wells, canals and other sources of irrigation in sample villages. (From IWMI–Tata Survey.)

Table 2.4. Comparison of farms with and without tube well water supply, Pakistan. (From Ministry of Agriculture, 1988.)

Item	Unit	Type	Sugarcane	Rice	Cotton	Wheat
Cropped area	Percent farm area	With TW	8	13	8	60
		Without TW	3	3	7.5	50
Yield per acre	Tonnes	With TW	23.6	1.3	0.40	1.10
		Without TW	12.6	0.9	0.38	0.76
Gross value per acre	Pak Rs.[a]	With TW	23,800	14,188	8,624	56,808
		Without TW	4,725	2,910	5,060	33,300

[a]$1 = Pak Rs. 65 in September 2001.
TW = Tube wells.

estimate, the annual benefits in the form of agricultural production of the order of Pak Rs.150 billion ($2.3 billion) accrue to over 2.5 million farmers, who either own tube wells or hire the services of tube wells from their neighbours. The best farm level productivity performance of course is obtained by those who can use a judicious combination of surface and groundwater. Table 2.4 reports physical and value productivity on 521 canal-irrigated farms in the Indus system in Pakistan Punjab and shows that farmers with wells obtain 50–100% higher yield per acre and 80% higher value of output per acre compared with canal irrigators without wells. Groundwater users in South Asia often use only a small fraction of scientifically recommended water requirements; rather than aiming at fully irrigated yields, they use sparse, life-saving irrigation to obtain substantial increases over rain-fed yields (see Fig. 2.11). This is because of the high marginal cost of groundwater use; some of the poorest irrigators in arid parts of South Asia – who purchase pump irrigation from well owners – commonly pay 10–14 cents/m^3 of water compared to a fraction of a cent paid by canal

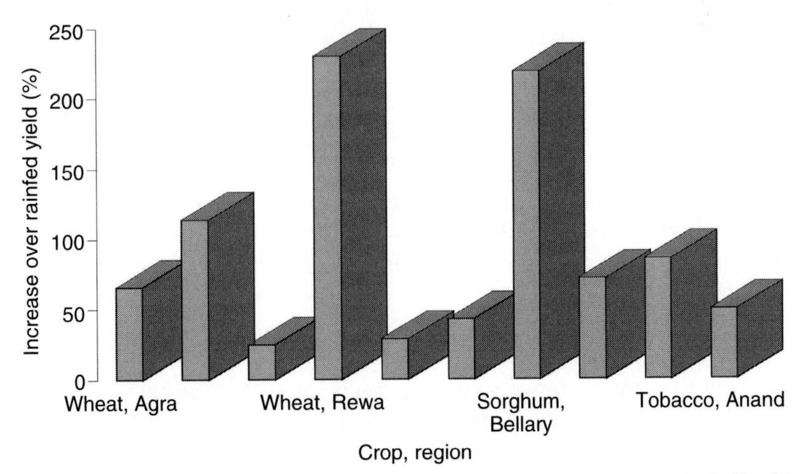

Fig. 2.11. Yield impact of life-saving 5 cm irrigation on rainfed crops in India. (From Dhawan, 1989, after Singh and Vijaylakshmi, 1987.)

irrigators. Finally, compared to large surface systems whose design is driven by topography and hydraulics, groundwater development is often much more amenable to poverty targeting. No wonder, then, that in developing regions of South Asia, groundwater development has become the central element of livelihood creation programmes for the poor (Kahnert and Levine, 1993, for the GBM basin; Shah, 1993, for India; Calow *et al.*, 1997, for Africa).

Socio-economic vs. socio-ecological impacts

All in all, as a purely socio-economic phenomenon, South Asia's groundwater irrigation boom has been an unalloyed success. By all accounts, it has served the purpose of a massive programme of strengthening rural livelihood. It has made the region food-secure at macro-level. It has done more to alleviate rural poverty than most public interventions expressly designed to that end. In scale and depth, its socio-economic impacts are comparable to some of the world's most successful development programmes such as the dairy cooperative movement of India that revolutionized India's dairy economy.

However, overall socio-ecological returns to the boom have long since been declining on the margin. In many regions, groundwater depletion that manifests in secular decline in water tables is beginning to take its toll. Pumping costs are rising; well failures and abandonment are evermore frequent. All the resource management challenges we outlined in Fig. 2.3 are in full play; and there are few regions left apart from pockets of the eastern Gangetic basin, where further groundwater development can be had more or less as a 'free lunch'.

The Pathology of Decline of a Groundwater Socio-ecology

A few years ago, David Seckler, the then director general of IWMI, wrote alarmingly that a quarter of India's food harvest is at risk if she fails to manage her groundwater properly. Many people today think that Seckler might have well underestimated the situation, and that if India does not take charge of her groundwater, her agricultural economy may crash. Postel (1999) has suggested that approximately 10% of the world's food production depends on overdraft of groundwater to the extent of 200 km^3; most likely, 100 km^3 out of this occurs in western India. In the lower Indus basin in Pakistan and the Bhakra system in northern India, groundwater depletion is not a problem but soil and groundwater salinization is. IWMI's past research to understand the dynamics of groundwater socio-ecologies indicates some recurring patterns. In much of South Asia, for example, the rise and fall of local groundwater economies follow a four-stage progression outlined in Fig. 2.12. This highlights the typical progression of a socio-ecology from a stage in which unutilized groundwater resource potential becomes the instrument of unleashing an agrarian boom to one in which, unable to apply brakes in time, it goes overboard in exploiting its groundwater.

The four-stage framework outlined in Figure 2.12 shows the transition that South Asian policymakers and managers need to make from a resource

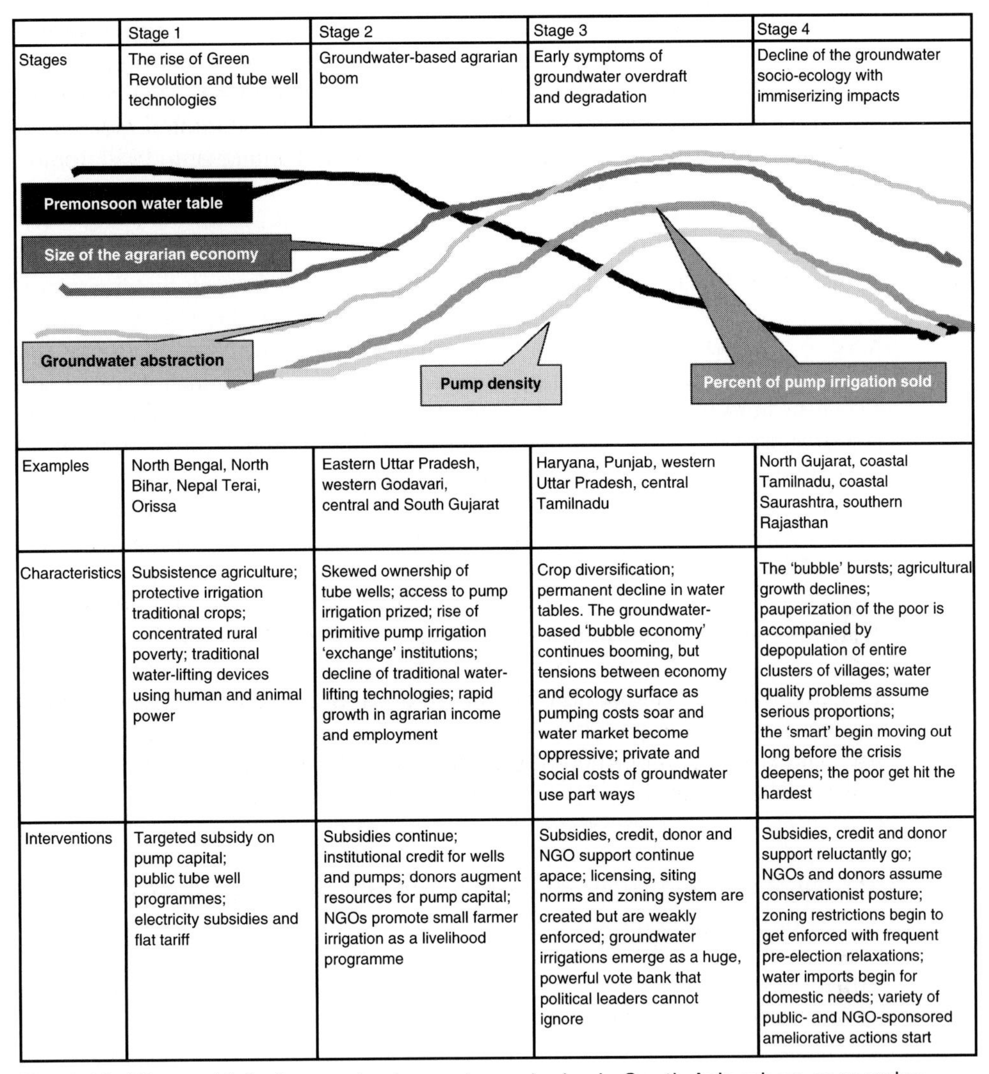

	Stage 1	Stage 2	Stage 3	Stage 4
Stages	The rise of Green Revolution and tube well technologies	Groundwater-based agrarian boom	Early symptoms of groundwater overdraft and degradation	Decline of the groundwater socio-ecology with immiserizing impacts
Examples	North Bengal, North Bihar, Nepal Terai, Orissa	Eastern Uttar Pradesh, western Godavari, central and South Gujarat	Haryana, Punjab, western Uttar Pradesh, central Tamilnadu	North Gujarat, coastal Tamilnadu, coastal Saurashtra, southern Rajasthan
Characteristics	Subsistence agriculture; protective irrigation traditional crops; concentrated rural poverty; traditional water-lifting devices using human and animal power	Skewed ownership of tube wells; access to pump irrigation prized; rise of primitive pump irrigation 'exchange' institutions; decline of traditional water-lifting technologies; rapid growth in agrarian income and employment	Crop diversification; permanent decline in water tables. The groundwater-based 'bubble economy' continues booming, but tensions between economy and ecology surface as pumping costs soar and water market become oppressive; private and social costs of groundwater use part ways	The 'bubble' bursts; agricultural growth declines; pauperization of the poor is accompanied by depopulation of entire clusters of villages; water quality problems assume serious proportions; the 'smart' begin moving out long before the crisis deepens; the poor get hit the hardest
Interventions	Targeted subsidy on pump capital; public tube well programmes; electricity subsidies and flat tariff	Subsidies continue; institutional credit for wells and pumps; donors augment resources for pump capital; NGOs promote small farmer irrigation as a livelihood programme	Subsidies, credit, donor and NGO support continue apace; licensing, siting norms and zoning system are created but are weakly enforced; groundwater irrigations emerge as a huge, powerful vote bank that political leaders cannot ignore	Subsidies, credit and donor support reluctantly go; NGOs and donors assume conservationist posture; zoning restrictions begin to get enforced with frequent pre-election relaxations; water imports begin for domestic needs; variety of public- and NGO-sponsored ameliorative actions start

Fig. 2.12. Rise and fall of groundwater socio-ecologies in South Asia where economies follow a four-stage progression.

development mindset to a resource *management* mode. Forty years of Green Revolution and mechanized tube well technology have nudged many regions of South Asia into stages 2–4. However, even today, there are pockets that exhibit characteristics of stage 1, but the areas of South Asia that are at stage 1 or 2 are shrinking by the day. Many parts of western India were in this stage in the 1950s or earlier, but have advanced into stage 3 or 4. An oft-cited case is North Gujarat where groundwater depletion has set off a long-term decline in the booming agrarian economy; here, the well-off farmers who foresaw the impending doom forged a generational response and made a planned transition to a non-farm, urban livelihood. The resource-poor have been left behind to pick up the pieces of what was a booming economy barely a decade ago. This drama is being re-enacted in ecology after groundwater socio-ecology with frightful regularity (Shah, 1993; Moench, 1994; Barry and Issoufaly, 2002).

In stage 1 and early stage 2, the prime concern is to promote profitable use of a valuable, renewable resource for generating wealth and economic surplus; however, already by stage 2, the thinking needs to change towards careful management of the resource. Yet, the policy regime ideal for stages 1 and 2 has tended to become 'sticky' and to persist long after a region moves into stage 3 or even 4. IWMI's recent work in the North China Plain suggests that the story is much the same over there. The critical issue to address is: Does stage 4 always have to play out the way it has in the past? Or are there adaptive policy and management responses in stage 2 that can generate a steady-state equilibrium, which sustains the groundwater-induced agrarian boom without degrading the resource itself? In the remainder of this chapter, we review the prospects and opportunities for forging such steady-state equilibrium.

In Search of Sustainability

Challenge of demand-side management

The South Asian debate on creating effective groundwater management regimes has been swayed by the success stories of groundwater regulation in Australia and the USA where the number of users is small, and their average size very large (see Table 2.5); or from Europe, which has a large number of small users but where the state has capacity to deploy huge financial and technological resources to mend its natural resources problems. The South Asian situation is different; as a result, the debate continues but the policy alternatives commended come unstuck. Enacting and enforcing a groundwater law, establishing clear tradable property rights on water, pricing groundwater as an economic good, installing and enforcing a licensing and permit system have all been discussed *ad nauseum* in South Asia as desirable policy interventions to regulate groundwater overdraft (see e.g. Arriens *et al.*, 1996, pp. 176–178,

Table 2.5. Structure of national groundwater economies of selected countries.

Country	Annual groundwater use (km^3)	No of agricultural groundwater structures (million)	Extraction/ structure (m^3/year)	Percent of population directly or indirectly dependent on groundwater irrigation
India	150	19	7,900	55–60
Pakistan Punjab	45	0.5	90,000	60–65
China	75	3.5	21,500	22–25
Iran	29	0.5	58,000	12–18
Mexico	29	0.07	414,285	5–6
USA	100	0.2	500,000	<1–2

239–245). Nobody seems to disagree with the need for these; yet, no Asian country has been able to deploy any of these interventions effectively even as the groundwater situation has been turning rapidly from bad to worse. The scale of the groundwater threat is long recognized; but viable strategies for dealing with it are not forthcoming; indeed, governments are still busy promoting more groundwater development, as if they were in stage 1. This is true for South Asia, but it is also true for North China.[7]

In principle, the groundwater threat can be met, provided national administrations can build a tight resource management regime well in time that focuses on both demand- and supply-side interventions. The catch is that nowhere in the world – barring in very rich countries – do we find such an ideal regime actually in operation. Worldwide, then, there is some action by way of a response to groundwater degradation, but it is too little, too late, too experimental, too curative, and too supply-side-oriented. There is precious little done to reduce demand for groundwater or on approaches to economizing on its use. The only examples we can find that combine demand- and supply-side interventions are in western USA, which has suffered amongst the most extensive groundwater depletion problems anywhere in the world, and before anyone else did.[8] The examples of western USA provide important pointers to the rest of the world about where to direct ameliorative action (see Peck, Chapter 14, this volume). A major problem in transferring these lessons wholesale to the developing country context, however, is the numbers involved: in a typical groundwater district in the USA, the total number of farmers is probably less than 1000; in an area of comparable size, Asia would have over 100,000 farmers (see Table 2.5). The average stakes per farmer too would vary by a factor of a thousand or more. As a result, spontaneous collective action by groundwater users to protect and manage the resource is far less likely – and more difficult to sustain – in Asia. In the Murray–Darling basin in Australia, widely held as a model for integrated river basin management, obtaining a permit is mandatory for all groundwater users, but small users extracting water for domestic or livestock needs, or for irrigating small plots of 2 ha or less, are exempt (see Turral, Chapter 15, this volume). If this exemption were to be applied in South Asia or the North China Plain, more than 95% of groundwater irrigators would be exempted (Shah *et al.*, 2006).

Legal/regulatory initiatives tried worldwide

The differing rules for obtaining a permit for groundwater irrigation is perhaps why Asian and other developing country governments tend to rely more heavily on enacting laws to regulate groundwater use and abuse. Although South Asia is yet to embark on this path, there is little evidence to suggest that water laws deliver the desired regulation, either in Asia or elsewhere in the developing world. China is way ahead of South Asian countries in legislative and regulatory measures to rein in groundwater withdrawals. Its new water law requires that all the pumpers get a permit; but the law is yet to be enforced. Only in deep tube well areas of the North China Plain are tube well owners obliged to get individual permits; elsewhere, the village as a whole holds a permit to use groundwater, which has little operational meaning. China's water administration is able to extract close to an economic price from canal irrigators; but groundwater is still free (Shah *et al.*, 2004a). South Africa's new water law and water policy enshrine the principles of 'user pays; polluter pays'; they work well in the commercial farm economy dominated by large-scale white farms but would fail to impact areas of 'black irrigation' in the former homelands. India has been toying around with a draft model groundwater bill for more than 30 years; but is not able to make it into a law due to doubts about enforcing such a law on more than 19 million irrigation pumpers scattered across a vast countryside. The establishment of Aquifer Management Councils called COTAS (*Consejos Técnicos de Aguas*) in Mexico as part of its water reforms and under the new Mexican water law is a notable development of interest to South Asia's groundwater policymakers. However, IWMI researchers in Guanajuato, Mexico are skeptical and hopeful at the same time:

> [S]everal factors bode ill for their (COTAS) future effectiveness in arresting groundwater depletion. Most importantly, their main role will be advisory in nature and they will not have the mandate to resolve conflicts between water users or restrict groundwater extractions. Moreover, there is an unclear division of tasks and responsibilities between COTAS, irrigation water users' associations, the federal and state water management agencies and the river basin council. On the other hand, the COTAS provide a vehicle for groundwater users to engage in self-governing, collective action and to find innovative solutions to the vexing problem of groundwater depletion. (Wester *et al.*, 1999)

A recent assessment of what COTAS have achieved is even gloomier. Mexican attempts to nationalize water, and create groundwater rights by issuing concessions to all users who are working in organized industry and with municipal users – sectors where these reforms are the least needed for effective regulation; however, in the farming sector, groundwater concessions have come unstuck. A major problem is the high transaction costs of enforcing the terms of the concession on 70,000 tube well owners and a similar number of farmers who impound rainwater in private *bordos* (ponds) in the highlands of Northern Mexico (Shah *et al.*, 2004b). South Asia is often advised to draw a leaf out of the book of Mexican water reform; but it is easy to imagine how difficult it would be to enforce such a regime on 19 million tube well owners when Mexico has been finding it difficult to enforce it on 70,000 groundwater irrigators.

Equitable control

Institutional solutions to sustainable groundwater management that have a chance to work may pose complex issues of equity and political economy. Some of these became evident in the tiny and experimental World Bank–supported Taiz project in the Habir aquifer of Yemen with the objective to develop a partnership between rural and urban groundwater users to transfer water from the countryside to a town on equitable terms and ensure the sustainability of the resource. The project – which affected a small group of 7000 rural residents on the Habir aquifer – failed to either transfer water or ensure its sustainability, but suggested important lessons about why it failed. Taking an egalitarian stance, the project tried capacity building of all the 7000 residents to assume rights over the aquifer and manage the transfer of water to the city; however, the real stakeholders were 22 irrigation pumpers – who used over 90% of the aquifer – and not the 7000 residents. The practicalities of achieving the project aims required that the de facto rights of these 22 users were recognized, and incentives created for *them* to sustainably manage the resource. The pumpers, however, opposed, got frustrated and sabotaged all institutional efforts that infringed their de facto rights and failed to provide *them* incentives for sustainable management – which meant that sustainability could be possible only by reinforcing existing inequalities. The report on a World Bank Consultation that analysed the lessons of the Taiz project concluded: 'In our judgment, "the egalitarian option" is not viable and ultimately counter productive since it is unlikely to work' (Briscoe, 1999, p. 12).

Indirect levers

There are potentially powerful *indirect* demand-management strategies that are not even part of the academic discussion on groundwater management in the developing world. For example, it has been suggested that India Punjab's groundwater depletion problems could be easier to resolve if its export of 'virtual' groundwater in the form of rice could be reduced or stopped. IWMI researchers have suggested that in the North Indian plains, using earthen canals for recharging with flood water of monsoon rains can help counter groundwater depletion (IWMI–Tata Water Policy Briefing 1). Water-saving irrigation research – such as Alternate Wet and Dry Irrigation (AWADI) for rice in China or the System of Rice Intensification, which has found enthusiastic following in scores of countries including India and Sri Lanka (Satyanarayana, 2005; Sinha and Talati, 2005) – can help reduce groundwater use; but it needs to be examined if these technologies would work as well in dry areas. In many developing countries, pricing and supply of electricity to tube well owners can offer powerful levers for agricultural demand management for groundwater. Since levying a price on groundwater itself may entail high transaction costs of collection, energy price can serve as a useful 'surrogate' (Scott and Shah 2004; Shah *et al.*, 2004c).

Energy-irrigation nexus

Another key area in the groundwater economy of South Asia, especially India, is the perverse energy subsidies for tube well irrigation. In the populous South Asian region, there seem no practical means for direct management of groundwater; laws are unlikely to check the chaotic race to extract groundwater because of the logistical problems of regulating a large number of small, dispersed users; water pricing and/or property right reforms too will not work for the same reasons. However, electricity supply and pricing policy offer a powerful tool kit for *indirect* management of both groundwater and energy use. Since electricity subsidies have long been used by governments in this region to stimulate groundwater irrigation, the fortunes of groundwater and energy economies are closely tied. India is a classic example. Today, India's farmers use subsidized energy worth $4.5–5 billion/year to pump 150 km^3 of water mostly for irrigation; the country's groundwater economy has boomed by bleeding the energy economy. With the electricity industry close to bankruptcy, there are growing demands for eliminating power subsidies; but governments are unable to do so because of stiff opposition from the farmer lobby. Recent IWMI research (Shah *et al.*, 2004c) has argued that sustaining a prosperous groundwater economy with a viable power sector is feasible, but it requires that the decision makers in the two sectors jointly explore superior options for energy–groundwater co-management. IWMI studies recognize that switching to volumetric electricity pricing may not be politically feasible at present. However, they advocate a flat tariff accompanied by better management of high quality but carefully rationed power supply to maintain at once the financial sustainability of energy use in agriculture and the environmental sustainability of groundwater irrigation. They argue that such a strategy can curtail wasteful use of groundwater in irrigation to the extent of 15–18 km^3/year.

Supply-side responses

Where the problem has begun to pinch hard, the Asian response to groundwater depletion has been supply-side rather than demand-side. The standard reasoning is that even after building 800,000 big and small dams around the world, the reservoirs can capture and store no more than one-fifth of the rainwater, the bulk of the remainder still running off to the seas. In India, which has built more than its share of the world's dams, 1150 km^3 of the rainwater precipitation still runs off to the seas annually in the form of 'rejected recharge' (INCID, 1999). If a fraction of this could be stored underground by reducing the velocity of the runoff and providing time for recharge, groundwater supplies could be enhanced significantly. But this presumes *active* aquifer management where planned drawing down of the water table in the premonsoon dry months is an important element of the strategy for enhancing the recharge from monsoon rainwater as well as from irrigation return flows. Such proactive aquifer management is an established practice in many industrialized countries; for instance, the share of artificial groundwater recharge to total groundwater use is 30% in western Germany, 25% in Switzerland, 22% in the USA, 22% in Holland, 15% in Sweden and 12% in England (Y. Li, 2001).

Mega projects for interbasin transfer of water from surplus to deficit basins are increasingly talked about in groundwater irrigation areas of Asia. China is already executing a mega project for trans-basin diversions of approximately 25 km^3/year of water from the Yangtzi river in the water-surplus South to the water-scarce Yellow River basin in the North (Keller *et al.*, 2000). India has for a long time talked about a garland canal to link Himalayan rivers with Cauvery and other South Indian rivers; these have so far remained at the ideas level but with the passing of every drought, these seemingly impractical ideas acquire new appeal and credibility. In 2002, the Supreme Court of India enjoined the central government to undertake such linking of rivers on a war footing partly to alleviate the pressure on groundwater in western and peninsular India. Gujarat, the western Indian state chronically dependent on groundwater overdraft for its agriculture, has already started using interbasin transfer of water from the controversial Narmada project to counter groundwater depletion in parts of Saurashtra and North Gujarat.

The economics of interbasin transfer are deeply influenced by the groundwater economy. In Gujarat, for example, it has been argued that the overall economics of the Narmada project become far more favourable when we include into the cost–benefit calculus the beneficial impact of Narmada waters in significantly countering groundwater depletion in North Gujarat where farmers are using subsidized electricity to pump groundwater from 250 to 300 m. The saving of electricity subsidy required to sustain groundwater-irrigated agriculture and rural livelihood systems in such regions can tilt the cost–benefit ratios in favour of surface irrigation projects.

Reviving and improving upon forgotten traditions

Some of the water-scarce regions of Asia have age-old traditions and structures for rainwater harvesting, which have fallen into disuse and are now attracting renewed attention (see Sakthivadivel, Chapter 10, and Mudrakartha, Chapter 12, this volume). India's Central Ground Water Board has been harnessing support for a National Groundwater Recharge Programme. Tarun Bharat Sangh and Pradan, two local non-governmental organizations (NGOs) in the Alwar district of western Rajasthan whose work IWMI has been studying, have helped local communities to rehabilitate centuries-old tanks (known locally as *johads* or *paals)* with dramatic impact on groundwater recharge and revival of dried-up springs and rivulets in a 6500 km^2 area (Agarwal, 2000). In southern India, where centuries-old tanks are on a decline, wells are widely thought of as enemies of tanks. Until the 1960s, when modern tube well technology became available to farmers, tanks were preserved, maintained and nurtured as valuable common property irrigation structures. All those who benefited from a tank participated in its upkeep and the cleaning of its supply channels. Recently, better-off farmers have been able to increasingly privatize tank water by sinking tube wells in their surrounding. As a result, their stakes in maintaining tanks declined; and so did the age-old traditions of tank management.

However, in the western region of India, hit hardest by groundwater depletion, well owners have become great champions of tanks because they keep their wells productive (Sakthivadivel *et al.*, 2004). Catalysed first by spiritual Hindu organizations – such as the *Swadhyaya Pariwar* and *Swaminarayana Sampradaya* – and supported by numerous local NGOs, local communities have spontaneously created a massive water-harvesting and recharge movement based on the principle: 'water on your roof stays on your roof; water in your field stays in your field; and water in your village stays in your village'. As many as 300,000 wells – open and bore – have been modified by the people to divert rainwater to them; and thousands of ponds, check dams and other rainwater harvesting and recharge structures have been constructed on the basis of the self-help principle to keep the rainwater from gushing into the Arabian Sea (Shah, 2000). While systematic studies are still to begin of the impact of the movement and the popular science of rainwater harvesting and decentralized recharge that has emerged as a result of farmers' experiments, available indicative evidence suggests that for regions critically affected by groundwater depletion, only mass popular action on regional scale may be adequate to meet the challenge of depletion (Shah and Desai, 2002).

India has begun to take rainwater harvesting and groundwater recharge seriously at all levels. These are at the heart of its massive Integrated Watershed Development Programme, which provides public resources to local communities for treatment of watershed catchment areas and for constructing rainwater harvesting, and recharge structures. Trends during the 1990s also suggest a progressive shift of budgetary allocations from irrigation development to water harvesting and recharge. One indication of the seriousness assigned to the issue by Indian leadership is the message delivered by the prime minister to the citizens on 26 January 2004, India's Republic Day; the nation's prime minister and water resources minister went to the people with a full-page story espousing the benefits and criticality of groundwater recharge.

From Resource Development to Management Mode

In the business-as-usual scenario, problems of groundwater overexploitation not just in South Asia but throughout the region will only become more acute, widespread, serious and visible. The front-line challenge is not just supply-side innovations but to put into operation a range of corrective mechanisms before the problem becomes either insolvable or not worth solving. This involves a transition from resource 'development' to resource 'management' mode (Moench, 1994, see also Moench, Chapter 9, this volume). Throughout Asia – where symptoms of overexploitation are all too clear – groundwater administration still operates in the 'development' mode, treating water availability as unlimited, and directing their energies on enhancing groundwater production. A major barrier that prevents transition from the groundwater *development* to *management* mode is lack of information. Many countries with severe groundwater depletion problems do not have any idea of how much groundwater occurs, and who withdraws how much groundwater and where. Indeed, even in European countries, where groundwater is important in all uses, there is no systematic monitoring of groundwater occurrence and draft (Hernandez-Mora *et al.*, 1999). Moreover, compared to reservoirs and canal systems, the amount and quality of application of science and management to national groundwater sectors has been far less primarily because, unlike the former, groundwater is in the private, 'informal' sector, with public agencies playing only an indirect role.

Gearing up for resource management entails at least five important steps:

1. Recognizing that even as the bulk of the public policy and investments is directed at large government-managed irrigation programmes, in reality, South Asia's agriculture has increasingly come to depend upon small-holder irrigation based largely on groundwater; policy effort as well as resource investments need to adjust to this reality if these are to achieve *integrated* water and land resources management in the true sense.
2. Implementing information systems and resource planning by establishing appropriate systems for groundwater monitoring on a regular basis and undertaking systematic and scientific research on the occurrence, use and ways of augmenting and managing the resource.
3. Initiating some form of demand-side management by: (i) registering users through a permit or license system; (ii) creating appropriate laws and regulatory mechanisms; (iii) employing a system of pricing that aligns the incentives for groundwater use with the goal of sustainability; (iv) promoting conjunctive use of surface and groundwaters by reinventing main system management processes to fit a situation of intensive tube well irrigation in command areas; and (v) promoting 'precision' irrigation and water-saving crop production technologies and approaches.
4. Initiating supply-side management by: (i) promoting mass-based rainwater harvesting and groundwater recharge programmes and activities; (ii) maximizing surface water use for recharge; and (iii) improving incentives for water conservation and artificial recharge.
5. Undertaking groundwater management in the river basin context. Groundwater interventions often tend to be too 'local' in their approach. Past and forthcoming work in IWMI and elsewhere suggests that like surface water, groundwater resources too need to be planned and managed for maximum basin level efficiency. A rare example where a systematic effort seems to have been made to understand the hydrology and economics of an entire aquifer are the mountain aquifers underlying the West Bank and Israel. The actual equity effects of shared management by Israelis and Palestinians here are open to controversy; however, this offers an early example of issues that crop up in managing trans-boundary aquifers (Feitelson and Haddad, 1998). Equally instructive for the developing world will be the impact of the entry of large corporate players in the business of using aquifers as interyear water storage systems for trading of water.

As groundwater becomes scarce and costlier to use in relative terms, many ideas – such as trans-basin movement or surface water systems exclusively for recharge – that in the yesteryears were discarded as unfeasible or unattractive, will now offer new promise, provided of course that Asia learns intelligently from these ideas and adapts them appropriately to its unique situation.

Conclusion

South Asia has experienced a veritable boom in groundwater irrigation over the last 35 years. This boom is a manifestation of the struggle of the region's peasantry to survive in the midst of inexorable increase in population pressure on farmland. Because small pumps and boreholes have proved one of the most potent land-augmenting technologies, smallholders in India, Bangladesh, Nepal Terai and Pakistan have taken to bore well irrigation with great enthusiasm.

Our analysis suggests that this enthusiasm has proved to be well founded, and that farmland productivity through Green Revolution technology has experienced a quantum jump thanks to the spread of groundwater irrigation. Wells have also brought greater spatial, social and interpersonal equity in access to irrigation, especially when compared to large public canal irrigation systems that have created islands of agrarian prosperity. Indeed, it can be safely said that the groundwater boom has been amongst the best things that have happened for South Asia's rural poor in the past few decades, and the size and dispersion of the livelihood benefits of this boom can arguably outcompete some of the best-known poverty alleviation programmes in the region.

The key concern in South Asia is managing this boom for socio-economic as well as environmental sustainability. Evidence is mounting that this runaway economy is taking its toll on wetlands, lean-season river flows, groundwater levels as well as quality. Evidence is also mounting that, unless effectively regulated, further indiscriminate expansion of bore well irrigation – except in pockets like the eastern Gangetic basin – will undo all the good it is doing to South Asia's poor. The sense of urgency about building effective mechanisms for governing the groundwater economy is already being felt. The challenge for the region's decision makers is to evolve a strategy unique to its peculiarities rather than blindly adopting approaches tried in groundwater economies with a totally different architecture.

Even if South Asia experiments with direct regulation of groundwater abstraction – such as licensing of bore wells, withdrawal permits and water fees – it should not bank on these schemes. It should instead devise a tool kit of *indirect instruments* to regulate overall groundwater abstractions. This requires that water policymakers eschew hydrocentric vision and embrace a broader, strategic view of groundwater governance. It is also important to realize that for a long time to come, the most potent response to groundwater overdevelopment in South Asia would come from effective supply-side interventions. Therefore, South Asia should scale up its commitment of financial and scientific resources to groundwater recharge management to a level commensurate with the high and increasing dependence of the region on groundwater resource.

Notes

1 This is an official Government of India estimate. Independent researchers suggest that the proportion is likely much higher. An IWMI survey of 2629 farmers from 278 villages across India, Pakistan Punjab and Sind, Nepal Terai and Bangladesh showed that groundwater wells serve as sole or complementary sources in serving 75% of irrigated areas in the entire sample; this ratio was higher at 87% for the Indian sample (Shah *et al.*, 2005).

2 For example, Henry Vaux, a senior agricultural economist from the University of California at Davis asserts: 'Persistent groundwater overdraft is self-terminating' (Vaux, personal communication, El Escorial, 2005).

3 This was when oil prices were less than half of their cost in October 2005.
4 It uses a district-wise data-set compiled by Bhalla and Singh (2001) covering 273 districts of India and provides data for the value of 35 agricultural crops at 1990 base year price (in rupees, which has been converted to dollars according to the 1990 rupee/dollar exchange rate) for four decades – 1960s to 1990s. These 35 crops cover more than 90% of the crop output and area cultivated in India. We have worked out productivity figures by dividing the value of these 35 crops (in dollars) by the net cropped area in the district. Bhalla and Singh (2001) span data across 273 districts (1960 base), and include all states except Himachal Pradesh and the North-eastern states.
5 Similar evidence is available from other parts of the world as well (see Hernandez-Mora *et al.*, 1999, for a comparative study in Andalucia, Northern Spain).
6 1$ = Pak Rs. 65 in September 2001.
7 A scholar of the Chinese groundwater degradation problem recently wrote: 'For more than twenty years – since almost immediately after large-scale mechanized groundwater pumping began – Chinese scientists have observed, reported, and warned against the dangers of ground water declines. In 1978, a network of 14,000 observation wells was established in North China. Water levels in every well are measured once every five days. Ground water investigations on all scales, from county to regional levels, and from annual reports to huge research projects involving hundreds of hydrogeologists, have documented water-level declines, and without exception have pointed the finger at over-pumping. Decision-makers in the Land Use Bureau, the Planning Bureau, and the Water Conservation Bureau have been well informed of the problem for years. Official responses have come all the way from the highest level of the Central Government, the State Council, which in 1985 issued 'the principles of determination, calculation, collection and use of water charge for water conservancy works' expressly to address water-shortage problems. Yet, policies continue to encourage unfettered water use. . . . Therefore, the most important question regarding sustainable water use in China is why policy makers ignore the groundwater crisis' (Kendy, 2000).
8 In the Santa Clara Valley south of San Francisco Bay, overdraft was estimated at 52,000 acre feet way back in 1949 when India was still on bullock bailers and Persian wheels. The response to sustained overdraft was for new institutions to be created, such as the Santa Clara Water Conservation District and a water user association. Ten dams were constructed to store flood waters for recharge; barriers of injection wells were created to prevent sea water intrusion; arrangements were made to import 100,000 acre feet of water annually. But, besides these supply-side interventions, there were also measures to restrict the withdrawals through the creation of groundwater zones and the levy of groundwater tax that varied across zones according to the cost of alternative supplies. As a result, in the mid-1980s, the groundwater table stabilized at 30 feet above the historic lowest, and land subsidence became a matter of the past (Coe, 1989).

References

Abderrahman, W.A. (2003) Should intensive use of non-renewable groundwater resources always be rejected? In: Llamas, R. and Custodio, E. (eds) *Intensive Use of Groundwater: Challenges and Opportunities*. Balkema, Rotterdam, The Netherlands, pp. 191–203.

Agarwal, A. (2000) Drought? Try capturing the rain. Occasional Paper. Center for Science and Environment, New Delhi, India, pp. 1–16.

Arriens, W., Bird, J., Berkoff, J. and Mosley, P. (eds) (1996) *Towards Effective Water Policy in the Asian and Pacific Region: Overview of Issues and Recommendations*, Vol. 1. Asian Development Bank, Manila, Philippines.

Barry, E. and Issoufaly, H. (2002) Agrarian diagnosis of Kumbhasan in North Gujarat, India. IWMI–Tata Water Policy Research Program, Anand, India and National Institute of Agricultural Sciences of Paris, Grignon, France.

Bhalla, G.S. and Singh, G. (2001) *Indian Agriculture: Four Decades of Development*. Sage Publications, New Delhi, India.

Briscoe, J. (1999) *Water Resources Management in Yemen: Results of a Consultation*. World Bank (Office Memorandum), Washington, DC.

Calow, R.C., Robins, N.S., Macdonald, A.M., Macdonald, D.M.J., Gibbs, B.R., Orpen, W.R.G., Mtembezeka, P., Andrews, A.J. and Appiah, S.O. (1997) Groundwater management in drought-prone areas of Africa. *Water Resources Development* 13(2), 241–261.

Coe, J.J. (1989) Responses to some of the adverse external effects of groundwater withdrawals in California. In: O'Mara, G. (ed.) *Efficiency in Irrigation: The Conjunctive Use of Surface and Groundwater Resources*. World Bank, Washington DC, pp. 51–57.

Daines, S.R. and Pawar, J.R. (1987) Economic returns to irrigation in India. Report prepared by SRD Research Group Inc., USA, for Agency for International Development Mission to India, New Delhi, India.

DebRoy, A. and Shah, T. (2003) Socio-ecology of groundwater irrigation in India. In: Llamas R. and Custodio, E. (eds) *Intensive Use of Groundwater: Challenges and Opportunities*. Swets & Zetlinger, The Netherlands, pp. 307–336.

Dhawan, B.D. (1989) *Studies in Irrigation and Water Management*. Commonwealth Publishers, New Delhi, India.

Döll, P., Lehner, B. and Kaspar, F. (2002) Global modeling of groundwater recharge. In: Schmitz, G.H. (ed.) *Proceedings of Third International Conference on Water Resources and the Environment Research*, Vol. I. Technical University of Dresden, Germany, pp. 27–31.

Feitelson, E. and Haddad, M. (1998) *Identification of Joint Management Structures for Shared Aquifers: A Cooperative Palestinian–Israeli Effort*. Technical Paper 415, World Bank, Washington, DC.

Food and Agricultural Organization (2003) Aquastat. Available at http://www.fao.org/ag/agl/aglw/aquastat/main/index.stm

GoI (Government of India) (1995) *Groundwater Resources of India*. Central Ground Water Board, Ministry of Water Resources, Government of India, Faridabad, India.

Hernandez-Mora, N., Llamas, R. and Martinez-Cortina, L. (1999) Misconceptions in aquifer overexploitation implications for water policy in Southern Europe. Paper presented at the 3rd Workshop SAGA, Milan, Italy.

INCID (Indian National Committee on Irrigation and Drainage) (1999) Water for food and rural development 2025. Paper presented at the PODIUM Workshop, Central Water Commission, New Delhi, India, 15–16 December 1999.

Kahnert, F. and Levine, G. (eds) (1993) *Groundwater Irrigation and the Rural Poor: Options for Development in the Gangetic Basin*. World Bank, Washington, DC.

Keller, A., Sakthivadivel, R. and Seckler, D. (2000) *Water Scarcity and the Role of Storage in Development*. IWMI Research Report 39, International Water Management Institute, Colombo, Sri Lanka.

Kendy, E. (2000) Countering groundwater depletion: achieving sustainable land and water use in a water-short region. Research Proposal, Cornell University, Ithaca.

Li, Y. (2001) *Groundwater Recharge*. Nanjing Institute of Hydrology and Water Resources, China.

Mainuddin, M. (2002) Groundwater irrigation in Bangladesh: 'Tool for Poverty Alleviation' or 'Cause of Mass Poisoning'? *Proceedings of the Symposiu m on Intensive Use of Groundwater: Challenges and Opportunities*, Valencia, Spain, 10–14 December 2002.

Martinez-Cortina, L. and Hernandez-Mora, N. (2003) The role of groundwater in Spain's water policy. *International Water Resources Association* 28(3), 313–320.

Michael, A.M. (2001) *Irrigation: Theory and Practice.* Vikas Publishing House Pvt. Ltd., New Delhi, India.
Ministry of Agriculture (1988) Report of the National Commission on Agriculture, Government of Pakistan, Islamabad, Pakistan.
Moench, M. (1994) Approaches to groundwater management: to control or enable? *Economic and Political Weekly*, September 24, A135–A146.
Palmer-Jones, R. (1999) Slowdown in Agricultural Growth in Bangladesh: Neither a Good Description Nor a Description Good to Give. In: Rogaly, B. Harriss-White, B. and Bose, S (eds) *Sonar Bangla? Agricultural Growth and Agrarian Change in West Bengal and Bangladesh.* Sage Publications, New Delhi, India.
Postel, S. (1999) *Pillar of Sand: Can the Irrigation Miracle Last?* W.W. Norton, New York.
Qureshi, R.H. and Barrett-Lennard, E.G. (1998) *Saline Agriculture for Irrigated Lands in Pakistan.* Monograph No. 50, Australian Centre for International Agricultural Research, Canberra, Australia.
Qureshi, A.S., Shah, T. and Akhtar, M. (2003) The Groundwater Economy of Pakistan. IWMI Working Paper 64 (Pakistan Country Series No.19), International Water Management Institute, Lahore, Pakistan.
Roy, T. (2004) *The Economic History of India 1857–1947.* Oxford University Press, New Delhi, India.
Sakthivadivel, R., Gomathinayagam, P. and Shah, T. (2004) Rejuvenating irrigation tanks through local institutions. *Economic and Political Weekly* 39(31), 3521–3526.
Satyanarayana, A. (2005) System of rice intensification: an innovative method to produce more with less water and inputs. Paper presented at 4th IWMI–Tata Annual Water Policy Workshop, Anand, India, 24–26 February 2005.
Scott, C. and Shah, T. (2004) Groundwater overdraft reduction through agricultural energy policy: insights from India and Mexico. *Water Resources Development* 20(2), 149–164.
Shah, T. (1993) *Water Markets and Irrigation Development: Political Economy and Practical Policy.* Oxford University Press, Bombay, India.
Shah, T. (2000) Mobilizing social energy against environmental challenge: understanding the groundwater recharge movement in Western India. *Natural Resources Forum* 24, 197–209.
Shah, T. (2005) Groundwater and Human Development: Challenges and Opportunities in Livelihoods and Environment. *Water, Science & Technology,* 51(8), 27–37.
Shah, T. and Desai, R. (2002) Decentralized water harvesting and groundwater recharge: can these save Saurashtra and Kutch from desiccation?" Paper presented at 1st IWMI–Tata Annual Water Policy Workshop, Anand, India, 13–14 February 2002.
Shah, T., DebRoy, A., Qureshi, A.S. and Wang, J. (2003) Sustaining Asia's groundwater boom: an overview of issues and evidence. *Natural Resources Forum* 27, 130–140.
Shah, T., Giordano, M and Wang, J. (2004a) Irrigation institutions in a dynamic economy: what is China doing differently from India? *Economic and Political Weekly* 39(31), 3452–3461.
Shah, T., Scott, C. and Buechler, S. (2004b) Water sector reforms in Mexico: lessons for India's new water policy. *Economic and Political Weekly* 39(4), 361–370.
Shah, T., Scott, C., Kishore, A. and Sharma, A. (2004c) *Energy-Irrigation Nexus in South Asia: Improving Groundwater Conservation and Power Sector Viability.* IWMI Research Report 39, International Water Management Institute, Colombo, Sri Lanka.
Shah, T., Makin, I. and Sakthivadivel. R. (2005) Limits to leapfrogging: issues in transposing successful river basin management institutions in the developing world. In: Svendsen, M. (ed.) *Irrigation and River Basin Management: Options for Governance and Institutions.* CAB International, Wallingford, UK, pp. 31–49.
Shah, T., Singh, O.P. and Mukherji, A. (2006) Groundwater irrigation and South Asian agriculture: empirical analysis from a large-scale survey of India, Pakistan, Nepal and Bangladesh. *Hydrogeology Journal* 14(3), 286–309.

Sharma, S.K. and Mehta, M. (2002) Groundwater development scenario: management issues and options in India. Paper presented at the IWMI–ICAR–Colombo Plan sponsored Policy Dialogue on 'Forward-Thinking Policies for Groundwater Management: Energy, Water Resources, and Economic Approaches' at India International Center, New Delhi, India, 2–6 September 2002.

Sinha, S.K. and Talati, J. (2005) Impact of system of rice intensification (SRI) on rice yields: results of a new sample study in Purulia district, India. Paper presented at 4th IWMI–Tata Annual Water Policy Workshop, Anand, India, 24–26 February 2005.

Wester, P., Marañón Pimentel, B. and Scott, C.A. (1999) Institutional responses to groundwater depletion: the aquifer management councils in the state of Guanajuato, Mexico. International Water Management Institute, Colombo, Sri Lanka.

Whitcombe, E. (1984) *Irrigation*. In: Kumar, D. (ed.) *The Cambridge Economic History of India, C 1757–C 1970*, Vol. II, Orient Longman, New Delhi, India.

World Bank and Ministry of Water Resources (1998) *India-Water Resources Management Sector Review, Groundwater Regulation and Management Report*. Washington, DC and Government of India, New Delhi, India.

3 The Development, Challenges and Management of Groundwater in Rural China

Jinxia Wang[1], Jikun Huang[1], Amelia Blanke[2], Qiuqiong Huang[2] and Scott Rozelle[2]

[1]*Center for Chinese Agricultural Policy, Institute of Geographical Sciences and Natural Resources Research, Chinese Academy of Sciences, Jia 11 Datun Road, Anwai, Beijing 100101, PR China;* [2]*Department of Agricultural and Resource Economics, University of California, Davis, CA 95616, USA*

Introduction

The history of groundwater in China is one of extremes, or apparent extremes. Before the 1960s, the story was one of neglect; only a small fraction of China's water supply came from groundwater (Nickum, 1988). Almost none of the Ministry of Water Resource's investment funds were allocated to the groundwater sector until the late 1960s. Certainly, to the extent that underground water resources were valuable, China was ignoring a valuable resource.

Since the mid-1970s, however, the prominence of the groundwater sector has risen dramatically. Over the last 30 years, agricultural producers, factory managers and city officials – far from ignoring groundwater resources – have entered an era of exploitation (Smil, 1993; Brown and Halweil, 1998). Arguably, there have been more tube wells sunk in China over the last quarter century than anywhere else in the world. As a share of total water supply, groundwater has risen from a negligible amount across most of China to being a primary source of water for agriculture, industry and domestic use in many of the nation's most productive regions. Unfortunately, the resulting fall in groundwater tables has been one of China's most serious environmental problems (World Bank, 1997).

Despite the rise in importance of the sector, and the threats to its continuation, relatively little systematic information is available about many key aspects of China's groundwater economy in rural areas. That is not to say that there is a shortage of scientific research studies that document some of China's groundwater-related problems, for example, land subsidence, salt water intrusion and overdrafts (Chen *et al.*, 2003; Sakura *et al.*, 2003). Moreover, there is recent work on groundwater usage and quality in China's cities (Tang, 1999). However, with the exception of a number of general summary pieces that are based primarily on anecdotes and secondary citations (e.g. Nickum, 1988, 1998; Lohmar *et al.*, 2003) and papers that look at groundwater use in relatively isolated agricultural areas (e.g. Kendy *et al.*, 2004; Wang *et al.*, 2005, 2006), there is little work based on original data that is sufficiently broad in scope to give the reader a general overview of the groundwater economy and its challenges and management, especially in rural areas.

The primary goal of this chapter is to overcome the absence of research on China's groundwater economy. We will pursue three specific objectives. First, we will characterize China's groundwater resources, briefly reviewing the main physical and geographic properties of northern China's groundwater resource development, describing the role of groundwater in the economy and examining the technology that producers are using to extract and utilize the resource. Second, we will examine the main problems that the sector is facing, including falling groundwater levels and deteriorating water quality. Finally, we will document responses of the major water stakeholders in China's agricultural sector – the government and agricultural producers – focusing primarily on the emergence of institutions as a response to some of these problems. Our findings draw primarily on two data-sets that we collected ourselves, covering nearly 450 communities in northern China.

Due to the broad nature of the issues dealt with in this chapter, we will narrow the scope of our analysis in several ways. First, we will limit our examination to northern China, the region that uses the majority of China's groundwater. In our study, northern China can be thought to include the following regions: north China (*huabei*), north-east China (*dongbei*) and north-west China (*xibei*). Our sample communities also represent all or part of four major river basins: the Hai River basin, the lower and middle reaches of the Yellow River basin, the northern bank of the Huai River basin and the Songliao River basin in the north-east. Although we use our data to extrapolate to the entire northern China region (12 provinces and 2 municipalities), most of our data come from six provinces – Liaoning, Hebei, Henan, Shanxi, Inner Mongolia and Shaanxi.[1] Since agriculture is the main water-using sector (68% in 2001; Ministry of Water Resources, 2002), our data-set was collected from rural communities, and we will focus on the use of water for agriculture.[2] We also explicitly exclude all wells in our sample villages that are used solely for drinking water.

Data

In addition to national statistics, our analysis is based on data that we collected as part of two recent surveys specifically designed to address irrigation practices and agricultural water management. The China Water Institutions and Management (CWIM) survey of September 2004 was the second round of a panel survey, the first phase of which was conducted in 2001 (Fig. 3.1). Enumerators conducted surveys of community leaders, groundwater managers, surface water irrigation managers and households in 48 villages in Hebei and Henan provinces. The villages were chosen according to geographic properties. In Hebei, villages were chosen from counties near the coast, near the mountains and in the central region between the mountains and the coast. In

Fig. 3.1. Fourteen counties surveyed in China in September 2004. (From China Water Institutions and Management (CWIM) survey.)

Henan, villages were chosen from counties bordering the Yellow River and from counties in irrigation districts at varying distances from the Yellow River.

We conducted a second survey, the North China Water Resource Survey (NCWRS), in December 2004 and January 2005 (Fig. 3.2). This survey of village leaders from 400 villages in Inner Mongolia, Hebei, Henan, Liaoning, Shaanxi and Shanxi provinces used an extended version of the community-level village instrument of the CWIM survey. Using a stratified random sampling strategy for the purpose of generating a sample representative of northern China, we first sorted counties in each of our regionally representative sample provinces into one of four water scarcity categories: very scarce, somewhat scarce, normal and absolutely scarce (mountain/desert).[3] We randomly selected two townships within each county and four villages within each township. In total, combining the CWIM and NCWRS surveys, we visited approximately 6 provinces, 60 counties, 126 townships and 448 villages.

The scope of the surveys was quite broad. Each of the survey questionnaires included more than ten sections. Among the sections, there were those that focused on the nature of China's rural water resources, the common types of well and pumping technology. There also were several sections that examined the most important water problems, government water policies and regulations as well as a number of institutional responses (e.g. tube well privatization). Although sections of the survey covered both surface and groundwater resources, we will focus mostly on those villages that have groundwater

Fig. 3.2. Fifty counties surveyed in December 2004 and January 2005. (From North China Water Resource Survey (NCWRS).)

resources (in some cases, even if they were not being used). The survey collected data on many variables for 2 years – 1995 and 2004 – asking about conditions back in time. By weighting our descriptive and multivariate analysis with a set of population weights, we were able to generate point estimates for all of northern China.[4]

China's Groundwater Resources

While China's water resources are substantial compared with those of many other countries, its population is even larger, and its water resources are not evenly distributed across the country or across important agricultural regions. China ranks fifth in total water resources among the countries of the world. On a per capita basis, however, its water resource availability is among the lowest. Moreover, the nation's water resources are overwhelmingly concentrated in southern China; northern China has only approximately 25% of the water endowment of the south and 10% of the world average (Ministry of Water Resources, 2000). The lower levels of rainfall in northern China are also much more seasonal than in the south, with more than 70% of the rain occurring between June and September. Northern China, however, remains an important agricultural region and the site for much of China's industrial production. Although it has only 24% of the nation's water resources, northern China contains more than 65% of China's cultivated land and produces roughly half of its grain (nearly all of its wheat and maize) and more than 45% of the nation's gross domestic product (China National Statistical Bureau, 2000; Ministry of Water Resources, 2000).

Groundwater resources in China are both unevenly distributed and unevenly used across regions. According to the latest estimates generated by the Ministry of Land Resources, the annual natural recharge of fresh groundwater resources in China is 884 billion cubic meters, about one-third of the nation's total water resources (Ministry of Land Resources, 2005). Of this, about 70% are in southern China, and only about 30% in northern China. However, the intensity of groundwater use occurs in a very different pattern. Of all the known groundwater resources, rural and urban users are using more than 70% in northern China, whereas less than 30% of the known groundwater resources in southern China are being used.

Despite the fact that most groundwater resources are located in southern China, it is fortunate that they exist across wide expanses of northern China's river basins and that these resources are relatively abundant and accessible. Alluvial deposits consisting primarily of sand, loess silt and clay extend to a depth of more than 500 m below the surface in some areas (Kendy *et al.*, 2003). These deposits comprise the aquifers that supply groundwater to regions in all major river basins in the North China Plain. The aquifers, however, vary greatly across northern China. For example, in the North China Plain, unlike the south where villages in mountainous areas can tap groundwater resources, mountainous areas are often groundwater-deficient.[5] In the flat plains, the aquifers are multilayered. The multilayered aquifers in the Hai River basin (NCP) typically have 2–5 layers; the first and third layers are the most water-abundant. The first layer is typically an unconfined aquifer made up of large-grained homogeneous sand and gravel. The other layers are typically confined aquifers. In some areas, especially in the eastern parts of the Hai River basin, there is a naturally occurring saline layer. Created during a previous ice age, the second layer often contains saline water, is confined and has a salt content high enough that it is typically unusable for agriculture without treatment.

Groundwater resources from the farmer's perspective

To obtain an understanding of how farmers view their water resources, we asked village leader respondents to describe the nature of the aquifers that are under their villages. Because most village leaders have not been a part of any hydrogeological surveys, they often were not able to answer questions concerning the existence, size or other geological properties of the aquifers. Instead, village leaders knew more precisely how many shallow and deep wells there were in their village as well as the depths of those wells. Although there is no complete correlation between the depth of the wells and the nature of the aquifer, in many cases, the existence of shallow or deep wells coincides with that of shallow or deep layers of village aquifers. Regardless of their exact hydrogeological properties, according to our data (and the perception of village leaders), 'deep wells' are almost always having a depth of at least 60 m. If a village needs to drill through an aquitard (a clay layer in most cases) to sink a well, the well is always defined as a 'deep well'. Shallow wells, in contrast, are mostly less than 60 m and do not penetrate an aquitard.

Whether deep or shallow, groundwater resources are extensive across regions of northern China. We asked village leaders if there were groundwater resources in the village. Most replied that there were, and the share of villages having these was almost 95% in 2004. However, not all villages having groundwater use this resource for irrigation. In 2004, more than 15% of irrigated villages with groundwater did not use it for irrigation. We further explore the reason behind this. According to the village leader respondents, there were two major reasons for not using it. Research results show that in 2004, the most important reason was that there were cheap and sufficient surface water resources (51% of villages). The second important reason was that there was no money to dig tube wells (37% of villages). Such findings suggest that there still may be potential to use greater volumes of groundwater resources in the future. With increasing water scarcity and rising water demands, more villages have begun to use their community's groundwater. For example, from 1995 to 2004, the share of villages using groundwater resources *for the first time* had increased by almost 12%.

Relying on the observations of our NCWRS respondents, one of our most prominent findings was the great diversity of aquifer development in northern China. Of the 238 sample villages that used groundwater for irrigation in 2004, 33% told us that they extracted groundwater only from shallow aquifers, 42% only from deep aquifers and the remaining 25% from both. Our data show that in some villages in northern China, the groundwater supply from shallow aquifers is sufficient to support current local water demand for irrigation. In other villages, maybe due to exhausted or unusable shallow aquifers, farmers extract groundwater only from deep aquifers.[6] In some villages (25%), both shallow and deep aquifers are being used. The groundwater supply from shallow unconfined aquifers is highly dependent upon precipitation, which supplies groundwater recharge. When rainfall is above average, as it was in 2004, water levels increase in shallow aquifers due to above-average recharge. This may be the reason that more villages extracted groundwater from shallow aquifers in 2004 than in 1995.[7]

According to our respondents, the depth to water also varied across northern China. Although the average depth to water in 2004 was 26 m, it varied sharply across our sample villages (Fig. 3.3). In fact, in most villages depth to water was fairly shallow. In 2004, the average depth to water for the villages from the shallowest quartile of villages was only 4 m and that for the second quartile was only 9 m. Villages in the third quartile were pumping from an average depth to water of more than 30 m. In only 4% of groundwater villages were villagers pumping from more than 100 m.

The contribution of China's groundwater

After the emergence of the tube well, and diesel and electric pumping technology, the role of groundwater rapidly grew in importance for all uses (Wang *et al.*,

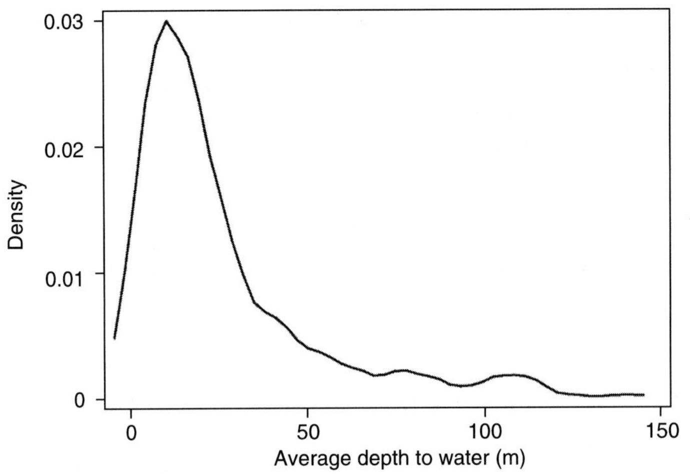

Fig. 3.3. Average depth to water in 2004. (From authors' survey in 2004.)

2005). In total, the use of groundwater rose from almost zero in the 1950s to 57 billion cubic meters annually in the 1970s. After the rural economic reforms in the late 1970s and early 1980s (which, among other things, shifted income and control rights from the collective to the individual household), groundwater use continued to rise, reaching 75 billion cubic meters in the 1980s and more than 100 billion cubic meters after 2000 (Ministry of Land Resources, 2005).[8] As the use of groundwater rose nationally, its share in the nation's water supply also rose (from almost nothing in the 1950s to a major fraction now). Across China, groundwater currently supplies about 20% of China's water. However, this amount is unevenly distributed. In southern China, groundwater comprises approximately 14% of water supply; in northern China it supplies 49%. From this point of view, the rise of the accessibility of groundwater has certainly played an important role in the emergence of northern China's regional economy.

Although the importance of groundwater has risen for all uses, it is likely that, as with water resources in general, groundwater resources are being increasingly allocated for non-agricultural uses.[9] Unfortunately, China does not systematically collect data on water allocation to economic uses by type of water resource. As a consequence, all we know are the shares of total water resources that are going for domestic, industrial and agricultural uses. Since much of the increase in water use over the last 20 years has come from groundwater, we believe it is safe to assume that the share of groundwater being allocated to domestic and industrial uses follows a somewhat similar pattern to that followed by water resource use in general. Although in 1978, only 1% of China's water use was allocated for domestic use, by 2002 about 11% of water went for domestic users (Table 3.1). The use of water for industry also rose from 14% in 1978 to more than 21% in 2002. Although the share of water used in agriculture has fallen (from 85% in 1978 to 68% in 2002), it is still the largest water user.

Table 3.1. Uses of water resources in China, 1978 to 2002. (From Ministry of Water Resources, 2002.)

	Domestic (%)	Industrial (%)	Agriculture (%)
Total water resources			
1978	1	14	85
1997	10	21	69
2002	11	21	68
Groundwater resources			
1997	26	20	54

Irrigation and the role of groundwater

Our data also demonstrate the importance of groundwater in supplying irrigation to northern China's agricultural sector. According to the respondents, nearly half (49%) of China's cultivated area is irrigated (slightly higher than the figure given in China National Statistical Bureau, 2004 – 42%). However, with our data we can understand the water economy more fully, since our survey covers more than what is available in official sources. For example, since national statistics do not collect irrigation data by type of irrigation water, we asked village leaders to carefully document the source of their irrigated area: either surface, groundwater or conjunctive use of both. On the basis of their responses, in 1995, of all of the cultivated land that is irrigated, only 40% came from surface water diversions (or was lifted from canals by pumps onto the fields). The remaining 60% came from groundwater sources. Between 1995 and 2004 the importance of groundwater has continued to grow. In 2004, 68% of irrigation in northern China was from groundwater.

Crop-specific incidences of irrigated area

Our data can also produce estimates of crop-specific sown area statistics by irrigated and non-irrigated portions. For example, major food grains in northern China are mostly irrigated (Table 3.2). Approximately 96% of rice and 80% of wheat are irrigated, levels that are above the national average (Table 3.2, column 1, rows 1 and 2). Hence, our data support the findings of Huang *et al.* (2006) that investment in irrigation has been central for China to maintain food security. Although it is well known that China's food crops are heavily irrigated and that this is an important factor in China being able to produce a large fraction of its own food, these crop-specific estimates are important because China's own statistical bureau does not report sown area by irrigated and non-irrigated portions.

In contrast to the case of food grains, a majority of feed grains and lower-valued staple crop area is not irrigated (Table 3.2, column 1, rows 3, 5 and 6). For example, despite its growing importance in China's agricultural economy, only 49% of China's maize is irrigated.[10] An even lower proportion of coarse grains and potatoes (including white and sweet potatoes) is irrigated. Although the proportion of irrigated area in cash crops also varies by crop, much of the

Table 3.2. Share of irrigated sown area by crop type in north China. (From authors' survey in 2004.)

Crop	Percent of cropland that is irrigated	Percent of irrigated sown area	
		Surface water	Groundwater
Rice	96	76	24
Wheat	80	28	72
Maize	49	30	70
Cotton	58	30	70
Potato	22	27	73
Soybean	24	32	67
Oil crops	47	38	62
Field vegetables	66	33	67

area of China's main cash crops is irrigated (e.g. 58% of cotton area, 47% of oil crop area and 66% of field vegetable area – Table 3.2, rows 4, 6 and 7).

Perhaps more importantly, in northern China irrigation for most crops mainly depends on groundwater resources (Table 3.2, column 3). For grains and other staple crops, except for rice, at least 70% of the producers in irrigated areas use groundwater resources (72% for wheat; 70% for maize; 73% for potatoes). For cash crops, groundwater is the major source of water for irrigation. For example, groundwater irrigates 70% of cotton area, 62% of oil crop area and 67% of field vegetable area.

Developing China's Groundwater

While the development of China's surface water resources has a long history and has played an important role in its growth as a state, the development of most groundwater resources has been compressed into less than 50 years. In this section, we briefly examine the way in which China has developed its groundwater resources by first describing the trends over the last 50 years in the installation of tube wells and pumps, focusing on the path of this development over time and across space. Because we have more detailed data from the last decade, much of the discussion will focus on the recent period. The second part of this section briefly introduces the technology that is being used.

The rise of tube wells

According to national statistics, the installation of tube wells began in the late 1950s and, although the number of wells has grown continuously, the pace of increase has varied from decade to decade (Ministry of Water Resources and Nanjing Water Institute, 2004). During the 1950s, the first pumps were introduced to China's agricultural sector. Although still fairly limited, the growth rate was fast. During the Great Leap Forward (the late 1950s and early 1960s), however, statistical reporting was suspicious and many irrigation projects that were started during the period were badly engineered and often abandoned. After the recovery from the Great Leap Forward and the famine that followed, statistical agencies recovered, and statistical series since the mid-1960s are relatively consistent.

Since the mid-1960s, the installation and expansion of tube wells across China has been nothing less than phenomenal. In 1965, it was reported that there were only 150,000 tube wells in all of China (Shi, 2000). Since then, the number has grown steadily. By the late 1970s, there were more than 2.3 million tube wells. After stagnating during the early 1980s, a time when irrigated area, especially that serviced by surface water, fell, the number of tube wells continued to rise. By 1997, there were more than 3.5 million tube wells; by 2003, the number rose to 4.7 million.

The path of tube well expansion shown in the official data is largely supported by the information we have from the NCWRS. During the survey we asked the village leaders to tell us about the initial year in which someone (either the village leadership or an individual farmer) in their village sank a tube well (Fig. 3.4). According to the data, we found that by 1960, less than 6% of villages had sunk their first tube well. Over the next 20 years, between the early 1960s and the onset of reform, the number of villages with tube wells rose to more than 50%. During the next 10 years, between 1982 and 1992, the number of villages with tube wells rose by only 7%. After the early 1990s, however, the pace of the expansion of groundwater accelerated, and by 2004 almost 75% of villages had wells and thus access to groundwater.

While the growth of tube wells reported by the official statistical system is impressive, we have reason to believe the numbers are far understated. According to the NCWRS, on average, each village in northern China contained 35 wells in 1995. When extrapolated regionally, this means that there

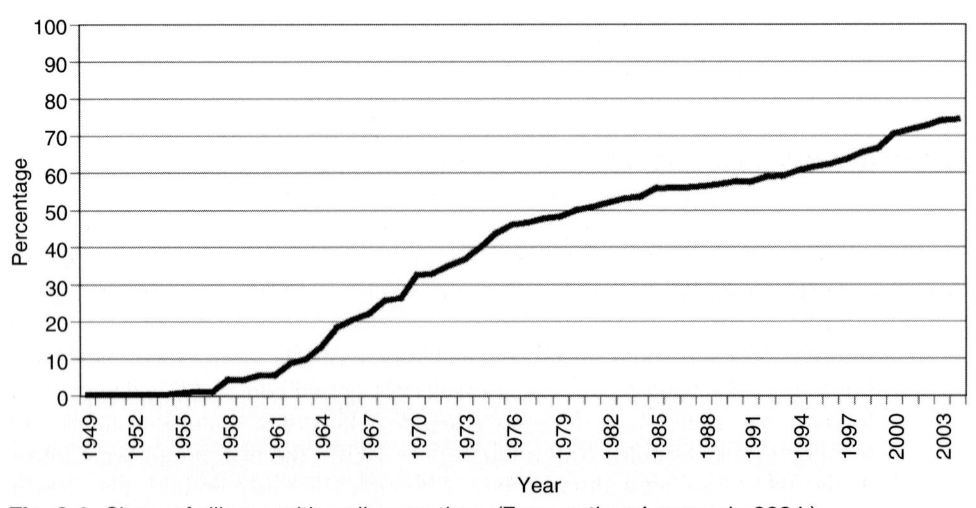

Fig. 3.4. Share of villages with wells over time. (From authors' survey in 2004.)

were more than 3.5 million tube wells in the 14 provinces in northern China by 1995. According to our data, there has been a rapid growth of wells.[11] By 2004, the average village in northern China contained 70 wells, suggesting that the rise in tube well construction since the mid-1990s has been even faster than indicated by official statistics. By 2004, we estimate that there were more than 7.6 million tube wells in northern China. At least in our sample villages, the number of tube wells has grown by more than 12% annually between 1995 and 2004.

The technology that pumps China's water

As China's groundwater usage has expanded, the characteristics of pumps used for shallow and deep wells have also evolved. In 1995, the average size of the pump used on a shallow well was 3 in., drawing 6.9 kW of power with a lift of 28 m. The average shallow well pump discharged about 32 m^3/h. By 2004, the average pump increased marginally in size (to 3.1 in.), power (to 7.2 kW), lift (31 m) and discharge (to 37.5 m^3/h).

The rate of change of deep well pumps was greater than that of shallow ones. In 1995, the average deep pump was 3.9 in. in diameter, drew 13.5 kW of power, had a lift of 53 m and discharged 61.2 m^3/h. By 2004, both power and lift had increased (to 14.1 and 58, respectively). Both diameter and discharge, however, decreased slightly (to 3.7 and 60.9, respectively).[12]

The evolution of pump technology was mostly being driven by new technologies that were coming on the market and the demand for more powerful pumps. When we asked villages if they had changed the pumping technology type between 1995 and 2004, more than one quarter of villages using groundwater in 2004 responded that they had. Interestingly, although pumps are generally getting bigger and more powerful, they are not necessarily increasing in price. In fact, there is evidence that the price of pumps in China is falling. While we cannot pinpoint the reason why, it is likely due to an increase in economy of scale over the last decade. Our data show a general trend in purchasing location from government (state-owned) to private pump dealers.

Groundwater Problems and Challenges

As with most periods of rapid economic growth and intensive resource use, many problems arise. In the case of northern China, however, because of the importance of water, much attention has been focused on the sector's problems (Smil, 1993; Brown and Halweil, 1998; Ministry of Water Resources and Nanjing Water Institute, 2004). In fact, we believe there are many misperceptions about the nature of China's water problems – especially as they relate to the rural economy. In many cases, problems, although serious regionally, are not national in scope. Other problems are often confined to urban or rural areas, but not both. Of course, most of the misperceptions are not intentional, but a result of poor information. The goal of this section is to try to provide a brief assessment of the main problems facing China's groundwater economy. Given the fact that most of the work in the past has had an urban focus, our work centers on those problems affecting the rural sector.

Overdrafting China's groundwater resources

According to a comprehensive survey completed by the Ministry of Water Resource in 1996, the overdraft of groundwater was one of China's most serious resource problems (Ministry of Water Resources and Nanjing Water Institute, 2004). Although we do not know the exact way in which the survey was conducted, the results of the survey provide evidence that groundwater overdraft is a widespread problem and may be getting worse. According to the report, overdraft is occurring in more than 164 locations and affects more than 180,000 km^2. The areas of overdraft range from 10–20 km^2 to more than 10,000 km^2, and are in 24 of China's 31 provinces. Groundwater overdraft is affecting all types of aquifers: the shallow groundwater table (87,000 km^2), the deep groundwater table (74,000 km^2) and the aquifers that have two layers, both the shallow and the deep (13,000 km^2).[13] Since the 1980s, the annual overdraft of groundwater has averaged about 7.1 billion cubic meters. In the late 1990s, the annual rate of overdraft exceeded 9 billion cubic meters. More than one-third of the volume of overdraft is from deep wells, many of which may be non-renewable on a short timescale.

Although the problem of overdraft is usually discussed in general, it appears to be particularly acute in cities. The Ministry of Land Resources has recently finished an evaluation of groundwater resources in China (Ministry of Land Resources, 2005). According to the final report, groundwater resources in most large and middle-sized cities in northern China are either in overdraft (extractions exceed recharge) or in serious overdraft conditions (the fall of the groundwater table exceeds 1.5 m/year).[14] For example, in many cities the volume of water extracted from the aquifer is nearly double the volume of average annual recharge.[15]

Such dramatic numbers for all of China, especially for urban areas, are the cause of the concern that has appeared in the literature. However, when analysing the effect on rural areas, at least according to NCWRS data, a somewhat different picture arises. According to our data, there was no fall in the groundwater table in 25–33% of villages in northern China using groundwater in both 1995 and 2004.[16] In 8.5–16% of villages (between one-third and one-half of villages that reported no fall in the groundwater table) respondents told the enumerators that the groundwater was actually higher in 2004 than in 1995. In another 10–17% of villages, the average annual fall in the groundwater table was less than 0.25 m/year. In other words, in more than one-third to one-half of China's villages using groundwater over the last decade, groundwater resources have shown little or no decline since the mid-1990s. Although, (based on our data, most villages are in or nearly in balance) we are not arguing that groundwater problems do not exist. In fact, there are still a large number of villages in which the water table is falling. Before classifying these villages as being irrational groundwater resource exploiters (although some of them may be), it is important to remember that a village's water resources may not be over-exploited even if the water table is falling. Given the fact that many of China's aquifers are fossil, by definition, any meaningful extraction will result in declining water levels. Hence, even under the most rationally planned groundwater utilization strategy, there will be a share of villages in China in which we should expect the water table to be falling. In addition, if we follow the Ministry of Water Resources (MWR) definition of serious overdraft, only 10% of villages using groundwater in the last decade have water tables that are falling at a rate faster than 1.5 m/year. Such a decline rate is not only serious, but also a crisis.

In summary, then, the point we want to make is that in many places – indeed, in most places in northern China – it is possible that water resources are not being misused. However, we do not want to minimize the problems that are occurring in some places. There are a large number of rural areas in which the water table appears to be falling at a dangerously fast pace. Where the resource is being misused, steps will be required to protect the long-term value and use of the resource. However, it is important to realize that many of the required measures (discussed in the next section) will have associated costs – to obtain adoption and productivity, and to avoid reduced income. Because measures to counter overdraft are not needed in all villages, leaders should not take a one-size-fits-all approach so that they can avoid inflicting unnecessary costs on producers in communities where overdraft conditions do not exist.

Subsequent effects of overdraft

As the groundwater table falls, producers face a number of impacts; above all, of course, the cost of pumping rises. According to our data, for every meter by which the groundwater table falls, pump costs rise by 0.005 yuan/m^3 (or about 2% of the mean level of pumping costs in 2004). In addition, wells may have to be replaced and the costs of investment increased, although in many cases new wells have been sunk for reasons other than the falling water table. The average cost of drilling a deep tube well (90 m) was more than five times the cost of drilling a shallow tube well (37 m). According to our data, well owners in China have sunk an enormous number of new wells in the last decade. On average, from 2002 to 2004, the typical groundwater-using village sank about 22 new wells, 5 deep and 17 shallow. Although a percentage of new wells (6 of the 22) were being installed because olds wells were abandoned, it should be noted that, according to the opinions of our respondents, only 2 of the 6 wells were abandoned because of the falling water table. In many cases, wells were replaced for other reasons (e.g. when a well structure collapsed).

Beyond the increases in pumping costs and well installation, there are also a number of other potential consequences of overdraft (Ministry of Water Resources and Nanjing Water Institute, 2004). One of the most commonly cited consequences is land subsidence. For example, in Hebei province alone, by 1995 more than 5000 km^2 had subsided more than 600 mm. In Tianjin municipality, the total exceeded 7000 km^2.[17] Groundwater overdraft may also lead to the intrusion of seawater into freshwater aquifers (Ministry of Water Resources and Nanjing Water Institute, 2004). By the mid-1990s overdrafting allowed sea water to intrude and contaminate aquifers under more than 1500 km^2 of land, especially in the coastal provinces of northern China such as Liaoning, Hebei and Shandong. The MWR has also been concerned about the impact of groundwater overdraft on desertification and depletion of stream flow that was previously supplied by natural groundwater discharge.

Although the consequences of overdraft are widely discussed in the literature and equated with China's water problems in general, interestingly none of these problems appears to be in any way associated with rural areas. According to our survey of more than 400 villages, no village leader ever reported that there was any land subsidence problem. Likewise, in no case did a village leader report that his or her village's groundwater was contaminated by sea water intrusion. Finally, there also was no evidence that villages that were using groundwater – both those that were drawing down their water table and those that were not – experienced a fall in cultivated area due to desertification. Clearly, although the attention that these problems get in the literature means that they are serious and require addressing, there appear to be no rural area problems.

Other Problems with Groundwater

Groundwater pollution

Both the literature on groundwater and our survey report a number of other problems that are not directly related to groundwater overdraft. For example, it has been widely reported in the press and in academic journals (e.g. Kendy *et al.*, 2003) that pollution from municipal sewage has contaminated the groundwater of many villages in China. Part of the problem is created when farmers pump from effluent canals, using sewage-laced water on their fields. The recharge from irrigation with such water can affect the entire aquifer. Even when villages do not use the water for irrigation purposes, recharge from streams and riverbeds can contribute to groundwater pollution. According to the Ministry of Water Resources and Nanjing Water Resources Institute (2004), the groundwater resources of more than 60% of the 118 largest cities in China have contaminated groundwater.

Drawing on our survey (in which we asked leaders about their perception of pollution), we find that the scope of the problem is somewhat less and the main source of pollution is different than those reported in other sources; interviews with leaders in communities suffering from contaminated groundwater demonstrate that pollution is still a serious problem. According to our sample communities, the groundwater is polluted in 5.40% of the villages. However, unlike the villages around cities, which are mainly being affected by municipal sewage waste, respondents identified industrial pollution and runoff from mining operations as the most common source of pollution. In fact, of all the villages that reported contaminated groundwater, 95% (5.15% of the total number of villages) said that the main source of pollution was from industrial and mining waste water. Only 0.25% of all villages (or less than 5% of villages that report contamination) said that their groundwater was polluted by agricultural chemicals; none said it was due to urban sewage.

While the extent of the perception of rural groundwater pollution problem appears to be less serious than the urban/suburban problem, it is still serious. Extrapolating our results to all of northern China, we can estimate that more than 20 million rural residents living in 20,000 rural villages are using groundwater that has been contaminated by industrial runoff. Moreover, unlike their urban and suburban counterparts, most villages in China lack any type of drinking water processing facilities. In most cases, the pollution causing the problems in one rural community was created by the actions of industrial and mining facilities that belonged to some other community or economic agent. There is no clear advocate to force upstream communities either to stop polluting or to compensate downstream communities for the damage. Moreover, there is little funding for rural groundwater pollution abatement. In short, there is no incentive or means to address and/or curtail the activities that are polluting the groundwater of millions of rural communities.

Soil salinization

Across China, the appearance of salinized soil has been a widespread problem but, according to a number of sources, this problem has been improving in recent years, unlike many others. According to the Ministry of Water Resources and Nanjing Water Resources Institute (2004), more than 1 million square kilometers of China's land has become salinized over the past several decades. The majority of the most serious problems has occurred in the north-east, the north-west and in some places in the North China Plain. Despite the widespread nature of the problem, in recent years, the area affected by salinization has fallen. Ironically, it may be that the same forces diverting surface water away from agriculture and forcing producers to rely increasingly on groundwater may be the primary cause of such improvements. Without access to cheap and abundant surface water, which led to the salinized soil problem, the problem has gradually disappeared as farmers have turned to groundwater and the water table has fallen (Nickum, 1988).[18]

In our sample of villages, we find that the salinization of the soil is one of the most commonly reported problems, although, consistent with national statistics, it is improving over time. According to our respondents, in 2004, 16% of villages reported having some salinized soils. Since the process that caused the soil salinization does not affect all cultivated areas in a village, only 3.4% of cultivated area was reported to be affected. Moreover, the scope of soil salinization is improving over time. In 1995, 20% of villages reported salinized soils and 4.4% of cultivated area was affected. Hence, between 1995 and 2004, there was nearly a 25% reduction in the severity of the nation's soil salinization problem.

Managing China's Groundwater

In this section we first examine the response – or more accurately, the lack of response – of the government to groundwater problems. We then track the response of producers – those at the community and individual household levels. As we will see, in contrast to officials, producers have responded sharply in many different ways.

Regulating (or not) China's groundwater: the role of the government

Over the last 50 years, China has constructed a vast and complex bureaucracy to manage its water resources. To understand the functioning of this system, it is important to first understand that, until recently, neither groundwater use nor water conservation has ever been of major concern to policymakers. Instead, the system was designed to construct and manage surface water to prevent floods, which have historically devastated the areas surrounding major rivers, and to effectively divert and exploit water resources for agricultural and industrial development. Historically, when attention was paid to water conservation, the emphasis was on surface water canal networks. Therefore, many of the most severe groundwater problems have not been directly addressed.

Laws and measures

Water policy is ultimately created and theoretically executed by the MWR. The MWR has run most aspects of water management since China's first comprehensive Water Law was enacted in 1988, taking over the duties from its predecessor, the Ministry of Water Resources and Electrical Power. The policy role of the MWR is to create and implement national price and allocation policy, and to oversee water conservancy investments by providing technical guidance and issuing laws and regulations to the subnational agencies (Lohmar *et al.*, 2003).

In fact, officials in the MWR and in other ministries have spent time and effort in passing laws and regulations concerning groundwater management in rural areas. For example, according to China's national 1988 Water Law, the property rights of all underground water resources belong to the state. This means that the rights to use, sell and/or charge for water ultimately rest with the government. The law does not allow extraction if the pumping of groundwater is harmful to the long-term sustainability of the use of the resource.

Beyond formal laws, there have also been many policy measures set up in part to rationally manage use of the nation's resources. In most provinces, prefectures and counties there are formal regulations controlling the right to drill tube wells, the spacing of wells and the price of water when sold. The national government has also set up the necessary regulatory apparatus to allow for the charging of a water extraction fee (surface water and groundwater in urban areas).

Despite the plethora of laws and policy measures that have been created by officials, there has not been an equal effort put out in implementing them. Certainly, part of the problem is one of historic neglect. In fact, the delegation of groundwater management at the ministerial level is still relatively small. There are far fewer officials working on this division than in other divisions, such as flood control, managing surface water systems and water transfer. Moreover, unlike the case of surface water management (Lohmar *et al.*, 2003), there has been no effort to bring management of aquifers that span jurisdictional boundaries under the ultimate control of an authority that can control the government and private entities that use water extracted from different parts of the aquifer. According to Negri (1989), when there is no single body controlling the entire resource, it becomes difficult to implement policies that attempt to manage the resource in a long-term, sustainable, more optimal manner.

Whether for lack of personnel or other difficulties in implementing the measures, inside China's villages few regulations have had any affect. For example, despite the nearly universal regulation that requires the use of a permit for drilling a well, less than 10% of the well owners surveyed obtained one before drilling. Only 5% of villages surveyed believed their drilling decisions needed to consider spacing decisions. Although price bureaus in every county were supposed to regulate the price for which groundwater was sold from one farmer to another, in only 8% of villages did this occur. Even more telling was that water extraction was not charged in any village; there were no physical limits put on well owners. In fact, it is safe to say that in most villages in China, groundwater resources are almost completely unregulated.

Producer response

Although China's central and regional governments currently have little control over groundwater in most parts of northern China, groundwater governance is not stagnant. In fact, when assessing the way groundwater is managed, the way farmers gain access to water and the way technology is being used to conserve the resource, the sector can be considered to be extremely dynamic. In this section, we examine three sets of issues: the privatization of tube wells, the emergence of groundwater markets and the adoption of new, water-saving technologies.

Privatization

Among any individual features of northern China's groundwater economy, the privatization of tube wells is perhaps the most prominent. Before the rural reforms in the 1960s and 1970s, township governments and village leadership councils financed, owned and managed most tube wells. In most villages individual farmers at most contributed their labour for tube well construction. Financed primarily by collective retained earnings, commune, brigade and team cadres were largely responsible for arranging for well-drilling companies run by the water resource bureau to sink tube wells. Pumps in the pre-reform era all came from either the water resource bureau pump supply company or the state-run local agricultural inputs corporation.

Soon after the general economic reforms began in the early 1980s, however, the ownership of China's tube well began to shift sharply. According to our survey in Hebei province in the late 1990s, collective ownership accounted for 93% of all tube wells in the early 1980s. Throughout the late 1980s and 1990s, however, the collective ownership of tube wells diminished. During this period the share of private tube wells increased from 7% to 64%. Data from the NCWRS largely support these findings. Tube well ownership in our study area, representing all of northern China, has also shifted sharply from collective to private (Table 3.3). In 1995, collective ownership accounted for 58% of tube wells in the average groundwater-using village. From 1995 to 2004, however, the collective ownership of tube wells diminished and accounted for only 30% of wells in 2004. In contrast, during the same period the average share of private tube wells increased from 42% to 70%.

Our interviews also revealed that the rise of privately financed investment means that the shift of tube well ownership is the result of the establishment of new tube wells rather than ownership transfers of collective tube wells. Due to the fall of the groundwater table and lack of maintenance on pumps and engines, a number of collective tube wells became inoperable during the last two decades and the absolute number of collective tube wells fell. During this time, the number of private wells increased rapidly.

Groundwater markets

As tube wells and the accompanying pumping equipment have come under the control of private individuals, access to groundwater for those farmers who do not own and operate their own wells has become a new issue. In fact, these markets have not always existed. In the 1970s and 1980s, when most wells were owned and operated by collective ownership, in almost all villages simple rules governed water allocations; most of the rules were based on a system in which all individuals were provided with water in an equitable way. In some villages, the collective ownership provided water free or at a subsidized rate. In the early period after reform, however, for a number of reasons the traditional institutions began to break down (see e.g. Wang *et al.*, 2005). In today's world in which most wells are owned by some, but not all, farmers there must be some way to transfer water from those with wells to those without.

In response to the demand for water in an environment increasingly dominated by private and privatized wells, following a pattern similar to that observed in South Asia (Shah, 1993), groundwater markets have begun to emerge in recent years as a way for many producers in rural China to gain access to

Table 3.3. Changes of well ownership from 1995 to 2004.

	All wells		Private wells	
	Collective	Private	Shareholding	Individual
Share of wells (%)				
1995	58	42	53	47
2004	30	70	38	62

groundwater.[19] In the 1980s, groundwater markets were almost non-existent. Indeed, according to the NCWRS, only 21% of villages had groundwater markets in 1995. By 2004, however, tube well operators in 44% of villages were selling water. Across all villages about 15% of private tube well owners sold water. Although groundwater markets exist in less than half of northern China's villages, the numbers are still significant: farmers in more than 100,000 villages are accessing water through groundwater markets. Moreover, in villages that have groundwater markets, these markets play an important role in transferring large volumes of water to a large number of households.[20]

Household and village adoption of water-saving technology

Another possible response to perceived water shortage is the adoption of new cultivation techniques and technologies. Our survey covered three sets of technologies: traditional technologies (agronomic-based, highly divisible[21]); household-based technologies (highly divisible, low fixed cost, requiring little collective action), which were generally practiced by farmers in pre-People's Republic of China; and community-based technologies (high fixed costs, requiring collective action for adoption and maintenance). The adoption paths of these three different water-saving technologies trace three distinct sets of contours. Moreover, the general path of each technology within each major category – traditional, household-based and community-based – tends to follow the trajectory of the other similar technologies within its category. In this section, we track the adoption with one set of measures – a village-based set of measures in which a village is considered to have adopted a technology if at least one plot or farmer in the village uses the technology. In another study (Blanke *et al.*, 2005), we also examine a measure of area of adoption (which gives largely the same pattern of results).

As the name implies, according to our data, traditional water-saving technologies have been used for many years (Fig. 3.5, top set of lines). The strongest distinguishing characteristic of traditional water-saving technologies is that they were being used in a relatively large number of China's villages even in the early 1950s. For example, in 1949 farmers in 55% of northern China villages were already leveling their land. During the reform period, the adoption of traditional technologies grew slowly, in part because traditional technology adoption rates were already high in the pre-reform and early reform era.

In contrast, household-based technologies have taken a different technological adoption path over the last 50 years (Fig. 3.3, middle set of lines). Although it is difficult to distinguish exact levels of adoption from Fig. 3.2 (the paths are too tightly bunched), household-based water-saving technology adoption rates were all low in 1949, ranging from 1% (surface pipe) to 10% (retain stubble/low till). Unsurprisingly, due to the relative abundance of water and the nature of farming at the time (collective-based with few incentives to maximize profits), household-based technology adoption rates at the village level remained low over the next 30–40 years. It was not until the early 1990s that these adoption rates soared. By 2004, farmers in at least 45% of villages were using each type of household-based water-saving technology mentioned in the Appendix.

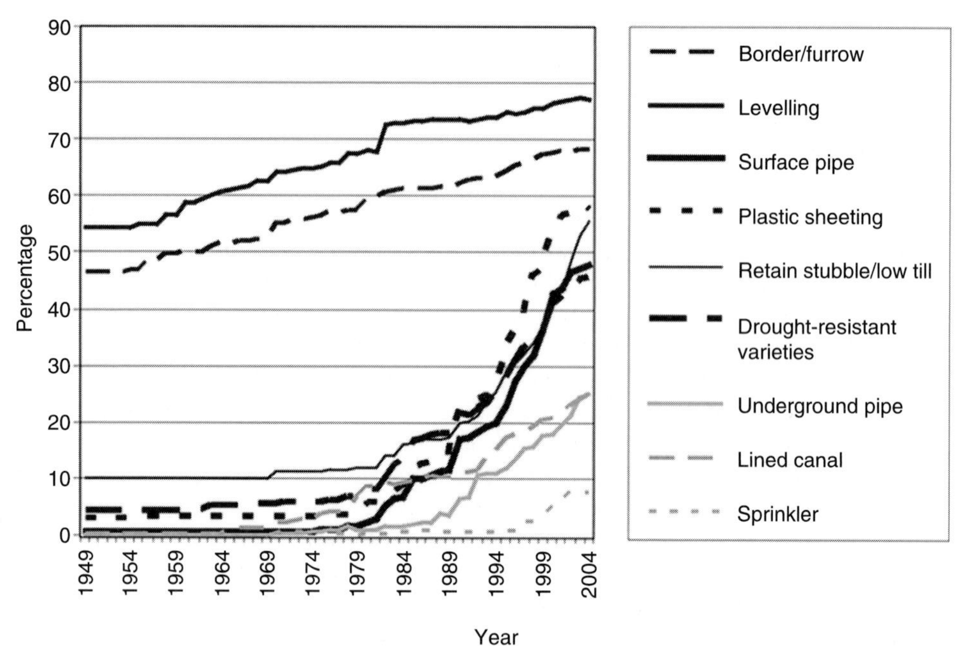

Fig. 3.5. Share of villages adopting water-saving technologies over time.

Finally, although the basic pattern of community-based technology adoption follows the same fundamental trend as household-based technologies, these paths start lower and rise at a slower rate (Fig. 3.3, lowest set of lines). Between the 1950s and 1980s, like household-based technologies, adoption rates were low. By the beginning of the reforms in the mid-1980s, the highest village-level adoption rate of a community-based technology (lined canals) was only 10%; on average the level of adoption of community-based technologies during the mid-1980s was around 5%. By 2004, as in the case of household-based technologies, the rate of adoption rose sharply relative to previous years. Because community-based technologies started from a lower level and rose less by 2004, the village-based measures still show that, on average, only about 20% of communities had adopted community-based technologies.

Although it is unclear, based on these descriptive contours, what is driving the adoption path of community-based technologies, it is likely that there are two sets of forces that are at once encouraging and holding back adoption. On the one hand, the rising scarcity of water resources is almost certainly pushing up demand for community-based technologies. On the other hand, the predominance of household farming in China (Rozelle and Swinnen, 2004) and the weakening of the collective ownership's financial resources and management authority (Lin, 1991) have made it more difficult to gather the resources and coordinate the effort needed to adopt technologies that have high fixed costs and involve many households in the community. In contrast, household-based technologies may be more widely adopted due to relatively low fixed costs, divisibility and minimal coordination requirements.

Conclusions

The primary goal of this chapter was to sketch a picture of China's groundwater water economy, with a focus on rural areas. Indeed, in our efforts to do so, we have generated a number of empirical-based findings that, at the very least, may help to clarify a number of misperceptions on which past discussions of China's water resources were sometimes based. China has some of the most abundant groundwater resources in the world. Over the last 2–3 decades in a large portion of China's localities, these resources have begun to be tapped. According to our results, however, there are still a significant number of areas that have undeveloped groundwater resources, even in the north-eastern areas, commonly believed to be generally overexploited.

In areas that have begun to use their groundwater resources, we have been able to paint a somewhat unorthodox picture. While there are serious groundwater problems (in around 10% of villages, the groundwater table fell by more than 1.5 m/year), including groundwater overdraft in some areas of northern China, in many other areas – indeed in more than one-third to one-half of China's villages using groundwater in northern China – groundwater resources have not diminished at all levels or are declining at less than 0.25 m/year over the last decade. In other words, the groundwater economy is heterogeneous, and as such, in dealing with policy in the future, considering the differences is important.

We also believe that we have been able to lay out a clear pattern of actual responses by major actors that will help to clarify the challenges for managing groundwater in the coming years. In short, government officials have done little to control the extraction of groundwater in rural China. Producers, especially individual farmers, on the other hand, have been responsive. Farmers have taken over control of most of the well and pump assets; they are increasingly taking on responsibility of transferring water from those who have wells to those who demand water. They are also figuring out ways to conserve the scarce resources.[22] Hence, the policy implication of all these results is clear. There needs to be a multistep response by officials. First, they need to determine where serious overdraft is occurring and where it is not. Attention then needs to be paid to the areas in which there is a problem. Policy must recognize that, with proper incentives, farmers will respond by saving water and transferring the resources from those who have it to those who need it.[23] Hence, if formulas can be designed to implement price-based policies or some other set of policies that make the scarcity of water more evident, farmers will respond. Such policies will not be easy to implement as they require a lot of information on the nature of the resource. In order to avoid negative income effects on those farmers who would have to pay more for water, it may also require complex transfer schemes in which farmers who are being charged for water and are being forced to cut back at the same time can be compensated in some way to try to minimize or offset the higher water fees. The transaction costs in such a system must also be considered. In some areas, it is possible that quantity control could work more efficiently than price-based control.

Acknowledgements

The authors would like to thank Mark Giordano, Tushaar Shah and Lijuan Zhang for their insights and helpful suggestions. We acknowledge the financial support from the National Natural Sciences Foundation (70021001) in China, the International Water Management Institute, the Australian Center of International Agricultural Research (ADP-200-120), the Food and Agriculture Organization of the UN and the Comprehensive Assessment on Water Management in Agriculture. Scott Rozelle is a member of Giannini Foundation.

Notes

1 The 12 provinces are: Heilongjiang, Jilin, Liaoning, Inner Mongolia, Hebei, Shandong, Shanxi, Henan, Shaanxi, Qinghai, Ningxia and Gansu. The two municipalities are Beijing and Tianjin. In this analysis, because of the lack of information on provinces in the extreme western areas of China, we do not include Tibet or Xinjiang in northern China.

2 Groundwater is also very important in urban water supply in northern China.

3 In Hebei province, where county-level groundwater overdraft statistics are available, the scarcity categories were defined according to a Ministry of Water Resource publication that categorized provinces by scarcity (which almost certainly is related to the degree of annual overdraft). In the remaining provinces, all four scarcity indices were defined according to the percentage of irrigated area as follows: very scarce (between 21% and 40%), somewhat scarce (between 41% and 60%), normal (more than 61%) and mountain and desert (less than 20%). Within each of the scarcity strata, we sampled two or three counties; of all the counties in the mountainous and desert areas, we chose one county.

4 The information that we collected comes from estimates provided to enumerators from village leaders based on their experience during the survey. For some technical data (data on water levels, water quality, soil salinity, etc.), although the village leaders do not have access to scientific measurements, they are readily able to state their perceptions on these issues. We believe, in many cases, that the information is fairly accurate. Even in the cases when information on the level of a variable for a given year may not be absolutely accurate (e.g. the salinity level of the water), due to the fact that they have been living and working in the village for many years, we believe that they are able to provide accurate estimates on the trends of these variables. Because these are based on the experience of village leaders, their response rates were high. In fact, for most variables the response rate was 100%, meaning our data are not subject to dropout bias.

5 In north China, almost all provinces have both mountainous areas and flat plains; therefore, it is hard to describe which regions are mountainous and which flat.

6 Although we have not asked the reason that why farmers only extract groundwater from deep aquifers, based on our experience in the field, it should be due to exhausted or unusable shallow aquifers.

7 We need more investigation to explore the reason in the future.
8 One of the important characteristics of the rural economic reforms in China is that land was distributed evenly to individual farm households. After the rural reforms, although land ownership was still collective, land use and income rights were transferred to individual farm households. Before the rural reforms, communes and brigades/teams (i.e. village collectives) financed most tube wells. After reform, the fiscal revenue position of many villages declined. More importantly, after the early 1980s the policy constraints that originally limited the scope of private activities were gradually relaxed and this resulted in the development of private tube wells.
9 Compared with the allocation of total water use to non-agricultural sectors, groundwater allocation to these sectors appears to have increased more rapidly. In 1997, non-agricultural sectors used 46% of total groundwater resources, while they only used 31% of total water resources (see Table 3.1, row 4).
10 Although maize is grown during the rainy season, and so the crop generally does not require as much irrigation as wheat (which is grown mostly during the dry season), irrigation can still play an important role in increasing maize productivity (Huang *et al.*, 2006). In North China, irrigation is supplementary.
11 According to Wang *et al.* (2006), the expansion of tube wells does not necessarily mean that there is an expansion in water consumption. However, according to our data, a significant share of the new wells is located in areas that are allowing for the expansion of cropping area, increased intensity of cropping and rising yields. Hence, while not all of the rise in wells will result in increased consumption of water, a part of it will.
12 This indicates that due to the decline in the groundwater table, it requires more power to extract water.
13 There are also several other minor types of aquifers that are being overdrafted, which account for about 7000 km^2.
14 The definition of overdraft here is from MWR in China. It is important to note, however, that there are other definitions. For example, Kendy points out that the MWR does not accurately define 'overdraft'. In a sustainable system, groundwater recharge should equal discharge over time. Extraction (groundwater pumping) is only a small part of total discharge from an aquifer. Other parts include natural discharge to rivers (which explains why rivers flow even long after rain and snow stop falling), and natural discharge to wetlands, lakes and plants. If extraction (groundwater pumping) exceeds recharge, all those other components of groundwater discharge would cease. Overdraft is better defined by long-term water level declines.
15 According to a comprehensive survey completed by the Ministry of Water Resource in 1996, groundwater overdrafting is a widespread problem and may be getting worse.
16 In our survey we asked village leaders about the average level of groundwater depth during the year and the 'static' level of the groundwater. We explained that the static level of the water table is the level that exists immediately prior to the irrigation season (e.g. in the North China Plain this would be around the month of March). According to our respondents, there were differences in the statistics on the changes in the groundwater table when using average or static groundwater levels. According to our data, the static level produced numbers that suggested there were fewer villages in which the groundwater table was falling.
17 Land subsidence mainly occurs in urban areas.
18 Salinization is caused by different factors, and responds to different solutions in different settings. Over time, continued groundwater use is likely to increase soil and water salinization. Each time groundwater is 'recycled' through the pumping and reinfiltration process, it becomes more saline.

19 We define groundwater markets as localized, village-level arrangements through which owners of tube wells sell pump irrigation services to other farmers of the village (i.e. they sell water to other farmers from their wells for use on crops). In this chapter, we are only going to examine 'private' water markets. In other words, we will examine the nature of groundwater markets that are being driven by individuals and groups of individuals that sink wells. In making such a definition, we are assuming that when village leaders (the collective owners) provide water to villagers, this is being done under non-market conditions.
20 Groundwater markets in northern China are not necessarily 'competitive' and may be more accurately characterized by captive selling. When farmers buy water through groundwater markets, they not only pay for operating costs, but also pay a little service cost that contributes to profit for the operators or owners.
21 Here 'highly divisible' means that individual farmers can adopt the technology by themselves.
22 Some researchers (Kendy *et al.*, 2004) argue that farmers are figuring out ways to reduce pumping without reducing crop production. Thus, they are conserving electricity, but not water.
23 Some researchers (Kendy *et al.*, 2004) argue that policies must ensure that water is actually saved (i.e. irrigated area decreases). So long as crop production stays the same, no water will be saved and any 'transfers' will only exacerbate the problem.

References

Blanke, A., Rozelle, S., Lohmar, B., Wang, J. and Huang, J. (2005) Rural Water-Saving Technology Adoption in Northern China: An Analysis of Survey Results. Selected paper prepared for presentation at the American Agricultural Economics Association Annual Meeting, Providence, Rhode Island, 24–27 July 2005.

Brown, L. and Halweil, B. (1998) China's water shortages could shake world food security. *World Watch* 11(2), 10–18.

Chen, C., Pei, S. and Jiao, J. (2003) Land subsidence caused by groundwater exploitation in Suzhou City, China. *Hydrogeology Journal* 11, 275–287.

China National Statistical Bureau (2000) *China Statistical Yearbook*. China Statistical Publishing House, Beijing.

China National Statistical Bureau (2004) *China Statistical Yearbook*. China Statistical Publishing House, Beijing.

Hoffman, G.J. and Durnford, D.S. (1999) Drainage design for salinity control. In: van Schilfgaarde, J. and Skaggs, W. (eds) *Agricultural Drainage*. American Society Agronomy Monograph, pp. 579–614.

Hu, R. (2000) A Documentation of China's Major Grain Varieties. Working Paper, Chinese Center for Agricultural Policy, Chinese Academy of Sciences, Beijing.

Huang, Q., Rozelle, S., Wang, J. and Huang, J. (2006) Irrigation, agricultural performance and poverty reduction in China. *Food Policy* 31(1), 30–52.

Kendy, E., Molden, D.J., Steenhuis, T., Liu, C. and Wang, J. (2003) *Policies Drain the North China Plain: Agricultural Policy and Groundwater Depletion in Luancheng County, 1949–2000*. IWMI Research Report 71.

Kendy, E., Zhang, Y., Liu C., Wang, J. and Steenhuis T. (2004) Groundwater recharge from irrigated cropland in the North China Plain: case study of Luancheng County, Hebei Province, 1949–2000. *Hydrological Processes* 18, 2289–2302.

Lin, J. (1991) Prohibitions of factor market exchanges and technological choice in Chinese agriculture. *Journal of Development Studies* 27(4), 1–15.

Lohmar, B., Wang, J., Rozelle, S., Huang, J. and Dawe, D. (2003) *China's Agricultural*

Water Policy Reforms: Increasing Investment, Resolving Conflicts and Revising Incentives. Agriculture Information Bulletin #782, Economic Research Service, USDA.

Ministry of Land Resources (2005) Report on Groundwater Resources and Environment Investigation in China, finished by Ministry of Land Resources, Beijing.

Ministry of Water Resources (2000) *China Water Resources Bulletin.*

Ministry of Water Resources (2002) *China Water Resources Bulletin.*

Ministry of Water Resources and Nanjing Water Institute (2004) *Groundwater Exploitation and Utilization in the Early 21st Century,* China Water Resources and Hydropower Publishing House, Beijing.

Negri, D.H. (1989) The common property aquifer as a differential game. *Water Resources Research* 25(1), 9–15.

Nickum, J.E. (1988) All is not wells in North China: irrigation in Yucheng County. In: O'Mara GT (ed.) *Efficiency in Irrigation.* The World Bank, Washington, DC, pp. 87–94.

Nickum, J.E. (1998) Is China living on the water margin? *The China Quarterly* (156), 880–898.

Rozelle, S. and Swinnen, J. (2004) Success and failure of reforms: insights from transition agriculture. *Journal of Economic Literature* 42(2), 404–456.

Sakura, Y., Tang, C., Yoshioka, R. and Ishibashi, H. (2003) Intensive use of groundwater in some areas of China and Japan. In: Llamas, R. and Custodio, E. (eds) *Intensive Use of Groundwater: Challenges and Opportunities.* A.A. Balkema, Rotterdam, The Netherlands.

Shah, T. (1993). *Groundwater Markets and Irrigation Development: Political Economy and Practical Policy.* Oxford University Press, Bombay, India.

Shi, Y. (2000) Groundwater Development in China. Paper for the Second World Forum, The Hague, 17–22 March 2000.

Smil, V. (1993) *China's Environmental Crisis: An Inquiry Into the Limits of National Development.* M.E. Sharpe, Armonk, NewYork.

Tang, K. (1999) Strategic Options for the Water Sector, TA NO 2817-PRC, Vol. 3, Working Paper 7, July 1999.

Wang, J., Huang, J. and Rozelle, S. (2005) Evolution of tubewell ownership and production in the North China Plain. *Australian Journal of Agricultural and Resource Economics* 49(2), 177–195.

Wang, J., Huang J., Huang, Q. and Rozelle, S. (2006) Privatization of tubewells in North China: determinants and impacts on irrigated area, productivity and the water table. *Hydrogeology Journal* 14, 275–285.

World Bank (1997) *At China's Table.* Monograph, The World Bank, Washington, DC.

You, S. (2001) Agricultural adaptation to climate change in China. *Journal of Environmental Sciences* 13(2), 192–197.

Appendix: Types of Water-saving Technologies

During our survey of leaders and water managers in more than 400 villages, we discovered that there are many types of water-savings technologies being used in northern China. For the purposes of this chapter, the term water-saving technology encompasses a wide variety of irrigation techniques and agricultural production practices. For analytical convenience, we have divided the list of technologies into three groups: traditional, household-based and community-based. In the rest of the chapter, we are excluding any discussion of a series of novel water-saving technologies (e.g. drip, intermittent irrigation, and chemicals and drugs) because across our sample, they had very low levels of adoption (i.e. nearly zero).

Our use of the term 'water-saving' is limited to perceived field-level applied irrigation savings. We understand that in the case of many technologies that we are considering, their adoption may not save water when net water use is measured on a basin scale. The real, or basin-wide, water-saving properties of each technology depend not only on the technical features of the technology, but also on the hydrology of the system and the economic adjustments to production that are associated with adoption of the technology.

Traditional technologies include border and furrow irrigation and field levelling. We have grouped these technologies because they are widely adopted and because village leaders in a majority of villages report adopting these techniques well before the beginning of agricultural reform in the early 1980s. These irrigation methods have relatively low fixed costs and are separable in the sense that one farm household can adopt the practice independent of the action of neighbours.

Household-based technologies include plastic sheeting, drought-resistant varieties, retain stubble/low till and surface-level plastic irrigation pipe. We have grouped these technologies because they are adopted by households (rather than villages or groups of households), have relatively low fixed costs and are highly divisible. Typically, adoption of these technologies is more recent than adoption of the traditional technologies.

Community-based technologies include underground pipe systems, lined canals and sprinkler systems. We have grouped these technologies because they tend to be adopted by communities or groups of households rather than by individual households. In most applications, they have large fixed costs and often require collective action or ongoing coordination of multiple households.

4 Rural Economic Transitions: Groundwater Use in the Middle East and Its Environmental Consequences

J.A. Tony Allan

King's College London & SOAS University of London, SOAS, Thornhaugh Street, London, WC1H 0XG, UK

Renewable groundwater is a fatally attractive source of water, especially in political economies that have not yet devoted substantial political energy to developing norms and laws to establish ownership, or to regulating water use to achieve efficiency and environmental consideration.

Farm lobbies are the oldest of all lobbies. They are incredibly strong in a young economy such as that of the USA. They are even stronger in societies that have been coping with periodic 'lean years' for four or more millennia.

Introduction: Groundwater in Its MENA Context

The first purpose of this chapter is to identify and explain the importance of groundwater in the Middle East and North Africa (MENA) region in the second half of the 20th century. The second is to demonstrate that one of the most important roles of the MENA region's very limited groundwater (Table 4.1) has been as an enabler of an important rural socio-economic transition. Groundwater played this important transitional role by strengthening individual family economic circumstances in rural areas. These stronger family economies were able to provide a stable pattern of expenditure, which enabled the acquisition of skills by the younger family members. Thus equipped, significant proportions of whole generations have moved to urban centres. Many rural communities in the region experienced relatively high rural incomes for about three decades on the basis of groundwater use.

At the same time, there is no question that those running the MENA political economies presided passively over a water-managing system that overused environmental capital – groundwater – that underpinned the process. This socio-economic groundwater-related phenomenon has also been identified in South Asia where numerous strategies have enabled similar transitions (see Tushaar, Chapter 2, this volume; Burke and Moench, 2000). Llamas and Custodio

Table 4.1. A typology of MENA economies useful in analysing the management of groundwater in the MENA region. (From the author.)

Low-income economies	Yemen
Partially industrialized economies	Lebanon, Syria, Jordan, Palestine, Egypt, Tunisia, Morocco
Oil-enriched economies with different groundwater endowments:	
• With modest renewable groundwater	Algeria, Iraq
• Mainly with non-renewable (fossil) groundwater	Saudi Arabia, Libya
• Very poor groundwater endowment	Kuwait, UAE, Oman, Qatar
Industrialized diverse economies with poor water endowment	Israel

For the purposes of this overview Turkey and Iran in relatively hydrologically favoured parts of the region have not been included.

(2003a,b, pp. 13–26) have drawn the attention of water professionals and water policymakers to the importance of taking a balanced view of the drawdown of groundwater aquifers. They show that lowering aquifer levels for productive economic purposes can bring economic and environmental benefits. Using the water stored in aquifers leaves space for the recharge following major rainfall events. The MENA region is particularly subject to such erratic rainfall patterns. Lowering groundwater levels where they have been at, or close to, the surface reduces evaporation and also transpiration where vegetation has been supported by groundwater (Llamas and Custodio, 2003b, pp. 18–19). It will be shown that there have been remarkable experiments in managing groundwater storage in the MENA region. However, the most significant feature of MENA groundwater management has been the revelation that groundwater resources are never sufficient to underpin food self-sufficiency. At best they can be an element in a complex of resource management strategies that achieve overall water security. Economic instruments and processes beyond hydrology and hydrogeology are the locus of water security.

Background

The MENA region's farmers and governments are very aware of water and the potential constraints of encountering seasonal and systemic water shortages. Irrigated farming is a deeply entrenched social phenomenon because it is the sole livelihood provider for many communities. Irrigated farming, as a result, is disproportionately prominent in national water allocation policy discourse. In the non-oil economies of the region, irrigated farming is still the basis of the livelihood of the largest employed sector. Secure livelihood is pivotal for rural societies. Traditional irrigated livelihood is integral to a range of powerful ideas that tend to reinforce the notion that irrigated farming is worthy, essential and even holy. Farmers do Allah's and God's work. Unfortunately irrigated farming brings the lowest economic returns to water of any productive combination of factor inputs.

The water predicament of the MENA economies is globally significant. Their experience in coping with progressively more serious water scarcity – demographically driven – in the second half of the 20th century provides an important parable. The experience is especially relevant to economies in arid regions where communities depend on irrigated farming. Water scarcity across the MENA region has been exceptional by global standards. The challenges facing some of the MENA economies are unprecedented at least in modern history.

In order to develop a comparative analysis – within the region and with other regions facing similar problems – it is tempting to identify a typology of the economies of the MENA region based on endowments:

1. Environmental endowments:
   - rich renewable groundwater endowments vs. poor renewable groundwater endowments in general (with the exception of Morocco there are only poor groundwater-endowed economies in relation to demographic circumstances in the region);
   - rich fossil water endowments vs. little or no fossil water endowments (Libya and Saudi Arabia vs the other economies);
   - other renewable water endowments (at the surface and in soil profiles) vs. very limited other water endowments (Egypt, Iraq, Syria and Lebanon vs. the other economies).
2. Non-water endowments and circumstances:
   - economies with large numbers of water-challenged water users living at high elevations – all dependent on scarce groundwater – vs. economies with water users living at low elevations (part of Syria, Jordan and Yemen vs. the other economies);
   - rich economies with poor water resources (e.g. oil-enriched), diversified economies and responsive political systems vs. poor natural resources, non-reforming political systems and limited economic developmental capacity (Algeria, Libya, Iraq, Saudi Arabia, Kuwait, Qatar, UAE and Oman vs. the other economies; Syria, Egypt and Yemen have limited – but important in the short term – oil resources).

The economies of the region do not fit neatly into two or three categories according to groundwater endowments. A typology based on political economy outcomes rather than groundwater endowments provides much greater analytical insights, i.e. a typology based on what political economies have done with their water endowment rather than how they are endowed. Environmental determinism has everywhere been discredited. Recent MENA groundwater management experience confirms that analyses based on water resource determinism must be avoided.

There is very strong evidence in the region that poor water endowments, especially poor groundwater endowments, do not determine approaches to utilizing and managing them. Israel is worse off in its water endowment than a number of its neighbours (see Table 4.2). It has no oil resources either. But it has

Table 4.2. Data on the Nubian sandstone aquifer systems (NSAS) of northern Africa. (From CEDARE/IFAD Programme for the development of a regional strategy for the utilization of the Nubian sandstone aquifer. Cited in Bakhbakhi, 2002.)

	Nubian system Palaeozoic and Mesozoic sandstone aquifers		Post-Nubian Miocene aquifers		Total freshwater in storage	Recoverable groundwater	Present extraction from NSAS		
							Post-Nubian	Nubian system	Total from NSAS
	Area ('000 km²)	Volume ('000 km³)	Area ('000 km²)	Volume ('000 km³)	('000 km³)[a]	(km³)[b]	(km³)	(km³)	(km³)
Egypt	815	155	426	97	252	5,180	0.306	0.200	0.506
Libya	754	137	494	72	208	5,920	0.264	0.567	0.831
Chad	233	48	NA	NA	48	1,630	NA	0.000	0.000
Sudan	373	34	NA	NA	34	2,610	NA	0.840	0.833
Total	2,176	373	921	169	542	15,340	0.570	1.607	2.170

[a]Assuming a storability of 10^4 for the confined part of the aquifers and 7% effective porosity for the unconfined part.
[b]Assuming a maximum allowed water level decline of 100 m in the unconfined aquifer areas and 200 m in the confined aquifer areas.

combined scarce renewable groundwater and some other renewable waters with its other endowments to develop a diverse and effective economy – albeit in very controversial asymmetric local power relations. (Allan, 2001; Selby, 2005).

The MENA region is also a very useful groundwater management laboratory, which confirms that water problems are only partly solved. It is evident from the levels of food imports of all the economies of the region that in almost all cases the region's problems are not even basically solved in the water sector. There are two ways of approaching the need for more water. Both encounter thresholds where they can contribute no further. The first threshold of water insufficiency is reached when supply management measures cannot deliver more solutions. The second threshold is reached when the increased efficiency associated with demand management proves to be insufficient to achieve or maintain self-sufficiency. At this point, the deficit has to be addressed in the political economy rather than in the water sector.

All the economies of the MENA region are solving their current water shortage problems outside the water sector. Table 4.1 indicates the extent to which the individual economies were solving their water deficit problems and avoiding international conflict over water by resorting to imports of water-intensive commodities (Allan, 2002, 2003). Their future water problems will also be solved outside the water sector in international trade. The MENA economies can pay for imports through the development of their own political economies. The capacity to pay for imports is politically determined. Politics determines whether an economy diversifies and strengthens. Groundwater endowment is a minor factor in relation to the bigger water picture. But in some economies in the region groundwater has provided a crucial and timely resource to support rural economies as families move to earn their livelihood in the cities.

With this evidence that a version of water security can be achieved despite poor groundwater endowments, a political economy analysis will be adopted. The strength or weakness of the individual MENA economies are expressions of their political, including institutional, capacity to combine water endowments effectively with other environmental capital, and with human, social and financial capitals.

On the basis of the current political economy outcomes – reflecting the ability of water users and governments to manage scarce groundwater endowments with different levels of effectiveness – we can identify the following typology.

MENA Groundwater Resources

Before analysing the MENA region's experience in managing its water in the different types of political economy, a brief quantification of the region's water will be provided. Its other freshwater resources – surface waters – will also be shown to provide a context. The very limited soil water resources[1] have not been estimated for the individual economies; however, a rough estimate for this chapter is provided. This number is necessary to make it possible to provide an estimated water budget for the region that approximates to its water needs. Surface and soil water will not be analysed in detail. They will be referred to when they are the most important or very significant elements in an economy's water use.

Renewable groundwater

The majority of the MENA region's accessible renewable groundwater aquifers are located in the region's extensive coastal plains. Human settlement has been supported by coastal aquifers for millennia along the coasts of northern Africa and the eastern Mediterranean as well as in the Gulf. As populations rose and groundwater levels fell, through excessive use, all these coastal aquifers have been subject to sea water intrusion. The second half of the 20th century witnessed progressive subsurface seawater intrusions of more than 20 km, e.g. in Libya, and of even greater distances under the delta of the Nile. A feature of the groundwater management of delta Egypt is the cultivation of rice associated with high inputs of water. High levels of water use prevent or at least slow the advance of the saltwater–freshwater interface. There is evidence from Israel that coastal aquifers can be managed to avoid serious degradation by means of technology and regulation, but at present such technology and regulation have not been deployed elsewhere.

Alluvial aquifers and other aquifers exist in inland basins and in the uplands of the region – for example, east of Damascus in Syria, in the highlands of Jordan and Palestine and in the highland basins of Yemen in Sana'a, Tai'iz and Sa'adah. These renewable aquifers are without exception being used beyond their rates of recharge.

Fossil groundwater

The deserts of northern Africa and the Arabian peninsula have been extensively explored for oil in the second half of the 20th century. Oil exploration companies are contracted to record and report on the hydrogeology, especially the non-renewable fossil waters. These groundwater data have been systematically collated by the national governments of the region.

Past rainfall regimes over the Sahara and the Arabian peninsula have left in place substantial reserves of groundwater with estimates of their age of between 12,000 and 30,000 years (Wright and Edmunds, 1971; Wright, 1986). Much of the ancient water is of usable quality (Edmunds and Wright, 1979). This water is often at accessible depths in terms of pumping costs, but it is located hundreds of kilometres away from potential users. An exception is the rapidly expanding capital city of Saudi Arabia, Riyadh. It lies close to 'fossil' groundwater but perversely the national government decided to devote almost all the 'fossil' water to agriculture rather than to supplying the city. In 2005 the city was mainly supplied by desalinated water pumped 450 km from the Gulf and lifted 600 m, costing $1.4/m^3$, to deliver to Riyadh. In the first few years of the new millennium the city used only 145 million cubic metres of local deep groundwater at a cost of about $0.44/m^3$ and 261 million cubic metres of desalinated water at a higher cost (E. Elhadj, London, 2005, unpublished data).

The volumes of water listed in Table 4.3 are more interesting to water scientists than of relevance to water managers and policymakers. The estimates are in many cases preliminary although helpful in so far as they narrow the levels of uncertainty associated with such 'fossil' water resources. For water managers the estimates are not useful. The high estimates of supposedly recoverable volumes of fossil water beneath the Libyan Sahara would secure the Libyan economy and its future estimated populations for 500 years at current rates of use. The low estimates would underpin all the water needs of the 2005 Libyan population for 100 years if they could be technically mobilized. On the basis of these estimates it is possible for Libyan water planners to claim in some international dialogues that Libya is very water-secure. Nevertheless, Libya accesses net annual volumes of virtual water of about 1.3 km^3 (Hoekstra and Hung, 2002; Chapagain and Hoekstra, 2003) in food imports rather than mobilizing sufficient fossil water to meet all its water needs. Egypt is also rich in fossil water but its net imports of virtual water at more than 18 km^3 annually reflect a national water self-sufficiency of 78%. There is no attempt to bring the fossil water into the national water budget. Nor in the case of Egypt is there significant development of the 'fossil' water to address Egypt's current water deficit. Present technologies are inadequate. International market circumstances are such that it is impossible to get the 'fossil' water into MENA economies cost-effectively for agricultural use, especially in Egypt. The oil-rich economies can meanwhile indulge their inclination to be self-sufficient in food via big infrastructures. The Great Man-made River project in Libya is an experiment of global significance testing technologies and the institutional capacity of a political economy of the region to develop MENA fossil waters.

Table 4.3. An estimate of the MENA region's annual water budget in 2005 showing the limited role of groundwater. (From FAO AQUASTAT 2003; Chapagain and Hoekstra, 2003, pp. 68–73; authors' estimates.)

	Groundwater (km³/year)	Percentage	Required for food self-sufficiency (km³/year)
Freshwater sector provides limited and partial solutions			
Surface water	154	49	Overused in many basins
Renewable groundwater	28	9	Generally severely overused
Fossil water contribution	3	1	Expensive option in agriculture in MENA
Total	185	50	Used at rates that severely impair environmental water services
Soil water	40	13	Very approximate estimate by author
Non-water sector provides solutions with substantial future potential			
Net virtual water 'imports'	77	25	Rising; very attractive economically
Manufactured water	3	1	Rising; too expensive for most agriculture
Total	80	26	Easily expandable, economically acceptable, no regional impairment of environmental services of water
Total	305	100	272 needed, assuming 1100 m³/person/year for 247 million population

A feature of the fossil groundwater hydropolitics in the Middle East is that economies that have to share their surface and groundwaters with other countries do not want to draw attention to the fossil water endowment. Presumably it would be a complication in any attempt to resort to the principle of *equitable utilization* in any legal or quasi-legal process to settle an international water dispute. For example, Egypt and Israel are very reluctant to discuss their fossil water endowments.

MENA groundwater in relation to the total water resources of the region's economies

Renewable groundwater is an important resource but a minor one in relation to the surface waters enjoyed by a few of the region's economies – Syria, Iraq, Lebanon and Egypt. The fossil water resources of three of the region's economies – Libya,

Egypt and Saudi Arabia – are vast but not able to be used to meet water needs in all sectors in current technological and economic circumstances.

Table 4.1 provides estimates of the water resources of the economies of the region. It gives basic information on current estimates of the water resource status of the 18 economies of the MENA region that endure water resource problems. The general impression provided by the data-set is that the Tigris–Euphrates economies are still relatively well endowed with water resources. Lebanon is also relatively secure with respect to water. All the other economies are enduring serious water deficits with respect to their capacity to produce enough food for self-sufficiency. And the situation is worsening as a result of rising populations. Future demographic trends will be very significant vis-à-vis water resources in the MENA region during the 21st century. The population of the MENA region will probably double before the mid century. This rate of increase is higher than most economies in Asia and South America and similar to those in Africa. The MENA region has indirectly benefited greatly from the population policies of Asia's major economy, China. China has lifted out about 300 million of its population from poverty in the last three decades. The population of the 18 economies of the MENA region considered here is only 280 million at the beginning of the millennium. In this type of calculus the role of water can be seen to be minor compared with the scale of the global demographic shifts and in population policies in other regions.

Table 4.1 provides estimates of the very limited volume of renewable groundwater in the MENA region. The 18 economies listed have in total only about 28 km^3 of renewable groundwater annually – 15% of the total freshwater used and 9% of the total water needed for food self-sufficiency. Most of this renewable groundwater has been used without giving attention to the institutions, regulations and technologies that would match water withdrawal with regional hydrological regimes for over four decades. This volume is sufficient for the *domestic* and *industrial* water needs of about 250 million people – a number close to the total population of the 18 economies. Domestic and industrial water use is about 10% of the total water that an individual or an economy needs. The remaining 90% of water needs is covered partially by renewable surface and soil water, with the deficit remedied by virtual water.

These macro-level estimates of the elements of water availability and use are subject to poor precision, with estimates of soil water being the least precise. However, it is interesting that the author's estimated figure of 40 km^3 of annual regional soil water was entered before the numbers for virtual, surface and groundwaters and before all sources were added together. Apart from the estimates for soil water all the others are based on best practice in respected agencies such as the Food and Agricultural Organization (FAO) for surface and groundwaters, or in a research institute such as the Institute for Water Education (IHE) for the virtual water data. The estimates used are: 154 km^3/year for annual surface water use; 28 km^3/year for renewable groundwater and fossil water use; and 80 km^3/year for virtual water and desalinated water. The very approximate estimate of 40 km^3/ of annual soil water use for rain-fed crop production brings the numbers for use and for estimated total regional water needs to a reasonable convergence with the estimate of water needed to secure the food and job needs of the region's population. If the absence of a more reliable estimate for soil water troubles hydrologists and soil scientists, they are invited to provide it. Meanwhile water users, managers and policymakers will remain comfortably unaware of the role of soil water. Ignoring soil water in a national water budget is part of normal political behaviour. Politicians, and political processes more generally, have to deal with all sorts of uncertainties including the absence of knowledge on most issues of importance. Water scientists could help if they can provide (accurate) evidence on soil water and offer such data to political processes in a friendly language register.

A very approximate water balance for the 18 economies of the MENA region considered in this discussion is shown in Table 4.1.

Table 4.1 and Fig. 4.1 are helpful in putting the MENA region's finite groundwater into perspective. The volumes of groundwater available are small. Renewable and fossil water at the beginning of the millennium amounted to only:

18% of all freshwater use in the region;
11% of the total water needed by the region's peoples and economies;
35% of the unconventional water use – virtual and manufactured water.

In addition, the MENA history of intense groundwater use is very short, i.e. about four decades. The era spans from the initiation of intensive use in the 1960s to the end of the millennium, by which time most renewable groundwater users were reducing pumping because the aquifers were damaged.

There are several strategic implications of these circumstances for water policymakers in the MENA region:

- The spectacular contribution of renewable groundwater to the very important urbanization transition in the MENA region in the past four decades was an important, but brief and unsustainable, moment in the region's economic history. Renewable groundwaters were too easy to develop in a cultural setting that rejected any regulatory regime. Renewable groundwaters have been dangerously impaired in some economies.
- As the region's renewable waters, including groundwaters, have been severely mismanaged, the priority should be to remedy their poor water quality and to initiate measures that restore some of the essential environmental services of renewable surface and groundwaters. The priorities should be, first, to put water back into the environment and, second, to reduce pollution. Experience in the region indicates that reforms will only be possible when the economies are diversified and strong.
- The future water security of the region's much higher population will be mainly addressed by measures outside the region's water sector. These measures enable the import of water-intensive commodities from global production and trading systems. In the region itself these non-water-sector processes are socio-economic development and especially economic diversification. The manufacture of desalinated water will also be a very important remedy for those economies that do not have sufficient water to meet domestic and industrial needs as most of the region's population lives either near the coast or on major river systems.

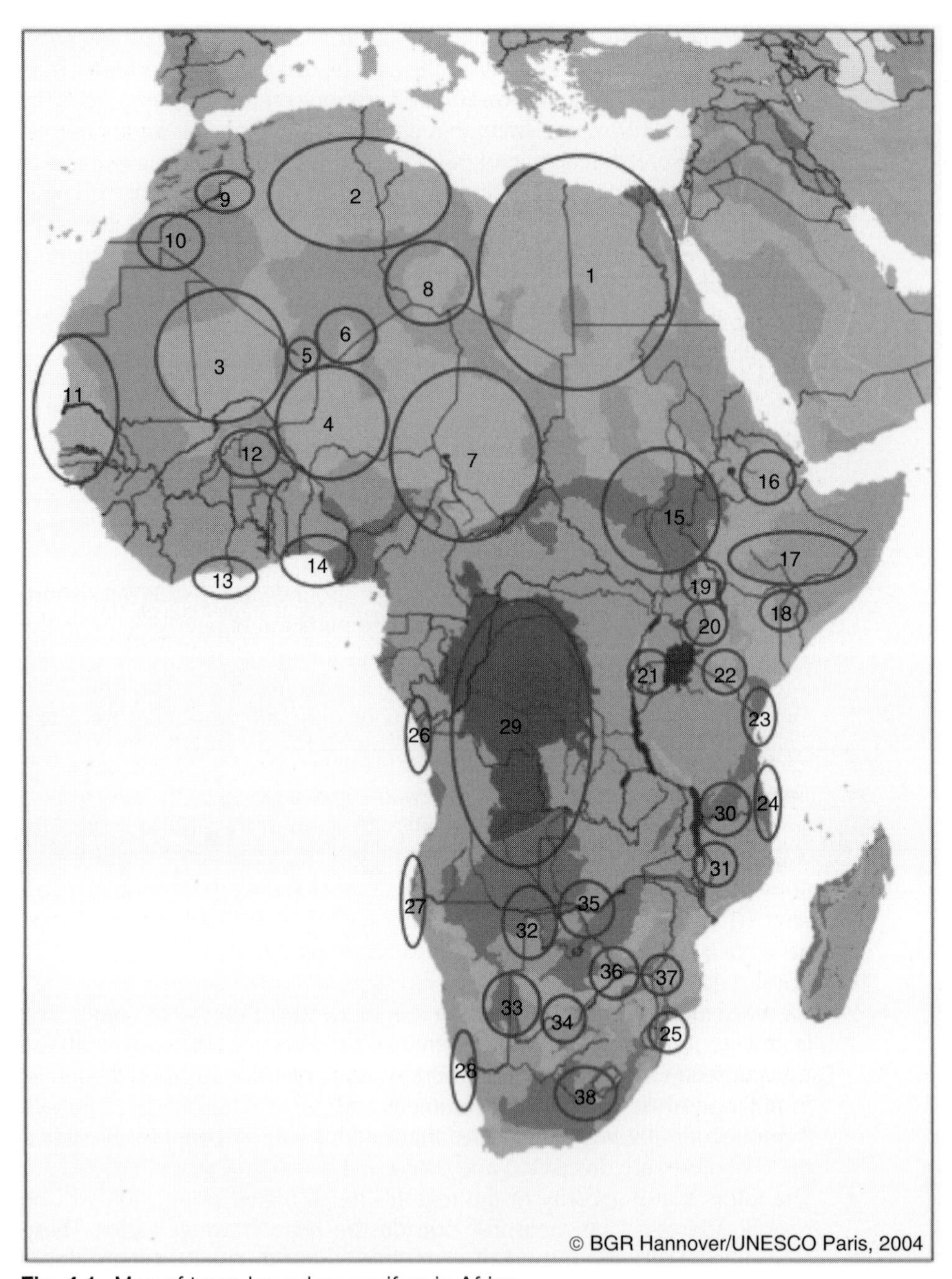

Fig. 4.1. Map of trans-boundary aquifers in Africa.

The rest of the discussion will examine the short history of intensive renewable groundwater use. It will show how a brief phase of intensive groundwater utilization has played a very important positive role because groundwater has a number of qualities for those facing immediate water scarcity. These qualities make renewable groundwater a fatally attractive source of water, especially in political economies that have not yet devoted substantial political energy to developing norms and laws to establish ownership, or to regulating water use to achieve efficiency and environmental consideration. The qualities of groundwater that make it fatally attractive are:

- Groundwaters can be beneath, or extremely close to, the projects and needs of water users.
- Groundwaters are often very close to the surface, at least when they are first developed, and can be developed at low cost.
- Groundwaters can be accessed by individual farmers and other individual users without the constraints of a regulated and bureaucratized water distribution infrastructure.
- Users can use water at will – provided only that they have the resources to acquire, operate and maintain the equipment and fuel it. This flexibility makes groundwater a very useful primary source of water for irrigation and an especially useful supplementary source in the extensive marginal rainfall tracts in the MENA region. A very important feature of the control that individual users have over groundwater is the capacity to address water scarcity in more than one annual cycle. The economic significance of this capacity to withstand droughts for more than 1 year is of immeasurable importance to those who have risked raising cash crops.

The qualities of groundwater that make it hard to monitor are technical, social and political:

- Users have little awareness of the impact of individual groundwater users on regional levels of use, especially in the early phases. The absence of awareness of the need for collective action is the norm.
- Renewable groundwater is regarded as a common pool resource – anyone who can access it is entitled to use it. This approach exists in all the economies of the region except Israel. Attempts to license wells and to limit groundwater use have generally failed.
- The impact of overuse is gradual and in the common pool circumstances of the MENA region a 'tragedy of the commons' has been accelerated (Handley, 2001; Lichtenthaler, 2003).

Mindsets and Sanctioned Discourse: Managing Groundwater in the MENA Region

More important than knowledge of the volumes and rates of use of renewable groundwater and fossil groundwater is the knowledge constructed by political classes and by the major users of water in the region – namely the irrigation communities. There are two main lessons to be learned from the MENA region about the unquestioning determination of such interests when in coalition to damage scarce renewable groundwater.

The first lesson is, as already mentioned, that the flexible accessibility of groundwater makes it a very easy and often cheap resource to mobilize. At the same time renewable groundwater has the capacity to enable rural communities to achieve higher incomes. Average yields of staple grains can be much more than doubled with supplementary irrigation, and more valuable crops, with higher returns to water, can also be raised.

The second lesson is that new users of renewable groundwater believe in their entitlement to the groundwater and will only stop using it when it runs out. No collective measures have been put in place to regulate groundwater use for the collective good – except in Isarel – at least certainly not in the last 40 years during which the MENA region's renewable groundwaters have been utilized at unprecedented rates.

Groundwater has nevertheless played a very important role in the economies of MENA since the mid 20th century. It will be shown in the following discussion that the mindset that has driven the development of groundwater at the level of government as well as at that of the individual farmer has been fixed on increasing agricultural production to meet national food needs and improve irrigators' incomes.

Supply management approaches, such as pumping more water, have dominated the water management of the last 50 years. A second goal has been productive efficiency, which means achieving more crop per drop via technological interventions. Renewable groundwater has been mobilized at progressively higher levels for irrigation use. Notions of economic efficiency and of the importance of economic returns to water have been evident only by the 1990s, and only in a minority of the economies – in Tunisia, Morocco and Jordan. Israel had embarked on the demand-managing measures of allocative efficiency a decade before in the early 1980s (Arlosoroff, 1996). Demand management requires that the allocation of water in an economy be informed by hydrological, hydraulic and economic science.

Knowledge-based groundwater policy has been rare in the MENA region. The norm is that users of groundwater pump water at rates that contradict the advice of groundwater scientists until the resource is exhausted. These practices reflect the assumptions of the region's water users and governments. These assumptions include a preference for water to be treated as a public good, even as an entitlement. Water should be provided freely and without restriction if possible or with little restriction otherwise. Where water is assumed to be privately owned, which is the case for most renewable groundwaters, the owner of the land assumes that he or she is entitled to use it without constraint other than the cost of drilling and pumping. Long-established uses of water in agriculture are also thought to be more important than uses for recently established activities such as industry and services. The environmental services of water resources are not significant in the minds of water users and governments.

The assumption that food self-sufficiency is a proper goal chimes with the immediate livelihood interests of the large rural populations of the region. The alignment of these interests with the mindset of the leaderships of the region has led to the overuse of the renewable aquifers in every MENA economy. The alignment of farmers' interests and those of the political class is the international norm. It is evident across the economies of the world including those of the USA and Europe. Farm lobbies have disproportionate influence in northern economies. Production and export subsidies are issues that are becoming increasingly politicized internationally. It should not be surprising that the realities of economic life and the imperatives of sustaining farmers' livelihoods in the south are elemental in the economies of the MENA region. The farm lobbies are the oldest of all lobbies. They are incredibly strong in a young economy such as that of the USA and in the recently established European Union (EU). That they are even stronger in societies that have been coping with periodic 'lean years' for four or more millennia should not be surprising.

It is argued next that it would have been strange if groundwater had been managed otherwise in the circumstances of the MENA region in the late 20th century. Even in economies that had the political and economic space to pursue knowledge-based groundwater management policies, both renewable and non-renewable aquifers have been seriously depleted. Overuse of the aquifers of the High Plains of Texas is a sorry tale (Rainwater *et al.*, 2005a). The political pressures are captured (Rainwater *et al.*, 2005b) in the following paragraph:

> The resistance [to reform] is not purely interest-based, but is generated in some measure by considerations of identity: we are not people who treat the sacred as something to be bought and sold. Similarly, for families who have worked the land for generations, even if only at the level of subsistence agriculture, telling them that it is irrational to farm in their location because climate change is producing extended drought conditions, or that it would be wiser for the government to import virtual water in the form of grain, rather than supporting irrigation projects, is not likely to be well received.
>
> The examples are not limited to the developing world. Consider the worldwide industry that is golf. Why is it so hard to convince people that planting golf courses in Arizona and Dubai is not rational or environmentally sustainable? The wealthy are by no means immune to constructed identities: we are people of leisure who have worked hard and deserve both sun and golf. The purely rational often succumbs to the powerfully normative or the radically political.

MENA farmers and municipalities benefited from advances in technology in accessing and distributing groundwater from the 1950s. In addition, by the 1960s half of the MENA population was oil-enriched. They were able to combine oil-rent-derived financial resources to develop accessible groundwater rapidly and more intensively – especially fossil water in a few economies. MENA groundwater users, like all others in water-scarce regions, found renewable groundwater to be a particularly useful and flexible resource since its use did not require major infrastructures to get the water to the points needed by irrigators. At least this is true for unregulated use, which has been the norm in the region. The development of non-renewable aquifers beneath the deserts of the region after 1980 has in the case of Libya required major pipelines, but this has been the exception.

Concluding Comments

This chapter has shown how MENA groundwater users, water professionals and politicians have managed renewable and fossil groundwater resources during three decades. The potential demand for water doubled during this period with the doubling of the region's population. There are four main conclusions. First, renewable groundwater aquifers are too easy to utilize and to damage in the absence of a regulatory culture. The ease with which they can be turned on and off to comply with the users' needs makes renewable groundwater very popular indeed. But nowhere is there a balanced approach to achieving collective interests.

Second, fossil water resources, of which the region has a significant volume both in northern Africa and in the Arabian peninsula, are expensive to develop, and the pumping and delivery infrastructures are also expensive to maintain. The oil-rich economies that have developed them have been expensively addressing a fantasy of self-sufficient food security without recognizing a much more hazardous technological dependency.

Third, the region has not developed the institutions and the political culture to install regulatory measures that would address the collective interests of the populations of an individual state. At the interstate level international customary law is very poorly developed with regard to groundwater shared by more than one state. With only minor exceptions, there have been no formal negotiations over trans-boundary groundwater despite the urgency of the problems facing managers. The issues have been discussed at scientific conferences on water resources, focusing on the shared North African aquifers. Data are beginning to be shared and published by agencies such as United Nations Educational, Scientific and Cultural Organization (UNESCO) and Centre for Environment and Development for the Arab Region and Europe (CEDARE), but as the issues are not urgent it is understandable that progress over cooperation is slow.

This chapter has emphasized that institutional development is slow as a result of the lack of diversity and strength of the MENA economies. This lack of diversity and strength is in turn the result of the patrimonial political and governance regimes that characterize the region. Israel, a non-patrimonial state, has demonstrated that water problems are easily manageable within a diverse and strong economy. Diversity and strength come when the political circumstances enable the productive combination of the factors of production. Sound water management is associated with economic strength and especially economic diversity. These socio-economic virtues are a consequence of political processes that combine and manage resources effectively.

The MENA region has too many examples of the social and cultural conditions that determine short-term water-using practices. Renewable groundwater is too easily developed. Regulation cannot be installed. Everywhere water policy is made by officials, politicians and water users with mindsets established in the demographic and water-using practices of the past. Water-managing policies evolve that molest as little as possible the users of big volumes of water in irrigation. This is especially the case in the use of groundwater, which has indeed proved to be a very popular water resource with farmers.

As a result of the easily initiated and impossible-to-stop forms of renewable groundwater use, individual MENA economies have experienced the comfort of renewable groundwaters in rural areas for periods of 20–40 years. Renewable groundwaters has supported very important and timely rural transitions. Irrigation, as well as supplementary irrigation, has enabled rural families to enjoy a period of higher income than in the past. The period of greater prosperity has come about in combination with improved public services. Together these factors have enabled a couple of generations of young people to gain education and skills that have eased their transfer from rural areas with poor long-term prospects to the cities of the region.

Notes

1 Soil water here is defined as the water intercepted by the root systems of plants and crops. It is the water that exists in soil profiles after a period of rainfall. It is not the water provided by an irrigation system from surface or groundwater resources. This rain-fed soil water is extremely difficult to quantify. Soil water is, however, very important globally as most of the agricultural production of the world, and almost all the forest products, are raised with soil water. Engineers are not comfortable with the concept as soil water cannot be pumped. They are also unhappy with attempts to quantify soil water as it normally moves downwards in response to gravity. The significant soil water is that intercepted by plants and crops. Economists are disposed to ignore soil water as it is even more difficult to value than to quantify. For this chapter it has been estimated that soil water available annually in the 18 economies considered averages to about 40 billion cubic metre. This volume is about 30% bigger than the renewable groundwater available (28 billion cubic metre) in these economies.

References

Algehariani, S. (2004) Water transfer versus desalination in North Africa: sustainability and cost comparison. Available at: www.ckl.ac.uk/geography/staff/allan.html

Allan, J.A. (2001) *The Middle East Water Question: Hydropolitics and the Global Economy*. I.B. Tauris, London.

Allan, J.A. (2002) Hydro-peace in the Middle East: why no water wars? A case study of the Jordan River Basin. *SAIS Review* 22(2), 255–272.

Allan, J.A. (2003) Virtual water – the water, food, and trade nexus useful concept or misleading metaphor? *Water International* 28(1), 4–11.

Appelgren, B. (ed.) (2002) Managing Shared Aquifer Resources in Africa. IHP-VI, Series on Groundwater No. 8, Unesco, Paris (with International Hydrological Programme, General Water Authority of Libya).

Arlosoroff, S. (1996) Managing scarce water: recent Israeli experience. In: Allan, J.A. (ed.) *Water, Peace and the Middle East*. I.B. Tauris, London, pp. 21–28.

Bakhbakhi, M. (2002) Hydrogeological frameworks of the Nubian sandstone aquifer system. In: Appelgren, B. (ed.) *Managing Shared Aquifer Resources in Africa*, IHP-VI, Series on Groundwater No. 8, UNESCO, Paris (with International Hydrological Programme, General Water Authority of Libya), pp. 177–202.

Burke, J.J. and Moench, M.M. (2000) Groundwater and society: resources, tensions and opportunities. Themes in groundwater management in the twenty-first century. UNDP & ISET, New York.

Chapagain, A.K. and Hoekstra, A.Y. (2003) Virtual water flows between nations in relation to international trade in livestock and livestock products. Value of Water Research Report Series No.13, UNESCO-IHE.

Edmunds, M. and Wright, E.P. (1979) Groundwater recharge and paleoclimate in the Sirte and Kufra Basin, Libya. *Journal of Hyrdology* 40, 215–241.

FAO AQUASTAT (2003) Agriculture Data, Rome: Food and Agriculture Organization of the United Nations. Available at: http://www.fao.org/, http://apps.fao.org/page/collections?subset=agriculture

Handley, C. (2001) *Water Stress: Some Symptoms and Causes: A Case Study of Ta'iz, Yemen.* SOAS Studies in Development Geography. Ashgate Press, Aldershot, UK.

Hoekstra, A.Y. (2003) Virtual water trade between nations: a global mechanism affecting regional water systems. *Global Change Newsletter*, IGBP, No. 54, June 2003.

Hoekstra, A.Y. and Hung, P.Q. (2002) Virtual water trade: a quantification of virtual water flows between nations in relation to international crop trade. Value of Water Research Report Series No. 11, IHE, The Netherlands.

Khater, A.R. (2003) Intensive groundwater use in the Middle East and North Africa. In: Llamas, M.R. and Custodio, E. (eds) *Intensive Use of Groundwater: Challenges and Opportunities*. A.A. Balkema, Lisse/Abingdon, Exton, Tokyo, pp. 355–386.

Knesset (2002) Report of the Parliamentary Committee of Inquiry on the Israeli Water Sector, Headed by M.K. David Magen. The Knesset, Jerusalem, June 2002. (English translation)

Lichtenthaler, G. (2003) *Political Ecology and the Role of Water: Environment, Society and Economy in Northern Yemen*. King's SOAS Studies in Development Geography. Ashgate Press, Aldershot, UK.

Llamas, R. and Custodio, E. (eds) (2003a) *Intensive Use of Groundwater: Challenges and Opportunities*. A.A. Balkema, Lisse/Abingdon, Exton, Tokyo.

Llamas, R. and Custodio, E. (2003b) Intensive use of groundwater: a new situation which demands proactive action. In: Llamas, R. and Custodio, E. (eds) *Intensive Use of Groundwater: Challenges and Opportunities*. A.A. Balkema, Lisse/Abingdon, Exton, Tokyo.

Puri, S. (2002) Management of transboundary aquifers: contribution to the water needs of Africa. In: Appelgren, B. (ed.) *Managing Shared Aquifer Resources in Africa*. IHP-VI, Series on Groundwater No. 8, UNESCO, Paris (with International Hydrological Programme, General Water Authority of Libya), pp. 39–46.

Rainwater, K., Stovall, J., Frailey, S. and Urban, L. (2005a) Impact of political, scientific, and non-technical issues on regional groundwater modeling in Texas, USA. In: Alsharhan, A.S. and Wood, W.W. (eds) *Water Resources Perspectives: Evaluation, Management and Policy*. Elsevier Science, Amsterdam, pp. x–xx.

Rainwater, K., Toope, S. and Allan, J.A. (2005b) Managing and allocating water resources: adopting the integrated water resource management approach. In: Alsharhan, A.S. and Wood, W.W. (eds) *Water Resources Perspectives: Evaluation, Management and Policy*. Elsevier Science, Amsterdam, pp. x–xx.

Reisner, M. (1984) *Cadillac Desert*. Penguin Books, New York (reprinted 1993).

Riaz, Khalid (2002) Tackling the issue of rural–urban water transfers in the Ta'iz region, Yemen. *Natural Resources Forum* 26, 89–100.

Rouyer, A.R. (2000) *Turning Water into Politics: The Water Issue in the Israeli–Palestinian Conflict*. Macmillan, London.

Salem, O. and Pallas, P. (2002) Transboundary aquifers: scientific–hydrogeological aspects. In: Appelgren, B. (ed.) *Managing Shared Aquifer Resources in Africa*. IHP-VI, Series on Groundwater No. 8. UNESCO, Paris (with International Hydrological Programme, General Water Authority of Libya), pp. 19–22.

Selby, J. (2005) The geopolitics of water in the Middle East: fantasies and realities. *Third World Quarterly* 26(2), 329–349.

Wright, E.P. (1986) Review of the hydrogeology of the Kufra Basin, North Africa. Proceedings of the Workshop on Hydrology, organized by the Special Research Project Arid Areas at the Technical University, Berlin.

Wright, E.P. and Edmunds, M. (1971) Hydrogeological studies in Central Cyrenaica, Libya. Symposium on the geology of Libya.

5 Sub-Saharan Africa: Opportunistic Exploitation

MUTSA MASIYANDIMA[1] AND MARK F. GIORDANO[2]

[1]*International Water Management Institute, 141, Cresswell Street, Weavind Park 0184, Pretoria, South Africa;* [2]*International Water Management Institute, 127 Sunil Mawatha, Pelawatte, Battaramulla, Sri Lanka*

Introduction

Better control of water is often cited as one of the most important elements for improving agricultural performance and the livelihood of the rural poor in sub-Saharan Africa (SSA). A key reason for this contention is the high variability in the region's natural water supplies. In fact, the spatial and temporal unevenness of the area's water resources (rainfall, river flows and groundwater) is perhaps the greatest of any major region of the world and is of concern in both low and high rainfall areas. While the construction of surface water storage and irrigation could help to even out water distribution – making it easier to take advantage of the Green Revolution and other technologies that revolutionized agricultural landscapes and food supplies in much of Asia – physical, economic and political factors have often hindered their development in SSA. In such circumstances, groundwater would seem to have great potential for a variety of reasons.

Groundwater has been described as a perennial source of water (Calow *et al.*, 1997), a much needed buffer during times of drought (Carter, 1988, in Carter, 2003), and a resource that can be developed for localized use (Butterworth *et al.*, 2001). Carter (2003) even describes groundwater as the ultimate resource for use at local scale, both because it lends itself to incremental development at relatively low cost and because it is more resilient to interannual variability than surface water is. With reference to groundwater, availability where it is needed reduces the need for large-scale infrastructure investments and low variability obviously counters fluctuations in surface supplies – two key issues in the region. Despite these positive and potential attributes, especially in the SSA context, groundwater plays only a relatively limited role.

The reason for the modest groundwater use across all sectors is partly because the hydrogeologic formations underlying most of SSA are not of the type necessary to supply large-scale water resources development. However, the lack of similarity to traditional groundwater regions can lead to an underappreciation of the use that does exist in SSA. At the extreme, an estimated 80% of human (mostly rural) and livestock populations in Botswana depend entirely on groundwater (Chenje and Johnson, 1996, in Nicol, 2002), with groundwater contributing up to 65% of all water consumed (Noble *et al.*, 2002). Groundwater is also critical for livestock production in large parts of the Sahel and East Africa. Similarly it plays a critical role in supplying water for small-scale but highly valuable irrigation as well as in stabilizing water supplies in times of drought. Numerous reports highlight the major role groundwater plays in rural domestic supply (BGS, 2000; Carter, 2003). The data on groundwater use are often found distributed among many agencies – donor offices, central government departments and local governments. Other abstraction points are unknown as they are privately installed. It is therefore difficult to estimate actual numbers involved, but the majority of poor rural households depend on groundwater for domestic supply, livestock, crop production and other purposes. Thus an appreciation of the impact of groundwater use in SSA agriculture goes beyond simple calculations of irrigated area to include livestock maintenance, drought mitigation and broader rural livelihood support.

In the past, there have been few attempts at broad-scale research on the role of groundwater in agricultural livelihood in the SSA context and even fewer attempts to quantify that role. As knowledge on this subject is relatively poor, the goal of this paper is to develop as full a picture as possible based generally on published information so as to consolidate known information, highlight critical gaps and inform further research on groundwater and its potential role in solving Africa's water and poverty problems. The paper is divided into four parts: an overview of the known groundwater resources of SSA and their relationship to human population; an overview of agricultural groundwater use and extent, highlighting groundwater's various roles and their possible contribution to rural livelihoods; the state of groundwater governance; and a set of recommendations for development of, and research on, groundwater in SSA.

Groundwater Resources of Sub-Saharan Africa

Understanding the general distribution of water resources in SSA is made difficult by the paucity of data. According to the FAO (2003b, p. 51): 'The information available is uneven and very poor for some of the African countries.' In addition to basic data problems, the distribution of water within Africa is not equal and the continent has the greatest spatial, and temporal, supply variability of any region in the world (Walling, 1996), thus making broad overviews difficult. In general, though, rainfall is greatest on the Guinea coast and in the west-central regions, and drops as one moves east and away from the equator. Low rainfall regions also tend to have irregular rainfall, often leading to crop failures. The unequal rainfall distribution is offset to some degree by the prevalence of exotic rivers such as the Niger, Nile and Okovango. The rainfall and surface water patterns, along with underlying geology, determine groundwater availability, accessibility and its utility for agricultural use.

SSA is generally divided into four main hydrogeological provinces: crystalline basement complex, volcanic rock, consolidated sedimentary rock and unconsolidated sediments (Fig. 5.1). Of the four provinces, the basement complex is largest and occupies over 40% of the area including most of West Africa as well as Zambia, Zimbabwe, the northern belt of South Africa and northern

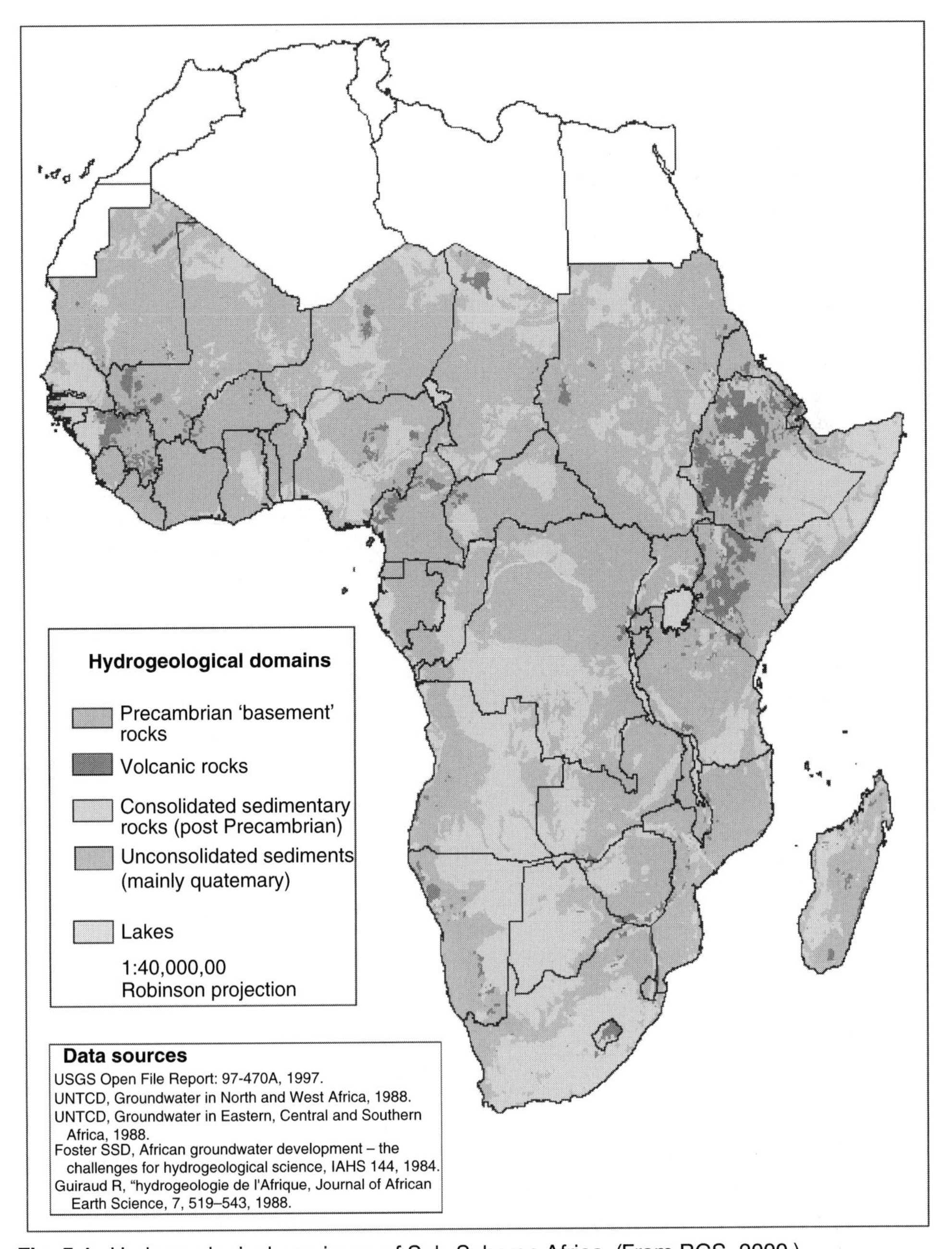

Fig. 5.1. Hydrogeological provinces of Sub-Saharan Africa. (From BGS, 2000.)

Mozambique. Basement complex aquifers have very little or no primary porosity and the groundwater in them is held in the weathered mantle and in fissure zones. These aquifers are characterized by poor storage and low yields, typically less than 1 l/s (Field and Collier, 1998; UNEP, 2003, p. 17).

The second largest aquifer complex, consolidated sedimentary rock, underlies 32% of the area and can hold substantial groundwater reserves (Walling, 1996). However, mudstone areas, which make up approximately two-thirds of this variety, store little groundwater. Most of South Africa, Botswana, southern Angola, eastern Namibia, eastern Democratic Republic of Congo, north-west Zimbabwe and western Zambia are underlain by this aquifer type with limited groundwater occurrence.

Unconsolidated sediments make up 22% of the region's area and often hold groundwater in unconfined conditions within sands and gravels. These aquifers are often found in river beds, and so their groundwater may be especially important for human use due to potential ease of access (MacDonald and Davies, 2000). Unconsolidated sediments are found along the Limpopo River and several of its tributaries, and also in coastal areas, such as the Cape Flats aquifer in South Africa and the coastal zones of the countries at the horn of Africa (Fig. 5.1). However, Purkey and Vermillion (1995) note that many African river systems are typified by fine to very fine sediments, rather than coarse sand and gravel, thus reducing extraction possibilities.

Volcanic rocks cover only about 6% of SSA. In paleosoils and fractures between lava flows they can produce high groundwater yields and supply springs (MacDonald and Davies, 2000). In Djibouti, where groundwater represents 98% of all water used, volcanic aquifers are an important source of water (Jalludin and Razack, 2004). However, in other volcanic areas, groundwater storage can be highly limited (Walling, 1996).

To exemplify the low-yielding aquifers in many parts of SSA, Table 5.1 shows the typical yields in the main aquifers found in South Africa where groundwater studies have been more rigorous than elsewhere in SSA. In Botswana, yields of up to 27 l/s (Table 5.2) have been reported, but generally yields are less than 5 l/s. Where high yields have been found, these have been unsustainable in the long term as they decline rapidly due to limited storage in lower layers of the aquifers (Water Surveys Botswana, Colombo, 2003, unpublished data). In

Table 5.1. Examples of favourable yield characteristics for major aquifers, South Africa (the hydrogeological provinces indicated here are the authors' inferences).

Aquifer type	Hydrogeological province	Typical yield[a] (l/s)
Alluvial deposits	Unconsolidated sediments	3–8
Coastal sands	Unconsolidated sediments	3–16
Karoo sediments	Unconsolidated sediments	1–3
Table mountain sandstone	Consolidated sediments	1–10
Dolomite (Karst)	Consolidated sediment	20–50
Granite (weathered)	Basement complex	5–10

[a]From DWAF (1998, p. 33).

Table 5.2. Borehole yields in selected well fields in Botswana. (From Department of Water Affairs, 2000.)

Wellfield	Hydrogeological province[a]	Average borehole yield 1998–2000[b] (l/s)
Palla Road	Unconsolidated sediments	11.11
Kanye	Unconsolidated sediments	10.47
Serowe	Unconsolidated sediments	1.75
Palapye	Basement complex	5.92
Gaotlhobogwe	Basement complex	27.78
Molepolole	Basement complex	6.47
Thamaga	Basement complex	4.17
Malotwane	Basement complex	3.39
Letlhakane	Unconsolidated sediments	6.22
Lecheng	Basement complex	2.53
Shoshong	Basement complex	1.75
Moshupa	Basement complex	1.58
Metsimotlhabe	Basement complex	1.53
Mochudi	Basement complex	1.39
Chadibe	Basement complex	0.94
Sefhare	Basement complex	2.67
Pitsanyane	Unconsolidated sediments	1.66

[a]From WMA Report to IWMI (2003).
[b]From Department of Water Affairs (2000).

the basement complex aquifer in Burkina Faso, yields are typically less than 1 l/s, whereas in the sedimentary aquifers yields reach 27 l/s (Obuobie and Barry, forthcoming). Planning for the use of groundwater in basement complex aquifers is further complicated by large seasonal variation in groundwater levels. These have been observed to range from 1 to 5 m in basement complex aquifers (Chilton and Foster, 1995). Depth to extractable groundwater appears to be another limiting factor for its use in SSA. In the Limpopo basin in South Africa depth to groundwater is highly variable, and borehole depths range from 50 to more than 100 m. In Lesotho, groundwater occurs mostly at depths of more than 50 m; in Zambia most boreholes are drilled to 44 m depth (Wurzel, 2001); in Zimbabwe borehole depths range from 25 to more than 100 m (Interconsult, 1986). In Mozambique, depth to extractable groundwater is up to 35 m in some areas, but can be up to 100 m in others. The high costs of abstraction associated with groundwater use including costs of unsuccessful drilling are seen as a major drawback to the use of groundwater in SSA.

The relationship between population distribution and SSA's groundwater provinces provides some insights into current agricultural groundwater use patterns and potential future development. Around three quarters of the SSA population lives in areas of poor groundwater availability, with 220 million people in low-yielding crystalline basement complex areas and about 110 million in areas of consolidated sediment. In these areas dwell most of the rural population, the socio-economic group often affected by problems of water access and who could potentially benefit from groundwater use. But because of the limiting factors alluded to above, there is a limit to how much groundwater they can use and the extent to which groundwater can impact their livelihood. Another 15% (60 million) of the population lives in areas with unconsolidated sediment, though most are not near areas with easy access to productive alluvial aquifers. The remaining 10% of the population (45 million) lives in volcanic rock zones with high but variable groundwater potential.

The most comprehensive water resource availability and use database for SSA to date is the FAO AQUASTAT. Although this database was originally designed with reference to agricultural use, it remains the most complete source of data for SSA. This database shows that for many of the SSA countries, groundwater is a small component of overall renewable water resources, suggesting limited contribution of groundwater to overall water requirements. Only 11 out of the 45 SSA countries listed in the AQUASTAT database have at least 10% of their renewable water resources made up of groundwater (Table 5.3),[1] and only 6 of these countries have per capita groundwater availability above 1000 m^3. Per capita water availability of surface water in Africa is generally much higher than the groundwater availability indicated here (see Savenije and van der Zaag, 2000),

Table 5.3. Groundwater availability and use in sub-Saharan Africa. Source: AQUASTAT, literature.

Country	Groundwater produced internally[a] (km^3/year)	Groundwater/ total renewable water resources[a]	Per capita groundwater availability[b] (m^3/year)	Information on use available[c]
Angola	2	0.01	179	No
Benin	0.3	0.03	40	No
Botswana	1.2	0.41	732	Yes
Burkina Faso	4.5	0.36	323	Yes
Burundi	0.1	0.03	16	No
Cameroon	5	0.02	305	No
Cape Verde	0.1	0.33	239	No
Central African Republic	0	0.00		No
Chad	1.5	0.10	153	No
Comoros	1	0.83	1490	No
Congo	0	0.00	–	No
Democratic Republic of Congo	1	0.00	17	No
Cote d'Ivoire	2.7	0.04	156	No
Djibouti	0	0.00	–	Yes
Equatorial Guinea	1	0.04	1866	No
Eritrea		0.00	–	No
Ethiopia	0	0.00	–	Yes
Gabon	2	0.01	1440	No
Gambia	0	0.00	–	No

Table 5.3. *Continued*

Country	Groundwater produced internally[a] (km^3/year)	Groundwater/ total renewable water resources[a]	Per capita groundwater availability[b] (m^3/year)	Information on use available[c]
Ghana	1.3	0.04	62	Yes
Guinea	0	0.00	–	No
Guinea-Bissau	4	0.25	2825	No
Kenya	3	0.15	89	Yes
Lesotho	0	0.00	–	No
Liberia	0	0.00	–	No
Madagascar	5	0.01	277	No
Malawi	0	0.00	–	No
Mali	10	0.17	814	Yes
Mauritius	0.2	0.09	163	No
Mozambique	2	0.02	103	No
Namibia	2.1	0.34	1034	Yes
Niger	2.5	0.71	214	Yes
Nigeria	7	0.03	54	Yes – limited
Rwanda	0	0.00	–	No
Senegal	2.6	0.10	234	No
Sierra Leone	10	0.06	1662	No
Somalia	0.3	0.05	35	No
South Africa	1.8	0.04	41	Yes
Sudan	2	0.07	50	Yes – limited
Swaziland	–	0.00	–	No
Tanzania	2	0.02	54	No
Togo	0.7	0.06	123	No
Uganda	0	0.00	–	No
Zambia	0	0.00	–	Yes
Zimbabwe	1	0.07	78	Yes

[a]Derived from AQUASTAT.
[b]From http://www.geohive.com
[c]From literature.

though in regions without surface water, groundwater becomes the only source available. Thus, to the region as a whole, groundwater will only play a relatively small role in agriculture because of the absolute levels of resource availability and the size of the resource relative to surface water. However, such generalizations can be misleading in national or even subnational contexts because of the great spatial and temporal variability in both ground and surface supplies.

Agricultural Groundwater Use in SSA

As shown in Table 5.3, national statistics on water use are not readily available for most countries. As such it is clear that one must consult multiple and often inconsistent data sources to paint even a rudimentary picture of its use at a continental scale. This approach is of course fraught with problems. For example, most use appears to be in small rural villages, where boreholes and wells have been installed by multiple agencies: government, individuals, non-governmental organizations (NGOs) and relief agencies. This use is scattered and individually quite small and therefore both difficult to measure and seemingly inconsequential. As a result, it frequently goes unreported (UNEP, 2003, p. 2), and total use tends to be underestimated. The cumulative impact of even small-scale uses of groundwater can be significant as the many scattered boreholes in Burkina Faso (Fig. 5.2) or the many shallow wells and deep tube wells in wadi systems (alluvial aquifers) in Djibouti (Jalludin and Razack, 2004) illustrate.

Examining the history of groundwater development in the region also highlights the difficulties in collecting meaningful statistics for groundwater use. For example, in southern Africa some of the literature relating to groundwater use is project-based or localized. Often, one has to consult multiple sources (government, consultants, NGOs and even individual water users) to construct a meaningful database relating to use. Here, an attempt is made at classifying groundwater use according to development objective and the agents responsible for installation of boreholes or wells (Table 5.4). One of the challenges arising from such a model of groundwater development is poor coordination of agents and the difficulty in trying to establish the actual extent of groundwater use or the number of boreholes drilled and used. In this scenario, it is also very difficult to capture the extent of groundwater use for livelihood and other purposes and its overall contribution to the economy.

Despite the problems associated with lack of data or incomplete data, some of the data available do present a picture that agricultural groundwater use is important at local scales in parts of SSA. For example, in the Limpopo province in northern South Africa there are reportedly more than 35,000 boreholes mostly used for domestic water and irrigation of small gardens, and Asian-style growth rates (see Wang *et al.*, Chapter 3, and Sakthivadivel, Chapter 10, this volume) in development have been documented (Tewari, forthcoming). In semi-arid Botswana, water supply is largely groundwater-based (Brunner *et al.*, 2004). Groundwater in Botswana is mostly used for rural, domestic and livestock purposes and this has steadily increased over the last 30 years, as shown by the number of registered boreholes in the country (Fig. 5.3). The increase in groundwater use in Botswana has been accompanied by overdraft as the abstraction is presumably greater than recharge (Kgathi, 1999). Such use of groundwater is mirrored in several other countries. Pockets of small-scale groundwater irrigation are found in Tanzania where reportedly 200 ha are irrigated using diesel and electric pumps; and in Malawi and Zimbabwe where collector wells are used to abstract water from weathered basement complex aquifers (FAO, 1997). In Cameroon, groundwater makes up only 2% of renewable water resources (Table 5.3). Yet, in the north of the country, where reservoirs are limited and precipitation is lower than the national average, groundwater is the most widely available water resource and is used for domestic, agricultural and industrial purposes (Njitchoua *et al.*, 1997). Similarly, in Borno and Yobe of Nigeria's Lake Chad basin, groundwater is the predominant source of domestic

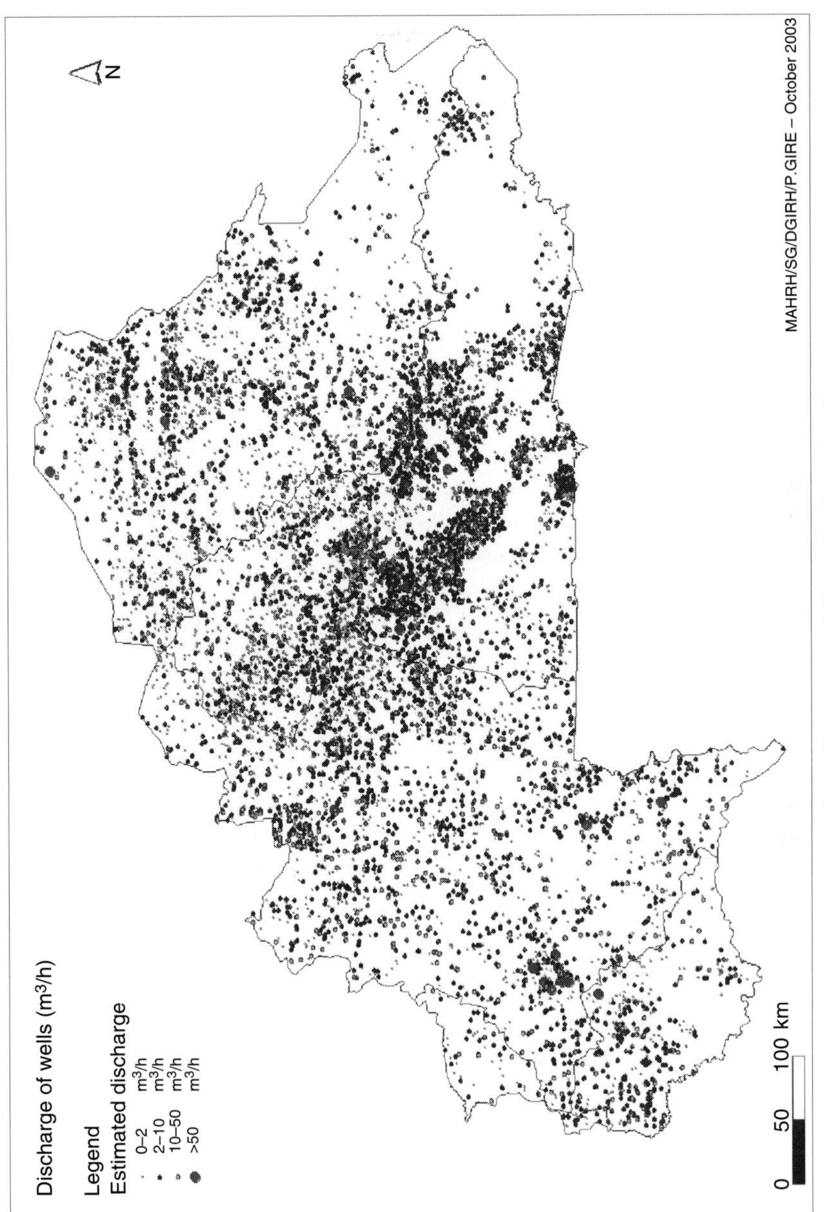

Fig. 5.2. Wells across Burkina Faso. (From Obuobie and Barry, forthcoming.)

Table 5.4. Types of groundwater use in SSA.

Type of groundwater use	Purpose	Responsible agent for borehole/well installation
Drought mitigation	Livestock watering Agriculture (crops)– bridging mechanism so that crops do not fail to mature Domestic water supply	Individuals Government
Normal supply	Domestic water Commercial irrigation	Individuals NGOs Municipalities Government in the case of rural communities (both central and local)
Emergency relief	Domestic water during drought years Stock water	NGOs CBOs Governments Churches
Social responsibility activities	Boreholes installed as part of ongoing aid and development activities	NGOs CBOs Governments Churches

NGO – non-governmental organization.
CBO – community-based organization.

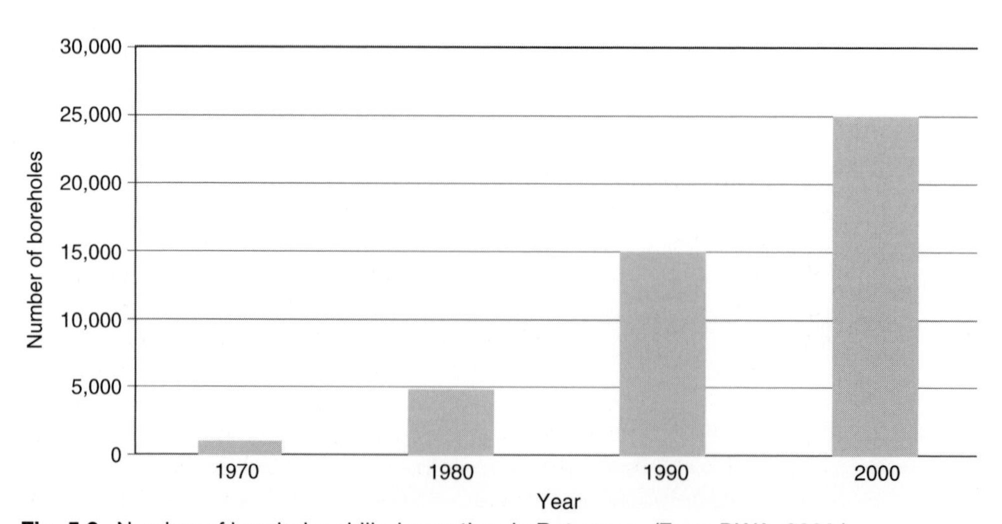

Fig. 5.3. Number of boreholes drilled over time in Botswana. (From DWA, 2002.)

water and for other non-irrigation uses, and more than 2000 boreholes are used in the two states alone (Bunu, 1999).

In addition to the numerous small-scale groundwater uses in SSA such as those mentioned above, large-scale commercial irrigation occupies the largest usage of groundwater, especially in South Africa in the Karst aquifer region of the upper Limpopo River basin in the north-west province where about 77 million cubic metres are abstracted annually for irrigation (IUCN, 2004), and in the wider Limpopo Water Management Area where about 850 million cubic metres are abstracted annually for irrigation (Basson *et al.*, 1997). Also, the Karst aquifers in the Lomagundi area (central Zimbabwe) and the Nyamandhlovu aquifer (western Zimbabwe) are exploited for commercial irrigation (Masiyandima, forthcoming). Irrigation officials in Zimbabwe estimate that more than 17,000 ha are irrigated commercially using groundwater.

Combining AQUASTAT figures with the results of a set of county surveys and some assumptions, Giordano (2006) estimated that there were perhaps 1 million hectares of groundwater irrigation in SSA. Although this is a rough estimate, it gives some indication of the possible direct role of groundwater in agricultural production in SSA. In an effort to measure the value of groundwater in other regions where use is more widespread and forms part of broader irrigated settings, irrigated area or the volume of water applied can be a reasonable measure of agricultural impact. By such measures, the value of groundwater in SSA is clearly small given the region's physical size and rural population. Yet groundwater is still considered the resource of choice in many, particularly rural, areas. The importance accorded to groundwater in parts of SSA is reflected by the number of site-specific studies on certain aspects of groundwater use such as recharge (Taylor and Howard, 1996; Njitchoua *et al.*, 1997; Brunner *et al.*, 2004). There have been many other groundwater recharge studies: in the Kalahari in Botswana (de Vries *et al.*, 2000), Ghana (Asomaning, 1992), Kenya (Singh *et al.*, 1984), Uganda (Howard and Karundu, 1992), Zambia (Houston, 1982) and Zimbabwe (Houston, 1990). Most studies try to quantify available groundwater resource from recharge.

Given the general belief that groundwater has been relatively undeveloped in SSA, it is not surprising that most studies focus on increased use. Yet, there are indications from a number of regions that the 'development' stage discussed by Shah and Kemper (respectively Chapter 2 and Chapter 8, this volume) has already been passed and overabstraction is now the issue. For example, farmers in the Dendron area in the Limpopo province of South Africa have experienced declining water levels over the last two decades in the aquifers that supply all of their irrigation (Masiyandima *et al.*, 2001). Similar problems have been reported in the Nubian aquifer system (which is admittedly a fossil system with no recharge) in northern SSA (Ulf and Manfred, 2002) and in other arid and semi-arid environments such as in Botswana. Abstraction of groundwater from Botswana's aquifers generally exceeds annual recharge (Kgathi, 1999). This is manifested by the declining water levels in several well fields. According to the Department of Water Affairs (Botswana), in some well fields groundwater levels are declining by as much as 2.6 m/year. Clearly there is little scope of additional groundwater development in such areas.

Livestock production

Large areas of savannah, semi-desert and desert areas in SSA are typified by livestock, rather than crop, production. While cattle tend to dominate the livestock economy, sheep, goats and, especially in deserts or near-desert environments, camels can also play important roles. In general, cattle density is highest in the Sahel region and roughly along the line from Ethiopia along the rift valley to South Africa and Lesotho (Thornton *et al.*, 2002). Livestock production is also pronounced in the drier areas of southern and eastern Africa, particularly in Botswana and Kenya.

In these arid areas, groundwater plays a critical role in the maintenance of the livestock economy, which is itself the basis of human survival of the poorest segments. In Somalia, for example, the only agricultural use of groundwater is for livestock watering (Ndiritu, 2004, unpublished data). In Botswana, a major livestock-producing country in southern Africa, groundwater is the main source of stock water. For Ghana, it is estimated that 70% of cattle and 40% of other livestock production account for 4.5% of agricultural gross domestic product (GDP) and all depend entirely on groundwater use (Obuobie and Barry, forthcoming). As a general indication of the role of livestock in rural livelihood and the role of groundwater in sustaining those livelihoods, the FAO (1986, p. 137) states that 'groundwater is more widespread than surface water in the Sahel, although it is at present exploited mainly for domestic and livestock purposes, from traditional wells with yields too low for irrigation'. As with irrigation, quantification of the contribution of groundwater to SSA's total livestock economy, based on published sources, is problematic. The World Bank has estimated that 10% of SSA's population is directly dependent on livestock production (McIntire *et al.*, 1992). Thornton *et al.* (2002) estimated that there are more than 160 million poor in SSA, and roughly one-third of the total population keep livestock. Given that a large share of livestock production is likely groundwater-dependent, the value of groundwater in SSA's overall livestock economy and in the livelihood of its poorest residents is clearly substantial.

Drought mitigation

Since groundwater supplies are less correlated with rainfall than surface supplies, one of groundwater's key functions can be its ability to mitigate the effects of erratic rainfall or drought on agricultural production. While this function is of global importance, it may be especially so in SSA where temporal rainfall variability, as outlined earlier, is amongst the highest in the world. In fact, African pastoral societies have taken advantage of groundwater to mitigate the impact of temporal variation in rainfall supply for centuries. The focus is now on the role of groundwater in moderating the impacts of drought on domestic water supply to rural communities (Gillham, 1997) and on crops. A case in point is the considerable expansion of irrigation in general, including wells, following the 1968–1973 droughts in Sahel (Morris *et al.*, 1984, p. 14). There are also numerous papers that highlight that role (Amad, 1988; Calow *et al.*, 1997).

In contrast to valuations of groundwater supply in crop and livestock production where relatively straightforward estimates can be made based on total area, number of animals or value of output (if data are available), estimating the drought mitigation value of groundwater is complicated by two primary reasons. First, the knowledge that groundwater is available as an alternative to surface or rainwater reduces risk and makes farming and livestock production possible in areas where it would otherwise not occur. Thus the value of some production based on non-groundwater sources, especially in marginal lands, can in fact be attributed to groundwater. Second, the role of groundwater in drought mitigation highlights the issue of *marginal*, as opposed to average, valuation of water resources.

Rural domestic supplies

Groundwater plays a role in providing domestic supplies to the rural population in many countries in SSA. According to the Southern African Development Community (SADC) statistics, groundwater is the primary source of drinking water for both humans and livestock in the driest areas of SADC, and it is estimated that about 60% of the population depends on groundwater resources for domestic water. In the Limpopo River water management area in South Africa rural domestic supply accounts for 55 million cubic metres of groundwater abstracted, or just more than 20% of all groundwater abstracted. Admittedly, groundwater resources in SSA are modest, but are sufficient and important at local levels as they are the main resource for water supply for rural populations, e.g. in parts of rural Zimbabwe, Mozambique (Juizo, 2005), Zambia and Botswana). Although precise numbers are lacking, it is likely that most domestic water supply in rural SSA is currently from groundwater and that expansion in rural supplies in the near future will likely be from groundwater sources. Further, within the rural sector, domestic use, rather than agriculture or livestock, appears to account for the vast majority of demand. This was true, for example, in all cases examined in SSA with the exception of South Africa (Obuobie and Giordano, forthcoming). Groundwater thus provides the foundation for rural livelihood whether or not it is directly used in agricultural or livestock production.

Mining

In Botswana and South Africa, groundwater is particularly important for mining. In Botswana, mining accounted for more than 60% of all abstractions at the turn of the century. In South Africa, the mining operations for platinum, diamond, tin, chrome, fluorspar, graphite, granite, silicon, vanadium, copper, manganese and coal in the Limpopo province depend largely on groundwater. In 2002, more than 70% of the water used for mining in the province was groundwater. In South Africa, the mining demand is overshadowed by both irrigation and domestic water demand but is expected to grow as the mining sector promises to remain a strong economic driver in the Limpopo province. The problem with groundwater use for mining may not be related as much to the volumes extracted as to the contamination of both surface and groundwater resources associated with it.

Urban use

In addition to small-scale use of groundwater in rural areas, there is pronounced use in many urban centres. The large cities that are groundwater-dependent in SSA are shown in Fig. 5.4. Even in cases in which groundwater is a small fraction of total water use, it represents a stable source of water, which is one of its important characteristics, particularly in dry years. In addition to the large

Fig. 5.4. Groundwater-dependent cities in SSA. (From UNEP, 2003.)

cities shown in Fig. 5.4, there are several urban centres that depend on groundwater which are not included in this map. Many small towns are dependent on groundwater for water supply. This is the case in Burkina Faso (Obuobie and Barry, forthcoming) and in Botswana's so-called minor villages. In South Africa about 105 towns depend entirely on groundwater (Tewari, 2002). It is generally accepted that many people in SSA depend on groundwater for drinking water supply. However, the actual number of people using groundwater for this purpose is unknown (see UNEP, 2003, p. 3). The fact that urban use is widespread shows that it is not as much a question of availability and accessibility and economic feasibility as it is of economic means and political decision and will to develop it.

Groundwater, livelihood and poverty

Groundwater use in SSA clearly contributes to livelihood through agricultural production – in the form of irrigation supply, livestock support and drought mitigation and in domestic supplies as outlined above. In the context of SSA, the benefits of groundwater use likely accrue primarily to the poor, because they make up the vast majority of rural agricultural producers. While the general connections between groundwater, livelihood and poverty in SSA are clear, quantifying the role that agricultural groundwater use plays in poverty alleviation and livelihood support is difficult.

Small rural communities in many southern and east African countries make use of groundwater from shallow aquifer systems associated with wetlands to produce crops both for household consumption and sale. In surveys carried out in about 20 communities across the southern African region,[2] wetland crop production contributed up to 50% of household food, and more than 50% of total annual household income (Masiyandima *et al.*, 2004). If we assume that 20% of the wetland systems in Zambia and Malawi are cultivated for such uses, about 600,000 ha are under cultivation. At an average annual household gross income of about $200/ha, the total gross income from such groundwater use is estimated to be well over $100 million. This can be compared to a value of $50–55 billion for the irrigation economy of India (Shah, Chapter 2, this volume).

While recognition of groundwater use in SSA wetlands is generally low, it is in fact better recognized than other small-scale uses of groundwater. In general, data on this sector are often limited and data on groundwater use, in particular that related to small-scale uses by poor farmers, are often non-existent as already discussed. Even government departments responsible for groundwater sometimes do not seem to have accurate information regarding the groundwater situation. While information from some government agencies indicates that the area under smallholder irrigation in South Africa is quite small (Nel, 2004, Pretoria, South Africa personal communication),[3] Busari and Sotsaka (2001) found that there is at least one community garden in each of the 70 villages around the Giyani area in the Limpopo basin, with gardens ranging in area from 1 to 25 ha. Community gardens are also to be found in many other villages across the Limpopo province. In 2001, the Limpopo province Department of Agriculture had a database with some of the community gardens irrigated with groundwater.[4] On the basis of pumping hours and pump discharges detailed in the database, abstraction for irrigating community gardens is estimated at about 3 million cubic metres annually, less than 2% of the reported groundwater abstraction for irrigation in the Limpopo water management area.[5] While this use may appear extremely modest from a water-accounting standpoint, it plays a significant role in the lives of the farmers who use it, enabling them to produce food and reduce their dependence on government and donor agencies for food.

In considering the impact of groundwater on livelihood and poverty in SSA, it is important to consider the costs associated with the use of groundwater, particularly drilling and operation and maintenance of equipment. In comparison with India, and perhaps other regions, such costs are high in SSA. Drilling costs, though variable across the continent, are still largely prohibitive. Wurzel (2001) estimated the average drilling cost in Africa to be $100/m, more than tenfold that in India. In 1996, borehole drilling costs were approximately $37/m in Mozambique while in Lesotho it was $23/m (Wurzel, 2001). In Zimbabwe drilling costs were estimated to be about $40/m in 2004 (Masiyandima, forthcoming). Combining these costs with the poor drilling success rate for boreholes (common in hard-rock areas), the cost of development of groundwater may still be difficult to justify in many places, even for targeted use such as rural domestic water supply.

Groundwater Governance

In SSA, there are customary or traditional mechanisms to regulate groundwater use in some areas. However, there have been relatively few efforts to develop formal groundwater governance mechanisms in most of the continent. This may be in part because of the general belief that groundwater potential has not been fully exploited and so the need for governance has not generally arisen. The lack of formal groundwater governance mechanisms may also be related to the fact that formal water policy in general has not received much emphasis until recently. Examples of this can be observed in Burkina Faso and Ghana where national water policies are still to be put into practice. Whatever the case, in many countries in SSA, the mechanisms for water governance in general, at least formally, were weak or non-existent prior to the recent set of water policy reforms that sprouted across Africa since the late 1990s. If the situation for surface water is bad, mechanisms for groundwater governance are as bad or worse.

However, the past few years have been marked by significant reforms in the water sectors in a number of countries in SSA. The aims of the reforms are numerous and these are summarized by Van Koppen (2002) as:

- Better integrate the management of water resources (multiple-use sectors; quantitative and qualitative; beneficial and non-beneficial uses; surface and groundwaters; hydrological, legal and institutional aspects; water and other sectors; governments and other stakeholders).

- (Further) prioritize domestic water supply in rural areas usually through local government and in urban areas sometimes through new public–private partnerships for water supplies (Mozambique, South Africa, Uganda, Zambia).
- Harmonize fragmented pieces of formal legislation into new policy and legislation.
- Specify the role of the government – invariably the custodian of the nation's water resources – complementary to newly established decentralized basin authorities and in some cases national bodies, such as the Water Resources Commission in Ghana or parastatals like the Zimbabwe National Water Authority.
- Shift and decentralize the boundaries of lower-level water management institutions to basins in order to better match hydrological reality.
- Design and implement national water right systems, accompanied by water charges and taxing.
- Stimulate users' participation, especially in basin-level and lower-tier water management institutions.
- Protect water quality and environmental needs.
- Improve hydrological assessments and monitoring for surface and groundwaters and ensure public availability of data.
- Promote international cooperation in trans-boundary basins.
- Redress the race, gender and class inequities of the colonial past (in Zimbabwe and South Africa).

The limited use of groundwater has perhaps meant little need for governance structures in the past. The situation has changed in some areas, with problems of overabstraction arising. Potentially, such cases will benefit from some form of regulation, and the reforms in the water sector offer opportunities for better control and regulation.

Conclusions

Given the impacts of groundwater utilization on agriculture and livelihood in Asia and the many advantages of using groundwater, it is not surprising that groundwater is considered as an option for water supply for various uses and also as having an impact on poverty in SSA. However, this chapter has highlighted some of the reasons why agricultural groundwater use is, and will likely remain, relatively limited.

The main reason for the limited contribution of groundwater to overall water resources in SSA is the hydrogeology – low-yielding aquifers and depth of occurrence of the groundwater. This is compounded by the fact that the rural population that could benefit from the groundwater is located in areas with aquifers not suitable for large-scale abstraction of groundwater or with their supply not prioritized by national agents. However, groundwater has its role – for mitigating the impacts of drought, rural domestic supplies, stock water and irrigation at local scale. To obtain a better picture of current and potential future contribution in these areas, there is need for a shift from the traditional analyses focusing on national and regional scales to more local levels where the limited opportunities exist.

Even where groundwater is available, most of the rural poor who could benefit most from it are not in a position to pay the capital costs associated with developing the resource. We have seen in the case of South Africa that development costs are higher than in many other regions. Combining this with the fact that farmers are poorer than in many other regions means that groundwater does not lend itself to the fast development that has been seen elsewhere.

We will likely continue to see the benefits of groundwater for rural domestic use and livestock watering, as well as small-scale irrigation in SSA. Increase in use beyond these sectors is highly unlikely due to resource limitations and high costs associated with the use of groundwater. Groundwater use is best explored where such factors working against its use are minimal. This has happened in some cases – in Botswana and in agricultural regions in South Africa (where incomes are also relatively high) where groundwater has continued to expand despite the associated overdraft. Cases need to be evaluated on an individual basis, and opportunities exploited in the best possible way.

While it is likely that the groundwater resources of SSA can provide solutions to the problems of water accessibility faced by some of the region's agricultural and rural communities, the limitations highlighted in this report suggest that this role should be seen as strategic. Opportunistic use of groundwater should be followed. The major challenge in following strategic and opportunistic approaches is limited information. The focus of the effort on groundwater research in many of the SSA countries should be to consolidate available knowledge and begin to construct adequate data on availability and how then to foster finance to develop use in those strategic locations.

Notes

1 Derived from AQUASTAT statistics (FAO, 2003a,b).
2 Communities in South Africa, Swaziland, Tanzania, Zambia and Zimbabwe.
3 Jaco Nel is a geohydrologist with the Department of Water Affairs, Pretoria, South Africa.
4 Data obtained from Engineer Martinus Gouws, Limpopo Province Department of Agriculture (2001).
5 From the Limpopo Province Department of Agriculture (South Africa) community garden database.

References

Amad, M.U. (1988) Ground water resources: the key to combating drought in Africa. *Desertification Control Bulletin* 16, 2–6.

Asomaning, G. (1992) Groundwater resources of the Birmin basin in Ghana. *Journal of African Earth Sciences* 15(3/4), 375–384.

Basson, M.S., van Niekerk, P.H. and van Rooyen, J.A. (1997) Overview of water resources availability and utilisation in South Africa. Department of Water Affairs and Forestry, Pretoria, South Africa.

British Geological Survey (2000) A brief review of groundwater for rural water supply in Sub-Saharan Africa. Technical Report WC/00/33. Overseas Geology Series. Keyworth, Nottingham, UK.

Bunu, M.Z. (1999) Groundwater management perspectives for Borno and Yobe States, Nigeria. *Journal of Environmental Hydrology* 7, Paper 19.

Busari, O. and Sotsaka, T. (2001) Groundwater in the Olifants Basin: assessing viable alternatives for small-scale irrigation. Project Progress Report. Water Research Commission, Pretoria, South Africa.

Butterworth, J., Kogpotso, M. and Pollard, S. (2001) Water resources and water supply for rural communities in the Sand River Catchment, South Africa. 2nd WARFSA/WaterNet Symposium: Integrated Water Resources Management: Theory, Practice, Cases, Cape Town, 30–31 October 2001.

Calow, R.C. *et al.* (1997) Groundwater management in drought-prone areas of Africa. *Water Resource Development* 13, 241–261.

Carter, R.C. (1988) Groundwater Development for small-scale irrigation in Sub-Saharan Africa. Paper presented to the Hydrogeological group of the Geological Society, April 9.

Carter, R.C. (2003) *Water and Poverty: Redressing the Balance.* Institute of Water and Environment, Cranfield University, UK.

Chenje, M. and Johnson, P. (eds) (1996) Water in Southern Africa. SADC/IUCN/SARDC, Maseru/Harare.

Chilton, J.P. and Foster, S.D. (1995) Hydrogeological characterization and water-supply potential of basement aquifers in Tropical Africa. *Hydrogeology Journal* 3(1), 36–49.

Department of Water Affairs (2000) *Groundwater Monitoring Final Report,* Vol. 1. Geotechnical Consulting Services, Botswana.

FAO (1986) *Irrigation in Africa South of the Sahara*. Food and Agriculutre Organization of the United Nations, Rome.

FAO (1997) Potential for irrigation technology transfer and uptake. In: *Irrigation Technology Transfer in Support of Food Security*. Water Reports –14. Food and Agriculutre Organization of the United Nations, Rome. Available at: http://www.fao.org/docrep/W7314E/w7314e0g.htm

FAO (2003a) Aquastat on-line database. Food and Agriculture Organization of the United Nations, Rome. Available at: http://www.fao.org/ag/agl/aglw/aquastat/water_res/waterres_tab.htm

FAO (2003b) *Review of World Water Resources by Country*. Water Reports 23. Food and Agriculture Organization of the United Nations, Rome.

Field, W.P. and Collier, F.W. (1998) Checklist to assist preparation of small-scale irrigation projects in sub-Saharan Africa. Institute of Hydrology report to ICID, New Delhi, India.

Gillham, S.W. (1997) Drought relief in rural KwaZulu-Natal. 23rd WEDC Conference, Durban, South Africa.

Giordano, M. (2006) Agricultural groundwater use and rural livelihoods in ub-Saharan Africa: a first-cut assessment. *Hydrogeology Journal* 14, 310–318.

Houston, J.F.T. (1982) Rainfall and recharge to dolomite aquifer in a semi-arid climate at Kabwe, Zambia. *Journal of Hydrology* 59, 173–187.

Houston, J.F.T. (1990) Rainfall-Run off–Recharge relationships in the basement rocks of Zimbabwe. In: Lerner, D., Issar, A.S. and Simmers, I. (eds). Groundwater Recharge; a guide to understanding and estimating natural recharge, Hannover, 271–283.

Howard, K.W.F. and Karundu, J. (1992) Constraints on the exploitation of basement aquifers in East Africa – water balance implications and the role of the Regolith. *Journal of Hydrology* 139(4),183–196.

Interconsult (1986) National Master Plan for Rural Water Supply, Vol. 2: Hydrogeology. Ministry of Energy and Water Resources and Development, Republic of Zimbabwe.

IUCN (2004) Institutional arrangements for groundwater management in Dolomitic terrains – water demand assessment. WRC Project 1324. Water Research Commission, South Africa.

Jalludin, M. and Razack, M. (2004) Assessment of hydraulic properties of sedimentary and volcanic aquifer systems under arid conditions in the Republic of Djibouti (Horn of Africa). *Hydrogeology Journal* 12,159–170.

Juizo, D. (2005) African Water Development Report: Mozambique Country Report. UNECA.

Kgathi, D.L. (1999) Water demand, population and sustainability in Botswana: implications for development policy. A paper prepared for the Population, Development and Environment Project, International Institute for Applied Systems Analysis, Luxemburg.

MacDonald, A.M. and Davies, J. (2000). A brief review of groundwater for rural water supply in sub-Saharan Africa. BGS Technical Report WC/00/33.

McIntire, J., Bourzat, D. and Pingali, P. (1992) *Crop–Livestock Interactions in Sub-Saharan Africa*. World Bank Regional and Sectoral Studies 1. The World Bank, Washington, DC.

Masiyandima (Forthcoming) Agricultural groundwater use in Zimbabwe. In: Obuobie, E. and Giordano, M. (eds) *Groundwater and Agricultural Groundwater Use in Sub-Saharan Africa: Case Studies from 9 Countries*. International Water Management Institute, Colombo.

Masiyandima, M., van der Stoep, I., Mwanasawani, T. and Pfupajena, S.C. (2001) *Groundwater Management Strategies and Their Implications on Irrigated Agriculture: The Case of Dendron Aquifer in Northern Province, South Africa.* Proceedings, 2nd WARFSA/WaterNet Symposium, Cape Town, pp. 201–209.

Masiyandima, M., McCartney, M.P. and van Koppen, B. (2004) *Sustainable Development and Management of Wetlands: Wetland Contributions to Livelihoods in Zambia*. 1AG: FNPP/GLO/002/NET. Food and Agriculture Organization of the United Nations, Rome.

Morris, J., Thorn, D. and Norman, R. (1984) *Prospects for Small-scale Irrigation Development in the Sahel*. USAID, WMS Report 26, Utah University, Utah.

Nicol, A. (2002) Sustainable livelihoods in southern Africa: institutions, governance and policy processes sustainable livelihoods in Southern Africa. Working paper 3. Water Theme Paper. ODI.

Njitchoua, R., Dever, L., Fontes, J.C. and Naah, E. (1997) Geochemistry, origin and recharge mechanisms of groundwaters from the Garoua Sandstone aquifer, northern Cameroon. *Journal of Hydrology* (Amsterdam) 190(1–2), 123–140.

Noble, C., Tomohiro, K., Van Loon, G., Jamieson, H. and Bijkis, L. (2002) Current Water Issues in Botswana. Available at: www.cameronnoble.com/botswana.pdf

Obuobie, E. and Giordano, M. (eds) (Forthcoming) *Groundwater and Agricultural Groundwater Use in Sub-Saharan Africa: Case Studies from 9 Countries*. International Water Management Institute, Colombo.

Obuobie, E. and Barry, B. (Forthcoming) Agricultural groundwater use in Burkina Faso. In: Obuobie, E. and Giordano, M. (eds) *Groundwater and Agricultural Groundwater Use in Sub-Saharan Africa: Case Studies from 9 Countries*. International Water Management Institute, Colombo.

Purkey, D.R. and Vermillion, D. (1995) *Lift Irrigation in West Africa: Challenges for Sustainable Local Management*. International Irrigation Management Institute (IIMI), Colombo, Sri Lanka.

Savenije, H.H.G. and Vander Zaag, P. (2000) Conceptual framework for the management of shared river basins; with special reference to the SADC and EU. *Water Policy* 2, 9–45.

Singh, J., Wapakala, W.W. and Chebosi, P.K. (1984) Estimating groundwater recharge based on infiltration characteristics of layered soil: challenges in African hydrology and water resources. *Proceedings of the Harare Symposium*, July 1984. IAHS Publication No. 144, 1984.

Taylor, R.G. and Howard, K.W.F. (1996) Groundwater recharge in the Victoria Nile basin of east Africa: support for the soil moisture balance approach using stable isotope tracers and flow modelling. *Journal of Hydrology* 180(15), 31–53.

Tewari, D.D. (Forthcoming) Agricultural groundwater use in Limpopo Province, South Africa. In: Obuobie, E. and Giordano, M. (eds) *Groundwater and Agricultural Groundwater Use in Sub-Saharan Africa: Case Studies from 9 Countries*. International Water Management Institute, Colombo.

Thornton, P.K. *et al.* (2002) *Mapping Poverty and Livestock in the Developing World*. International Livestock Research Institute, Nairobi.

Ulf, T. and Manfred, H. (2002) *Groundwater Resources of the Nubian Aquifer System N-E Africa*. Technical University of Berlin, Germany.

UNEP (2003) *Groundwater and Its Susceptibility to Degradation: A Global Assessment of the Problem and Options for Management*. UNEP/DEWA, Nairobi, Kenya.

Van Koppen, B. (2002) Water reform in sub-Saharan Africa: what is the difference? Paper presented at 3rd WaterNet/Warfsa Symposium 'Water Demand Management for Sustainable Development', Dar es Salaam, 30–31 October 2002. Available at: http://www.waternet.ihe.nl/aboutWN/pdf/VanKoppen.pdf

Vries de, J.J., Selaolo, E.T. and Beekman, H.E. (2000) Groundwater recharge in the Kalahari, with reference to paleo-hydrologic conditions. *Journal of Hydrology* 238,110–123.

Walling, D.E. (1996) Hydrology and rivers. In: Orme, A. (ed.) *The Physical Geography of Africa*. Oxford University Press, Oxford.

Wurzel, P. (2001) *Drilling Boreholes for Handpumps*. Working Papers for Water Supply and Environmental Sanitation, Vol. 2. Swiss Center for Development and Cooperation, St. Gallen, Switzerland.

6 Groundwater in Central America: Its Importance, Development and Use, with Particular Reference to Its Role in Irrigated Agriculture

MAUREEN BALLESTERO, VIRGINIA REYES
AND YAMILETH ASTORGA

Coordinator of Global Water Partnership of Central America (GWP-CA), Oficial Técnico, GWP-Centroamérica, Correo electrónico

Introduction

Groundwater in the Central American region is currently being exploited mainly for human consumption and industrial activities. Utilization of groundwater for agriculture activities in Central America is still very limited when compared with that of other Latin American countries such as Mexico or Brazil, or, as highlighted elsewhere in this volume, with developing countries such as China or India. To date, agricultural activity in the region continues to rely on rainfall and, to a lesser extent, gravity irrigation. Nevertheless, during dry season in the Pacific region of Central America the exploitation of aquifers for irrigation in agriculture is increasing.

Unfortunately, almost no systematic data exist in any of the Central American countries about the potential volume of the main aquifers and of the existing demands on them. Still, in areas where use does exist, there are reports of a continual reduction in the water table levels, leading to concerns that the resource is already being used in a potentially inefficient and unsustainable manner. Likewise, although there have always been restrictions on certain high-risk activities in recharge areas and in important aquifers, a discussion is just beginning about protective measures to regulate urban expansion and limit the introduction of economic activities in these areas.

The objective of this chapter is to analyse the actual situation, using limited publicly available data coupled with interviews with key professionals, of groundwater in Central America, emphasizing the utilization of groundwater for agricultural production. The chapter is divided into four sections. The first contains a presentation of the Central American region and examines the availability of water in each of the seven countries, the levels of extraction and the amount of extracted water being used in various ways. The second gives a review of the use of water in agriculture in each of the countries of the region and of the irrigation techniques used. The third describes the existing institutional framework for groundwater management in Central America. The final section highlights the key issues for groundwater management in a region that has yet to experience a 'groundwater revolution' and the parallels with, and divergences from, regions where agricultural groundwater use is already more developed.

Background

Central America, the isthmus connecting the main body of North America with the South American continent, is made up of seven countries: Belize, Guatemala, Honduras, El Salvador, Nicaragua, Costa Rica and Panama (Fig. 6.1). The region has an area of over 500,000 km^2 and a population of more than 37 million, growing at an annual rate of 2.4% (GWP, 2005). Urbanization over the last 30 years has shifted the population balance, with nearly 53% of the people now living in areas officially considered to be urban, and the remaining 47% residing in rural areas.

The Central American region shares a volcanic chain that extends from the north to the south and serves to divide the region's waters into Atlantic and Pacific draining basins. The Atlantic region is approximately 2.3 times larger than the Pacific and drains 70% of the territory. The most abundant rivers, as well as those with the largest watersheds, are also found within the Atlantic region (Leonard, 1987).

The water issue

According to the World Meteorological Organization, all the countries of the Central American isthmus with the exception of El Salvador are classified as wealthy in terms of water resources. In other words, they use less than 10% of their available water resources (SG-SICA, 2001). The average per capita availability of water in the region is more than 28,000 m^3/year, with the maximum value in Belize of around 58,500 m^3/year and the minimum in El Salvador of less than 2,800 m^3/year (CEPAL, 2003).

Considering the overall abundance of water, it would seem that the Central American region would have no problems in meeting demands for its various water uses, but this is not the reality. Three main factors are responsible for this apparent contradiction: seasonality in supplies, quality and population distribution. In terms of seasonality, rainfall in Central America, like most other regions, is not distributed evenly throughout the year. There are heavy rains and river flows in some months (May to December) and little in others. Further, storage facilities that might mitigate the effects of seasonality are not generally developed.

Even when supplies are high, they are often of low quality with high degrees of turbidity and sedimentation caused by erosion. The main cause of

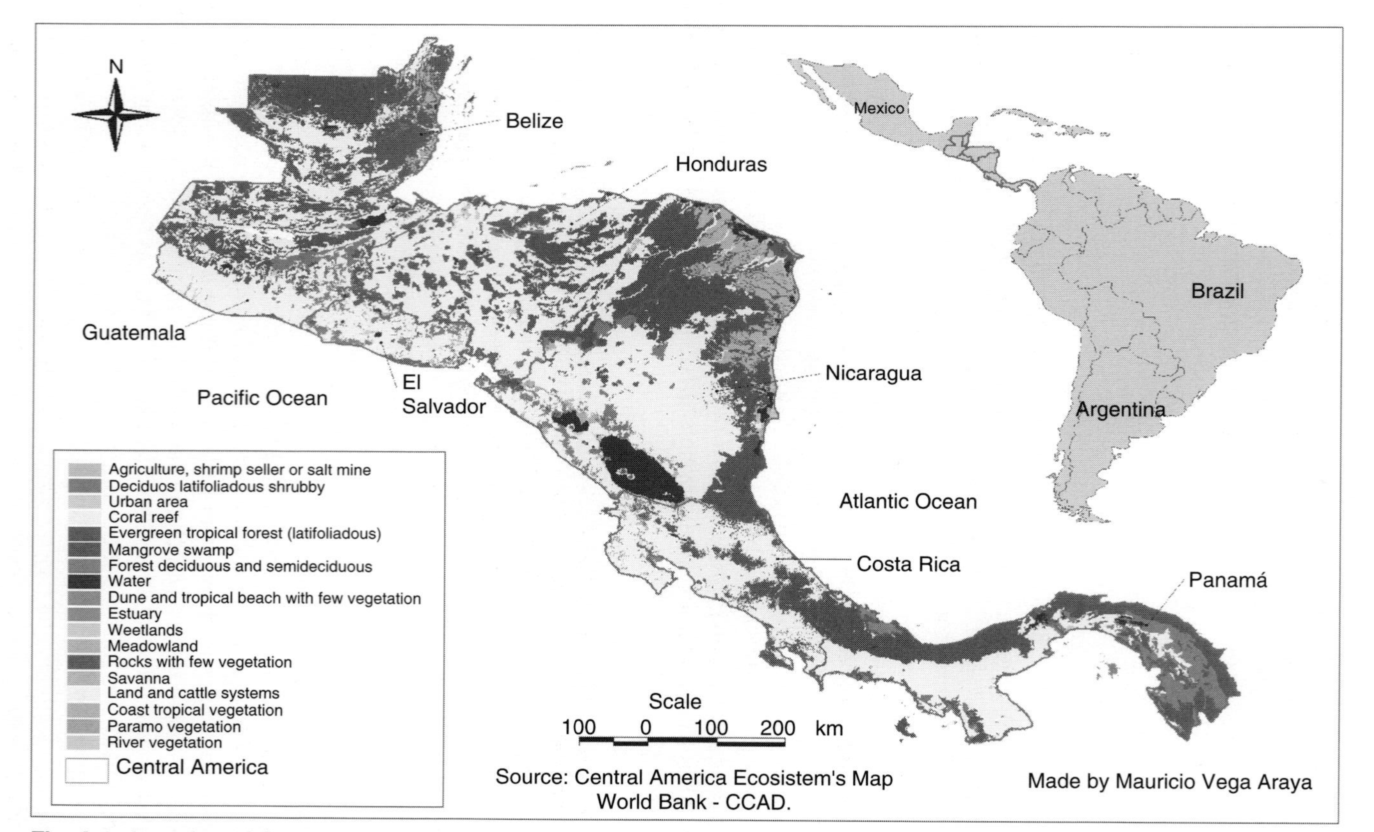

Fig. 6.1. Location of Central America.

the erosion in Central America is generally believed to be deforestation, which is itself a product of agricultural expansion. Water quality has also been heavily degraded in the areas surrounding many cities.

Adding to the supply problems from both these hydrologic factors is the location of human activity in the region. Interestingly, the population distribution in Central America is inversely related to the total potential availability of water. Almost symmetrically, 30% of the water is found in watersheds flowing towards the Pacific and 70% towards the Atlantic, whereas 30% of the population is located in the Atlantic zone and 70% in the Pacific. Because of higher population, the Pacific region also has the greatest economic activity. This, coupled with the seasonality and quality problems, has led to a water shortage in many areas despite what appears to be high average availability. Scarcity has now become an issue in places such as the peninsula of Azuero in Panama, the north-west of Costa Rica, Nicaragua's central and Pacific region, the entire country of El Salvador, western Honduras, as well as the high plateaus and Pacific coast of Guatemala.

Water uses and sources

Overall water extraction in Central America is about 19,000 million cubic metres or less than 3% of the total water availability (Table 6.1). Per capita extraction is estimated at 656 m^3/year. Costa Rica has the greatest extraction, both in total and per capita, but it still amounts to only 5% of total available supplies.

Agriculture is the main user of extracted water in the region, with a per capita consumption of approximately 2200 m^3/year (excluding Belize and El Salvador for which data are not available). Most of agriculture uses surface water as supplemental irrigation during the dry season (Losilla *et al.*, 2001). Agricultural use is followed in importance by domestic consumption and industrial demand. By the statistics in Table 6.1, agriculture accounts for more than 80% of use. However, figures for the industrial sector are likely underestimates. The primary reason for the underestimate is poor information on groundwater abstraction. The main source of water for industry is groundwater, and there is no system for concessions or inventories that would make it possible to measure, or control, consumption levels. Also omitted from the estimates are environmental water use and the generally non-consumptive use of water for tourism and hydroelectric power.

In addition to its use in industry, groundwater from springs and wells also accounts for an estimated 50–95% of the water being used in the public (domestic) supply system (Losilla *et al.*, 2001). One reason for the high utilization of groundwater is that the majority of surface supplies are of insufficient quality due to poor land-use practices and poorly planned urban expansion. Groundwater is particularly important for domestic supply in Belize, where it accounts for 95% use. Groundwater is similarly important for domestic supply in Costa Rica. In fact, groundwater accounts for almost 88% of Costa Rica's extraction to satisfy all consumptive demands, i.e. all uses with the exception of hydroelectric generation.

Table 6.1. Central America: water resources and use.

	Rainfall[a] (m/year)	Total water resources (million m^3/year)	Per capita water resources[b] (m^3/year)	Total extraction (million m^3/year)	Per capita extraction[c] (m^3/year)	Per capita extraction[c] (m^3/year)			Extraction as a percentage of resources
						Domestic	Industry	Agriculture	
Belize	1.5–4.6	15,258	58,458	101	389	–	–	–	0.7
Guatemala	2.2	111,855	8,857	2,702	214	33	36	145	2.4
Honduras	1.9	90,031	13,776	1,745	267	37	10	220	1.9
El Salvador	1.2	18,616	2,755	797	118	–	–	–	4.3
Nicaragua	1.0–4.0	195,238	34,672	1,759	312.3	59	2.4	250.9	0.9
Costa Rica	3.3	118,720	27,967	6,032	1421	158	76.4	1,187	5.1
Panama	3.0	156,259	49,262	59,316	1870	1453	14	403	38.0
Central America		705,976	27,965	19,069	656	–	–	–	2.7

[a]Information taken from each country report.
[b]World Bank (2005).
[c]CRRH-SICA, GWP-CATAC-UICN (2002).

For Costa Rica, projections for water demand for all uses by 2020 are estimated to reach 39 km^3, equivalent to 35% of the total water resources in the country. Even so, urban development continues to increase the pressure on water resources, regardless of the policies for conservation and protection adopted by the country. In some regions, signs of conflict and competition for water use are already being observed. In conclusion, the use of water, groundwater in particular, is becoming increasingly more complex every day.

Groundwater resources in Central America

Information on the location and availability of groundwater resources in the seven countries of Central American is both limited and variable. For example, while understanding is somewhat greater in Nicaragua and Costa Rica, in Belize existing aquifers and their annual discharge have hardly been studied. Still general information is available for most countries and is summarized in Table 6.2 and general patterns are discussed here.

In general terms, the higher parts of the watersheds in Central America are underlain by volcanic aquifers. In the lower river basins and inland valleys, aquifers of recent alluviums predominate, whereas in the middle parts of the river basins the aquifers are a mixture of volcanic materials, colluvial alluvials and, of lesser importance, aquifers of sedimentary rocks (Losilla *et al.*, 2001). Although important alluvial aquifers do exist throughout the region and some important sedimentary rock aquifers are found in Honduras, Guatemala and Belize, of particular importance to groundwater use are the highly porous soils throughout the Pacific volcanic chain that permit very high levels of rainwater infiltration to recharge the local aquifer systems (GWP, 2005). Their presence has been important historically for attracting settlements and population concentrations in the region's Pacific watersheds where, in most cases, the springs and eventual pumping from local wells have met the demand for water. Volcanic aquifers now provide potable water for most major Central American cities including Guatemala, Tegucigalpa, San Salvador, Managua and San Jose.

Unfortunately, the volcanic aquifers consist mainly of interstratifications of tuffs, gaps and quaternary as well as some tertiary lava, which present high permeability and fissure flows. In many cases, these make the aquifers highly vulnerable to human contamination from the cities they help to support. The heterogeneity of these aquifers, with differential horizontal and vertical flows, also makes them quite complicated to study and therefore to manage.

In general, the recharge of the main aquifers in Central America is accomplished by rainwater infiltration and to a lesser degree by a connection with surface water and excess of irrigation water application.

Threats to groundwater

The overexploitation of water is most clearly seen where the population is concentrated in metropolitan areas, which increases the demand for extracting

Table 6.2. Characteristics of the main aquifers in Central America. (From Losilla *et al.*, 2001.)

Country	Characteristics
Belize	No information is available.
Guatemala	Four very important aquifers are located in Guatemala: the upper and lower aquifers in the central highlands and in the valley of Guatemala. The upper aquifer in the high plateau is mainly formed by volcanic quaternary rocks, and the lower aquifer basically consists of lutitas, welded dacite and andesite tuffs, and basaltic andesite lava flows from the tertiary era, which has been fragmented locally. Although we have little information about the aquifers in the high plateaus, which have yet to be exploited, it is known that they are not confined, with a depth of 250 m and a production of 3–70 l/s. Land in the central plateau is now mainly for cultivating coffee and vegetables, as well as for pastures and brushland.
	The aquifers in the upper and lower valley of Guatemala are hydrologically connected. The upper aquifer has no quaternary formations, and its depth varies from 5 to more than 50 m; the lower aquifer is of tertiary formation, with a depth of 200–250 m, and extends over 550 km^2 with a water flow of up to 300 l/s. The recharge area of these aquifers is in the valley of Guatemala, with the exception of part of Lake Amatitlán, as well as other areas covered by urbanization.
Nicaragua	The Managua aquifer is located in the western central area of the country and extends for approximately 600 km^2; it has a saturated thickness that ranges from 200 to more than 450 m. There are approximately 160 excavated central wells and a total of 663 perforated wells with different uses, among which are domestic, municipal, industrial, agribusiness and irrigation. The depth of the perforated wells ranges between approximately 42 and 500 m. The average water flow from the wells is 3170 m^3/day, with a production capacity that ranges between 470 and 8500 m^3/day. In 1996, the production of water was 131.4 × 10 m^3/year. Nearly 1.5 million people were supplied from the groundwater in this aquifer, via the Managua aqueducts. According to a 1993 JICA/INAA study, west of the Managua aquifer there is large-scale irrigation for the main crops of maize, sorghum and beans. Central-pivot irrigation covers an area of approximately 247 ha. During 1993, the irrigated area was about 170 ha, and there was a total of 1700 h of pumping. There is an annual discharge of approximately 1.24 million cubic metres. The quantity of water required is approximately equal to the water consumed; when the planted area is 150 ha, consumption is 1275 million m^3, which is similar to the amount of water extracted annually.
Honduras	In Tegucigalpa, the capital of Honduras, there are more than 500 perforated wells, which yield 1–3 l/s; there are zones in the volcanic ash that have low yields, although some have greater yields (2–20 l/s).
El Salvador	In the department of San Salvador, a region of economic importance to the country, there is a group of aquifers that form a very complex water system due to the emergence of springs and of connections between surface and groundwater flows. To the west of the San Salvador aquifers, water for agricultural activity is extracted from wells located in the Zapotitán Valley, which is dedicated to agriculture. To the east is the San Salvador aquifer, which coincides with the metropolitan area of the capital city, and covers

Table 6.2. *Continued*

Country	Characteristics
	an area of approximately 185 km^2, with an annual yield of 42 × 10^6 m^3/year (1.35 m^3/s). The rest of the aquifers of this zone are mainly in the coastal plains or along the coastline. In general, they are very limited aquifers because of their proximity to the coast and because of the influence of salt water.
	Approximately 227 millior cubic metres of groundwater is extracted annually and 80% of the potable water supply comes from groundwater. The reliable yield of the existing groundwater deposits in the country are estimated at 83 m^3/s.
	The largest basins at the highest elevation with potential groundwater, in order of importance, are the Lempa and Jalponga rivers combined, Grande de San Miguel and Paz. However, the potential is not uniformly distributed. The aquifers in El Salvador were formed according to structural zones: the northern Sierras, the central depression, and the mountains and plains of the Pacific coast are formed by impermeable rocks. There are coastal aquifer formations with depths of more than 150 m and with average water flows of 16 l/s. The Santa Ana aquifer in the western zone has a high potential for exploitation, with a flow of 3.5 × 10^6 m^3/year/km.
Costa Rica	The characteristics of high permeability in the layers of fragmented and igneous lava, combined with high rainfall, favoured the formation of highly potential aquifers in the central and northern part of Costa Rica's Central Valley, where more than half of the population lives. These aquifers are called the Upper and Lower Colima and are separated by a low permeability layer that acts as an aquitard, which allows the descending and ascending vertical transfer of water.
	It has been estimated that the Lower Colima extends for approximately 230 km^2 and that the Upper Colima spreads over approximately 170 km^2. The maximum thickness is about 300 m. The outcropping of this lava is limited to the river canyons in the lower part of the valley. According to SENARA/BGS (1989), the Upper Colima aquifer recharges from the Barva aquifer through the tuffs of the unit known as the Tiribí formation and from the La Libertad aquifer by vertical percolation. The Upper Colima also receives a large part of its recharge from rain infiltration in those areas where there are no overlying layers. The Lower Colima is recharged from the Upper Colima by vertical percolation through the tuffs and ignimbrites of Puente de Mulas, or from the surface where the Upper Colima is absent. The average recharge in the aquifer system has been calculated at 8200 l/s (TAHAL, 1990).
	The flows extracted from the wells that collect from both aquifers are 50–120 l/s (SENARA/BGS, 1989). The depth of the water table level varies, depending on the surface topographical irregularities; but, in general terms, it ranges between 50 and 100 m. The direction of the underground flow is from north-east to south-west in both aquifers.
Panama	The hydrological characteristics of the geological formations in Panama are little known because of the lack of systematic studies. It is known that production from wells is generally acceptable. The majority of the aquifers that are exploited are of a type of fissure flow in volcanic rocks and of sedimentary and fissured conglomerates. The depth of the majority of the wells ranges from 20 to 110 m and production is 2–15 l/s.

groundwater at rates that exceed the capacity of the natural cycle to recharge the aquifers. Aquifers under virtually all of the metropolitan areas in Central America show signs of overexploitation. At the same time, urban expansion is covering the surfaces from which the aquifers would naturally be recharged. So while demands on the aquifers rise, their supply falls. The case of the aquifers in Guatemala, Managua in Nicaragua, and San Pedro Sula in Honduras exemplify this problem.

In Guatemala, a continuous decline of groundwater has been identified in the southern basin of the valley of Guatemala, as well as in the metropolitan area. In the case of El Salvador, the urbanized surface of the metropolitan area has increased almost exponentially, from 6.8 km^2 in 1935 to the current 91.5 km^2, and this has mainly taken place in the largest aquifer recharge areas. Because of this, the areas with the highest rate of infiltration have been reduced, whereas the areas with an infiltration rate of 0.05 (the rate assigned to areas of low impermeability) have increased by the same proportion. The same thing has happened in San Pedro Sula, Honduras.

The demographic projections of the United Nations Development Program (UNDP) indicate that by 2010 more than 60% of the population in all countries in the region with the exception of Guatemala will be concentrated in urban centres. All of these, with the exception of Honduras, are located within the Pacific region.

As mentioned above, groundwater in Central America is extremely vulnerable to pollution because the aquifers are relatively superficial and are covered by fractured or permeable materials. In areas of high precipitation, the infiltration of polluting agents potentially toxic to human health can be from 30% to 50% (Reynolds, 1992). The main sources of groundwater pollution are agricultural and industrial activities, along with domestic runoff.

A study conducted by the Food and Agricultural Organization (FAO) of the problems of pollution in 16 Latin American countries determined that in urban areas the main sources of polluting agents from agricultural activities are fertilizers, pesticides and food-processing industries, whereas in rural areas the contaminants are associated with pesticides and fertilizers of chemical origin. Intensive agriculture is one of the main sources of income in Central America. Great volumes of water are used as farmers seek the highest possible levels of performance, thus forcing irrational use in the dry season and uncontrolled use of pesticides and herbicides, which in turn lead to situations of risk.

A total of 10.1 million tonnes of chemical fertilizers were used in the region from 1980 to 2000. Interestingly, as shown in Fig. 6.2, the use of chemical fertilizers in Central America has revealed quite irregular tendencies, with use increasing by 20% in some years (1983 and 1997) and falling markedly in others (1982 and 1989). This use of agrochemicals and fertilizers in the region has contaminated some important aquifers.

Pollution caused from mercury and phosphates has been observed in Guatemala. In El Salvador, rivers and streams in the principal agricultural areas are highly polluted by pesticides, particularly by DDT in cotton cultivations in the south-eastern coastal plains. Concentrations of 3.15 mg of DDT per litre of water have been discovered in the Río Grande de San Miguel, which is triple the

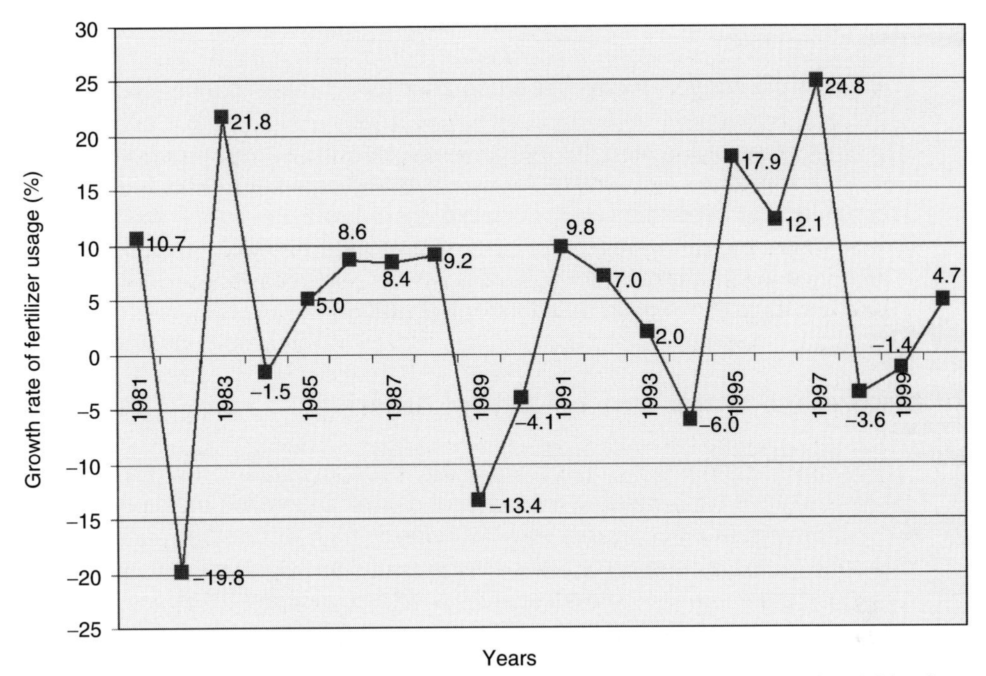

Fig. 6.2. Growth rate in the amount of chemical fertilizers used in agricultural activities for the period 1980–2000 in Central America.

lethal limit for fish. Toxaphene (non-biodegradable) pollution has been detected in Nicaragua in concentrations that exceed acceptable standards (Silvel *et al.*, 1997; CIEUA, 1998; Aquastat, 2001). In the case of Nicaragua, this problem is of great concern for the western aquifers (León-Chinandega) and the valley of Sébaco, as is the case in Guatemala for the sugarcane and banana plantations in the Pacific and Caribbean coastal regions (Choza, 2002).

Pesticides such as toxaphene were detected in the Siguatepeque aquifer in Honduras and in the León-Chinandega aquifer in Nicaragua, although in smaller concentrations to those permitted by the World Health Organization (WHO). There was intensive cultivation of cotton on top of these aquifers during the 1960s and 1970s. Likewise, insecticides such as clorado, carbofuran and 2,4-D, which are used for agricultural production, have been detected in rural areas of Honduras, although in concentrations that are lower than those permitted by the WHO.

In La Libertad spring in Costa Rica, on three occasions, concentrations of nitrate that exceed the norm of 45 mg/l NO_3^- have been detected, while a slight increase in the nitrate concentration of 2 mg/l was observed between 1986 and 1997. This indicates that in less than one decade the maximum permissible concentration would have been reached.

Nitrates have been detected in El Salvador, Nicaragua and Costa Rica. In the north-eastern sector of Managua's aquifer, high concentrations of almost 45 mg/l NO_3^- were reported, probably because of the use of nitrogenous fertilizers (Hetch, 1989).

The non-agricultural causes of contamination in urban areas are domestic runoff and industrial waste. For example, there is substantial data on faecal pollution in the aquifers located in urban zones. According to monitoring by the National Administration of Aqueducts and Sewers (ANDA) of the public wells in El Salvador, of 183 samples that were taken during a 10-year period from 31 wells, it was found that 10% of the wells showed high rates of coliform, which exceeded the permitted level. Likewise, there are zones in the metropolitan area where the percentage of wells contaminated by coliforms in excess of permitted levels is more than 50%; this is where there has been indiscriminate land use for both residential and industrial purposes.

Agriculture and Water Use in Central America

Although Central America has traditionally been an agricultural region, in the last 20 years it has witnessed a process of diversification and increased wealth generation from other sectors such as industry, high technology and services. In spite of this, agriculture continues to generate an important part (14.8%) of the total gross domestic product (GDP) and has continued to expand, at 2.3% between 1995 and 2002 (CEPAL, 2003). Nicaragua is the country most dependent on agriculture. There the sector contributes one-third of total GDP, with contributions coming particularly from the export of coffee ($98.3 million) and sugar ($33.4 million).

In Guatemala, where 75% of families live in extreme poverty, agriculture is the most productive sector and provides 25% of the GDP. It is also the most important activity for the people who live in the high central plateau, and creates jobs and income for 68% of the population. The export products are principally coffee and cotton, while maize, beans, wheat, vegetables and fruit are produced for domestic consumption.

Countries such as Costa Rica and Panama have diversified their exports by producing high-tech goods followed by seafood products. This has reduced the agricultural sector's proportion of the GDP from 11% to 7% in both countries, although they still export products such as bananas, coffee, sugar, pineapple and melon. However, this relatively low contribution of agriculture remains high, even in the case of Panama, when compared with that of developed countries. We must remember that the region's economies are still highly dependent on agriculture. This discussion and the figures shown in Table 6.1 highlight that even though agriculture may be declining as percentage of output, it is still important for the overall economy and for water use in particular.

Irrigation by country

Belize[1]

Because of land quality, only 6.1% of Belize's area is dedicated to agriculture. Still that area contributes 19% of GDP. Of total agricultural area, only about 5% or 3,500 ha is irrigated, supporting about 177 producers, as shown in Table 6.3.

Table 6.3. Belize: irrigated area, irrigation methods, supply sources and main crops by district, 2005. (From Ministry of Agriculture of Belize, 2005.)

District	Crop	Location	Estimated Number of farmers	Area (ha)	Method of irrigation	Water source	Average yield (gpm)	Quality of water
Corozal	Papaya Tainong/ Solo	Fruita Bomba	1	623.8	Drip	Well	120	Hard water
	Papaya Tainong	Little Belize	30	81	Drip	Well	100	Hard water
	Vegetables (onions, cabbage, hot pepper)	Small farms	70	31.6	Drip	Well	100	Hard water
Subtotal				736.4				
Orange Walk	Rice	Blue Creek	3	1417.5	Basin/flood	River	Unlimited	Fresh
	Papaya Tainong	Indian Creek	1	2	Drip	Well	50–100	Fresh
	Hot pepper	San Carlos	4	3.2	Drip	Well	50–100	Fresh
	Vegetables (hot pepper, cabbage, sweet pepper, tomatoes, onions, lettuce, sweet corn, broccoli, carrots, potatoes)	Small farms	28	20.3	Drip	Well	50–100	Fresh but saline Along coastal areas
Subtotal				1443				
Belize District	Rice	Singh Tut	1	81	Basin/flood	River	Unlimited	Fresh
	Vegetables	Small farms	15	14.2	Drip	Well	50–100	Fresh but saline along coastal areas
Subtotal				95.2				

Continued

Table 6.3. *Continued*

District	Crop	Location	Estimated Number of farmers	Area (ha)	Method of irrigation	Water source	Average yield (gpm)	Quality of water
Cayo	Papaya	D&L	1	14.2	Drip	Well	120	Fresh
	Vegetable (cabbage, carrots, lettuce, cucumbers)	Small farms	5	2.8	Drip	River/ stream	Unlimited for rivers but streams, ponds can run dry in summer	Fresh
Subtotal				17.0				
Stann	Bannanas for export	Big Creek	6	1215	Sprinkler/ under canopy	River		Fresh
Creek	Vegetables (hot pepper)	Small farm	2	2	Drip	Well	100	Fresh
Subtotal				1217				
Toledo	Rice	Farms	2	36.5	Basin/flood	River	Unlimited	Fresh
	Rice	Small farm	8	3.2	Basin/flood	River	Unlimited	Fresh
Subtotal				39.7				
Total			177	3548.3				

As much as 66% of the current irrigation is located in the mid-southern region and more than 94% is delivered by systems of low water use efficiency, sprinklers and surface water. Experiences have so far indicated that irrigation development has been successful only in monocrops of high-intensity production such as bananas, papayas and double cropping of rice. It played no role in the milpa system. The low input and low productivity levels that characterize other crops could not justify this additional input.

Contrary to the other countries in Central America, all of the irrigation systems in Belize are private and were developed with private funds or loans from international cooperation organizations such as the European Union (EU). State support through the Ministry of Agriculture is reduced to technical assistance for small producers and to facilitating the financing requests of large producers. Because these systems are private, there is no tariff system; the costs of development and maintenance are totally paid for by the producers.

There is a deficit of crop water during 4–9 months of the year, and the potential for irrigation increases from south to north. For the purpose of considering resources for irrigation, the country can be subdivided into three main areas as summarized in Table 6.3: (i) a southern high rainfall area (Stann Creek and Toledo); (ii) an intermediate rainfall area in the rolling lands of the central foothills (Cayo and the southern area of Belize); and (iii) a much drier northern plateau with diverse land systems (Corozal and the northern area of Belize).

The south has an abundant supply of good-quality surface water for use in the dry season and could support irrigation systems with low water use efficiencies, such as flood irrigation and, to a lesser degree, sprinklers or drip. Soil macro-structure poses a constraint to water use as the experiences of the largest irrigation project – 1215 ha of irrigated bananas – have illustrated. Southern communities in Toledo, where principally rice is grown, do not experience water problems in the dry season.

The central foothills, home to many small farming communities, face a water shortage during the dry season. Because the shallow and often stony soils present no real problems for soil water dynamics, the development of water storage facilities is most relevant to this area. However, the area is most conducive to irrigation systems with high water use efficiencies for the production of rice, vegetables and papaya. Water quality is good throughout the dry season although the supply is low.

The northern plateau, with its diverse land systems, is characterized by lagoons, creeks, swamps, subsurface storage in limestone aquifers and slow and sluggish flowing rivers. Availability of water for dry season use is good, but access to surface and groundwater sources poses problems for domestic and agricultural use in small farming communities. The water resources can support irrigation systems of low water use efficiency on the river banks, but is conducive to the more efficient water use system in other areas. Except for its high natural variability and shallow nature, the soil structure offers no constraints to soil water dynamics.

Guatemala[2]

As can be seen in Table 6.4, the total irrigated area in Guatemala is estimated at just more than 140,000 ha, of which surface water is used to irrigate 78.2% and

Table 6.4. Guatemala: area under irrigation by water source. (From Support Program for the Reconversion of Food and Agriculture Production (PARPA), based on PLAMAR-MAGA data, 2003–2004.)

	Irrigated area (ha)		Total irrigated
Department	Groundwater	Surface water	area
Alta Verapaz	252	479	731
Baja Verapaz	76.8	2,073.5	2,150.3
Chimaltenango	481.7	211.5	693.2
Chiquimula	644.2	1,498.6	2,142.9
El Progreso	325.7	1,722.6	2,048.3
Escuintla	23,456.9	28,205.4	51,662.3
Guatemala	170.5	1,036.5	1,207
Huehuetenango	–	1,035.8	1,035.8
Izabal	339.5	10,037.1	10,376.6
Jalapa	1,015.3	964.5	1,979.8
Jutiapa	700.5	2,067.3	2,767.8
Petén	353.1	258.3	611.4
Quetzaltenango	209.2	791.5	1,000.7
Quiche	–	413.5	413.5
Retalhuleu	864.5	7,113	7,977.5
Sacatepéquez	125.1	1,663	1,788.1
San Marcos	530.4	5,326.1	5,856.5
Santa Rosa	703	3,099.5	3,802.5
Sololá	0.7	308.2	308.9
Suchitepéquez	428.7	31,508.8	31,937.5
Totonicapán	43	61.1	104.1
Zacapa	290.1	11,612.8	11,902.9
Total	31,011	111,487.8	142,498.9

groundwater, 21.8%. The departments with the largest areas under irrigation are Escuintla (51,662.3 ha), Suchitepéquez (31,937.5 ha), Zacapa (11,902.9 ha) and Izabal (10,376.6 ha). This sector utilizes approximately 41 m^3/s, equivalent to 1.3 km^3/year.

The process of constructing community irrigation systems in Guatemala began in the early 20th century. Since the 1960s the government of Guatemala has supported the construction of public irrigation units that cover 15,229 ha and directly benefit 4402 families; another 64,000 ha is under private irrigation systems. This is a small percentage compared to the 2.5 million hectares that could potentially be irrigated. The government-constructed systems generally use surface water, whereas the majority of the private systems use pressurized systems (sprinklers or drip). The efficiency of the irrigation systems has been questioned; however, water utilized for irrigation accounts for scarcely 1.3% of the surface water sources in the country.

In 1982, the first attempts were made to transfer irrigation units to users. The process was intensified in 1988, and in the 1990s the transfer of the administration, operation and maintenance of the irrigation units took place. Currently, only five of these units remain to be transferred to the users: Llano de Morales in Sanarate municipality, El Progreso department; and Las Canoas and Rincón de la Pala in the department of Guatemala.

The state began a parallel process to reduce its attention to irrigation management so as to reduce the budget and personnel assigned. However, it was not successful in obtaining the conscientious and voluntary participation of the users. The result has been an accelerated deterioration of the infrastructure. Additional measures have been taken to solve this problem, in particular the 'remodelling' of irrigation infrastructure supported by the state, but problems remain.

In order to provide a short-term solution to the needs of the producers, 940 small irrigation systems were built between 1990 and 2004, at a cost of $20.3 million. These cover an area of 9049.5 ha and are administered by the users. In the areas where the majority of the projects are located, groundwater is used with the most efficient techniques such as drip, sprinkler or mixed irrigation, for example, in the production of sugarcane and plantains in Escuintla and Suchitepéquez, and of cantaloupe, mango and citrus fruit in Zacapa.

El Salvador[3]

In El Salvador, 77.4% of the land is dedicated to agriculture though only 4.4% (35,000 ha) is irrigated, mostly for crops such as pasture, maize, sugarcane, rice and vegetables. However, nearly 70% of the total water consumption in the country is for irrigation, in strong competition with other uses.

It is estimated that 7715 families benefit from the irrigation systems in the country, as can be seen in Table 6.5. However, in 2005, the Ministry of Agriculture and Livestock began taking a census to update the registry of irrigation users throughout the country, which would provide information about the exact number of hectares under irrigation, the techniques and the supply sources (surface or groundwater), as well as the number of families that benefit. The partial results indicate that in departments such as Sonsonate and Ahuachapán there was a 128% increase in the number of hectares irrigated and a 172% increase in the number of producers utilizing irrigation compared with the previous registry.

Table 6.5. El Salvador: land irrigated with groundwater and the number of beneficiaries by department. (From Department of Irrigation and Drainage, Ministry of Agriculture and Livestock, El Salvador, 2005.)

Department	Area irrigated with groundwater (ha)	Number of beneficiaries
Ahuachapán	1014	2535
Sonsonete	31.3	78
La Paz	7	18
Usulatán	1.5	4
La Unión	32.5	80
La Libertad	2000	5000
Total	3086.3	7715

More than 90% of the water utilized for irrigation comes from surface sources and gravity irrigation is used by more than 90% of the users, according to the results obtained in Sonsonate and Ahuachapán; techniques such as sprinklers, drip or mixed are used by few producers. This suggests high levels of inefficient water use despite the fact that El Salvador has the greatest water stress in the region.

Irrigation systems are organized in 43 irrigation associations that cover 50% of the irrigated area; 39 correspond to private systems and 4 to public systems (irrigation districts). Only 32% of the irrigated land in private systems is in associations, whereas 100% of public systems are in associations. Of all irrigation associations 51% are in Sonsonate.

The irrigation and drainage law authorizes the formation of federations. To date, the Federation of Irrigator Associations of El Salvador (FEDARES) and the Federation of Irrigators in the Sensunapán River basin have been formed. The first group brings together the associations of the irrigation and drainage districts of the southern Atiocoyo sector (ARAS), northern Atiocoyo sector (ARAN) and Zapotitán (AREZA); the second group is made up of seven private associations.

Investment in the public irrigation districts is financed by the state, to a maximum of $2,225,000. However, the users of the irrigation system are responsible for recuperating 50% of the investment. Each district establishes a tariff that will make it possible to recuperate the amount agreed upon in the time period that is negotiated between the beneficiaries and the Ministry of Agriculture and Livestock. In the same manner, the state finances the private irrigation associations; however, these groups must contribute 20% of the total investment, which cannot exceed $2000/ha.

In addition, by executive agreement, each district charges a minimum rate of $5.24/ha/year to operate and maintain the system. However, this amount does not cover the real needs of the irrigation systems, so the users have established voluntary tariffs of about $40/ha/year, which permits them to keep the systems operating. The private systems establish tariffs through a negotiation process among the members.

Honduras[4]

Agriculture is the economic base of Honduras. As much as 82% of the water exploited (national water balance) is directed to agricultural activities, supplying water to a total of 86,631 ha as shown in Table 6.6. Of this total, 92.3% is supplied by surface water and 7.7% is extracted from groundwater through wells. The majority of the irrigation systems are located in the departments of Yoro (30.5%), Choluteca (21.2%) and Cortés (20.8%), where, unlike in the rest of the region, mainly sprinkler or drip irrigation is used to cultivate bananas, cantaloupe and sugarcane.

Because of the continuous increase in water utilization, a 25-year master plan for irrigation and drainage is being promoted. Feasibility studies already exist for incorporating 14 projects that include 25,763 ha of irrigated agriculture, in accordance with the need to increase agricultural production for domestic consumption and for exportation. These projects, with an investment

Table 6.6. Honduras: irrigated area by source and by department. (From the authors, based on information provided by the Secretariat of Agriculture and Livestock (SAG), General Department of Irrigation and Drainage, 2005.)

	Water source			
Department	Surface	Groundwater	Total	Percentage
Comayagua-La Paz	6,684.2	995.9	7,680.1	9.4
Valle	173.3	69.2	242.5	0.3
El Paraíso	1,407.1	19.2	1,426.2	1.7
Choluteca	14,568.2	2,740.5	17,308.7	21.2
Olancho	194.8	162.3	357.2	0.4
Atlantida	1,718	200	1,918	2.3
Colon	2,613.5	–	2,613.5	3.2
Yoro	23,866.4	1,018.9	24,885.3	30.5
Cortes	15,920.1	1,058.3	16,978.4	20.8
Copan	5,099.8	–	5,099.8	6.2
Santa Barbara	1,286.9	–	1,286.9	1.6
Intibuca	209.4	–	209.4	0.3
Fco. Morazan	1,334.6	–	1,334.6	1.6
La Paz	235.7	–	235.7	0.3
Lempira	54.8	–	54.8	0.1
Total	75,366.6	6,264.3	81,630.9	100

Note: The classification of surface and groundwater is the responsibility of the authors and not of the source.

of more than $143 million, are intended to benefit 8224 families mainly for the cultivation of vegetables, grains and citrus fruits.

Through implementation of this plan, the government intends to promote and stimulate the private sector to develop secondary and tertiary productive infrastructure in large and small irrigation projects, by giving long-term credit incentives, effective technical assistance and investment guarantees. In addition, it contemplates supporting micro-irrigation and drinking water projects using the modality of co-participation between the community and the government, which would be administered by the users.

Nicaragua[5]

Irrigated agriculture in Nicaragua began in the 1950s and by the 1970s covered more than 70,000 ha. The best soils of the Pacific zone are used and groundwater is the main source of irrigation water. In the 1980s, a contingency plan was implemented for basic grains, using sprinkler systems with an automatic central pivot which have been nearly abandoned. The systems deteriorated mainly because of lack of maintenance, high cost of equipment and lack of technical assistance for operating the systems.

Given that 62.3% of the land in Nicaragua is dedicated to agriculture and that the potential irrigable area is estimated at 700,000 ha, the government, through the Ministry of Agriculture, created the Western Irrigation Program in 1998 as a mechanism of economic transformation and modernization of agriculture. This was accomplished by constructing new irrigation and drainage works, through soft loans from private banks and the Rural Credit Fund, with interest rates of 10–11.5%. The programme began in the departments of Chinandega and León and later were expanded throughout the rest of the country.

A $20.2 million loan from the government of the Republic of China was obtained to carry out the Western Irrigation Program. The majority of the funds were used by the Ministry of Agriculture for reconstruction and reposition of irrigation equipment; some were used for preliminary investment and technical assistance, and the remaining were put into investigative studies of surface and groundwaters.

Table 6.7 gives a summary of the funds spent from 1998 to 2005 by each department. The total number of projects constructed includes about 294 beneficiaries and a total investment of $6.8 million. The largest number of beneficiaries is concentrated in Matagalpa and León; however, the highest investment is concentrated in Chinandega.

Currently, the irrigation system covers only 4% of the potential area, or about 30,000 ha. However, even though this represents a small area, there have been conflicts about water use, above all in the central region where the flow of surface water is insufficient to cover the demands of the region and groundwater resources are very limited. The Las Canoas dam is one example. This was constructed by the Victoria de Julio Sugar Refinery to irrigate sugarcane and is causing water use conflicts between water users in the upper and middle basin of the Malacatoya River.

Costa Rica[6]

The National Irrigation and Drainage Service (SENARA) was created by Law 6877 on 18 July 1983. SENARA is given the authority and direct responsibility for developing the infrastructure, administration and operation of the system,

Table 6.7. Nicaragua: number of beneficiaries and investment by department by irrigation project between 1998 and 2005. (From Ministry of Agriculture and Forestry (MAG-FOR), Nicaragua, 2005.)

Department	Number of beneficiaries	Amount ($)
León	44	1,551,726.9
Chinandega	18	1,959,918.9
Matagalpa	62	1,169,695.6
Carazo	3	134,428.9
Estelí	39	286,423.9
Chontales	2	56,493.2
Rivas	25	392,749.3
Boaco	3	70,491.8
Managua	34	442,391.2
Madriz	52	420,983.6
Granada	9	306,486.9
Jinotega	1	25,234.6
Masaya	2	19,882.5
Total	294	6,836,907.3

for which it can establish a tariff system that must be approved by the Public Services Regulatory Authority (ARESEP), since irrigation is considered to be a public service.

Currently, SENARA is administering two irrigation systems: Arenal-Tempisque Irrigation District (DRAT) and Irrigation and Drainage of Small Areas (PARD).

ARENAL-TEMPISQUE IRRIGATION DISTRICT (DRAT) This system is located in Guanacaste province, the driest area of the country (during 5 months a year), and is nearly 100% supplied by surface water, utilizing water from the artificial Lake Arenal.

In the early 1980s, stage I of DRAT, which covered 6006 ha, was constructed at a cost of $15.1 million. Stage II (1986–1994) included the expansion to 12,170 ha at a cost of approximately $44.5 million. Currently, DRAT covers approximately 28,000 ha, of which 10,000 ha was included in the system of irrigation in 2003, when the Canal Oeste Tramo II was constructed at a cost of $2.5 million.

The total investment in the DRAT infrastructure is estimated at $67 million and benefits approximately 1125 families who produce mainly sugarcane, fodder, rice and fish from fish farms (400 ha of ponds), generating income of approximately $163.7 million in the zone. The producers in the area pay SENARA a fixed rate of $42.5/ha/year. The need to develop a tariff system based on volume used has been proposed; however, the social and economic conditions in the area have made it difficult to implement in the short term.

In addition, financial resources of $13.7 million are being negotiated to construct stage IV of DRAT, which consists in the continuation of the Southern Canal and the distribution network of the Lajas and Abangares subdistricts, which would benefit about 155 families and would irrigate an additional 8800 ha.

IRRIGATION AND DRAINAGE OF SMALL AREAS (PARD) This system is promoted by SENARA and corresponds to requests made by associations of producers, individual producers or state institutions. SENARA is in charge of facilitating the process and, in some cases, constructing the irrigation canal. However, these are not state property but belong to the producers, who are in charge of properly maintaining the irrigation system. Currently, as can be seen in Table 6.8, 95 projects are in operation using pressurized systems of irrigation (drip, micro-sprinkling or

Table 6.8. Costa Rica: irrigation and drainage of small areas in operation by region. (From SENARA, 2005.)

Region	Number of projects	Area (ha)	Number of families
Brunca	5	203.4	103
Chorotega	24	256.7	251
Central Occidental	29	1189	909
Central Oriental	12	337.5	360
Huetar Norte	7	445	226
Pacífico Central	18	254.8	174
Total	95	2686.4	2023

sprinkling), which include an area of 2686.4 ha and benefit 2023 families who mainly cultivate vegetables, root crops, tubers, decorative plants and prickly pears. The majority use is of surface water; it is estimated that less than 3% uses groundwater.

The areas where DRAT and PARD operate include approximately 30,686.3 ha and the total water demand is estimated at 35.2 m^3/s. Of this total demand, the Ministry of Environment and Energy (MINAE) has granted 1240 concessions for exploiting surface and groundwaters for agricultural use; less than 3% of the water in Costa Rica that is utilized for irrigation comes from groundwater.

In terms of short-, medium- and long-term investments, the expansion of DRAT and PARD is planned in the short term, which would be financed by the Central American Bank of Economic Integration (BCIE). In order to resolve future long-term water needs for domestic consumption, irrigation and tourism in the dry Pacific region of the country, studies are being conducted to consider building two multi-purpose dams in the Piedras and Tempisque rivers that would permit the utilization of rainwater and would reduce pressure on groundwater.

Panama[7]

Irrigation is relatively new in Panama. It is used in the production of three main groups of crops: traditional (rice, sugarcane and banana); vegetables and fruit for domestic consumption; and non-traditional export crops. Its purpose is to complement the rainfall conditions that prevail in the country, and its importance is growing along with economic liberalization, given the new opportuities in foreign and domestic markets.

Irrigated areas expanded from approximately 22,000 ha in 1970 to nearly 40,000 ha in 1990. A large part of this growth took place in the 1970s, with an approximate increase of 13,000 ha, principally in two state-owned companies (COBAPA and La Victoria Sugar Refinery). However, nearly 12,000 ha stopped being irrigated between 1990 and 1997, the majority of which was part of the state and public irrigation systems. This was a consequence of economic stagnation at the beginning of the 1990s and of the deterioriation and subsequent abandonment of agricultural projects on state farms and of public irrigation systems.

There is more than 270,000 ha of arable land in the country with soils and topography that could benefit from irrigation. Of this, approximately 71,500 ha has soil that is appropriate for irrigation and there is an adequate supply of water for irrigation, even in the dry season. All of this land could benefit from supplementary irrigation during the rainy season and would produce another cycle of crops if irrigation with surface water was practised during the summer. Despite this, Panama has not taken advantage of its irrigation potential. Currently, only 27,475 ha is under irrigation (Table 6.9), leaving dry about 44,000 ha of the land that is suitable for irrigation.

The majority of the 27,475 ha under irrigation is located in Coclé (12,963 ha), Veraguas (4478 ha) and Chiriquí (3288 ha). A system of gravity irrigation is used in 72% of the cultivated area (19,871.2 ha), mainly for rice cultivation, and

Table 6.9. Panama: land surface under irrigation by province and technique. (From National Department of Rural Engineering and Irrigation, Ministry of Agricultural Development, 2005.)

	Irrigated area (ha)				
Province	Gravity	Sprinkler	Drip	Micro-sprinkler	Total
Chiriquí	1,716.6	70	671.4	830	3,288
Veraguas	1,301.7	3,115.7	59.6	1	4,478
Herrera	50	–	877.7	6	933.7
Coclé	12,747.8	–	209.2	6	12,963
Panamá	1,642.5	97.5	28.5	109.6	1,878.1
Capira	83	96	16.5	68.59	264.09
Chepo	1,559.50	1.50	11.96	41	1,613.96
Los Santos	770.2	18.5	1,258.6	9	2,056.2
Total	19,871.2	3,399.2	3,133.4	1,071.2	27,475

other techniques are used to a lesser degree: sprinklers (12.4%), drip (11.4%) or micro-sprinklers (3.9%, in cantaloupe and watermelon crops). Only in the provinces of Herrera and Los Santos is groundwater used for irrigation, reaching some 1122 ha, which is 4% of the total irrigated area. Groundwater is used only for private and individual projects, and the National Environmental Authority (ANAM) must grant a concession for that activity. The remaining 96% of the area is irrigated by surface water.

Table 6.10 shows the new irrigation projects that were built between 1998 and 2004 (e.g. Boquete and Arco Seco projects) and the systems that have operated since 1972 and have been in a process of reconstruction since 1998 (e.g. public use irrigation). These systems cover approximately 5283 ha and benefit about 937 families in the provinces of Herrera, Los Santos, Chiriquí, Coclé and Veraguas. The irrigation systems for public use are administered independently by the users' associations who receive technical assistance and supervision by the Ministry of Agricultural Development. The associations that use gravity

Table 6.10. Panama: irrigation projects, number of beneficiaries and land area by province, 2005. (From National Department of Rural Engineering and Irrigation, Ministry of Agricultural Development, 2005.)

Project	Province	Land area (ha)	Number of beneficiaries
Agro-exportation of Azuero	Herrera and Los Santos	2113	383
Agro-exportation of Boquete	Chiriquí	270	114
Irrigation systems for public use	Coclé, Herrera, Veraguas and Los Santos	2900	440
Total		5283	937

systems charge their members a fee of $20/ha/season.[8] The systems that require pumping of surface water charge a fee of $30–40/ha/season. However, a revision of these fees is needed because they do not cover the necessary costs for maintenance and reinvestment in the irrigation systems.

Approximately 80% of the investment in Panama is from private funds and only 20% is invested by the state. The Ministry of Agricultural Development's expansion plans include the projected construction of six new irrigation projects and the conclusion of the reconstruction of the Irrigation Systems of Public Use located in the provinces of Coclé, Veraguas and Los Santos, which are being completely financed by the state with resources from the Development Trust Fund. The cost for remodelling of this system is $12.5 million. The estimated budget expense for five of the new projects is approximately $300 million.

In 2005, the government was to seek bids for building the Remigio Rojas Irrigation project in Chiriquí province, in the western region of the country. This project would incorporate 3200 ha into intensive irrigated production, basically directed to agro-exportation. The project comprises the construction of public and hydrological works; irrigation systems on farms, including local irrigation systems; postharvest plants, as well as the development of specialized technical assistance programmes; and technological transfer and implementation of marketing programmes through a 3-year programme of continual accompaniment by the company that is awarded the project.

In Panama, the selection of projects for state investment is based on the recommendations found in the National Irrigation Plan, a regulating, guiding and planning instrument for the development of the irrigation subsector. This plan was prepared through the initiative of the Ministry of Agricultural Development in 1977, in consultation with Utah State University, and was financed by the Inter-American Development Bank. The plan contains a database of projects for irrigation investment in which the most suitable areas for irrigated agricultural development have been prioritized, beginning with the parameters related to soil characteristics and water availability.

Institutional Framework

Within the framework of Central American integration, there were interesting initiatives in the late 1980s such as subscribing to the Central American Agreement for Environmental Protection, which was created by the Central American Environment and Development Commission (CCAD); it later became an organization of the Central American Integration System (SICA), which had been established previously in the Tegucigalpa Protocol.[9]

Various initiatives have been developed in Central America for the purpose of harmonizing the policies and legislation for water management in the region. In 1994, the Central American Water Agreement was signed, which sought the efficient use of water resources based on criteria of fairness and justice. In the agreement, water was considered the 'germ of life, source of development and peace, and a public good with economic value', and the interests of the involved actors must therefore be considered in its management. Also, in the same year, the Central American Ecological Summit for Sustainable Development was held in Nicaragua, where the Alliance for Sustainable Development (ALIDES) was signed in which the formulation of policies and legislation regarding water management and conservation was established as a priority (Aguilar, 2005).

In 1999, CCAD prepared the Central American Regional Environmental Plan (CAREP), which contemplates integrated management as one of its principle policies: social, economic and ecological, equitable access, and the promotion of shared responsibility in the management of water. Its objectives included an attempt to guarantee the protection of water sources and to assure the long-term provision of the adequate quantity and quality of water in order to define uses and to promote the total economic valuation of water resources.

In March 1997, CCAD, together with the Regional Committee on Hydraulic Resources (CRRH), prepared a proposal for the Central American Action Plan of Integrated Water Resource Management (PACADIRH). This proposal was understood to be a group of strategies and actions to 'direct and harmonize the joint development of the water-related wealth enjoyed by the Central American Isthmus, in harmony with the principles of sustainable development'. In 2004, the updated version of CAREP (2005–2010) included the topic of water within the theme of prevention and control of environmental contamination in addition to being considered a transversal theme for action.

In spite of the efforts made in the region, water management is still considered sectorially, depending on whether its use is for irrigation, domestic consumption, industry or energy production. No differentiation is made between surface water and groundwater. No specific law exists in any country about regulating the management of groundwater. Rather, the existing policies and laws have been established to regulate individual uses. At the moment, the Central American countries lack a policy for integrated water management. Only Costa Rica (1942), Honduras (1927) and Panama (1996) have a General Water Law (Table 6.11). These, however, contain no vision of integrated management.

In the countries of the region, with the exception of Panama and Belize, water administration is the responsibility of the Environmental Ministries. In the case of Panama, responsibility falls on the National Environmental Authority (ANAM) and in the case of Belize it is not defined. Although the administration is defined in almost every country, in practice it has not functioned. Due to the lack of clear laws and strong institutions to assume this role, administration continues to be sectorial and falls on the water users.

The institutional framework has been characterized as fragmented and dispersed, with badly defined roles and functions, and with overlapping responsibilities. In terms of groundwater, policies and laws in countries such as Costa Rica and Panama have focused on regulating its use through a system of concessions.

However, given the need to update the normative and legal frameworks, four countries – Costa Rica, Nicaragua, Honduras and Guatemala – are currently preparing proposals for new laws, which are being discussed or are about to be discussed in the respective congresses. In Panama, a process of public bidding is in process for the preparation of a new water law, and the Watershed Law is being regulated. In Belize and El Salvador, discussions have begun to prepare the necessary conditions for drafting a water law and national water plans.

Table 6.11. Central America: current water legislation and water law projects. (From Aguilar, 2005.)

Country	Current legislation	Water law project
Belize	Water and Sewerage Ordinance, Chapter 185, 1971	Not found
	Water and Sewerage Sanitary Instrument, No. 29, 1982	
	Environmental Protection Act, No. 22, 1992	
	Public Health Ordinance, Chapter 31, 1943	
	National Lands Act, No. 83, 1992	
	Water Industry Act, Chapter 222, 1993	
Costa Rica	Law No. 276, General Water Law, 1942	Water Resources Law Project, 2004
	Law No. 1634, Potable Water Law, 1953	
	Law No. 5395, General Health Law, 1973, and its reforms	
	Environmental Law No. 7554, 1995	
	Law No. 2726 to Create AyA, 1961	
	Law No. 7779 for Land Use, Management and Conservation, 1998	
	Regulation 25992-S for the quality of drinking water, 1997	
	Regulation 26042-S MINAE for the disposal and reuse of wastewater	
	Environmental Tax for Effluents, Decree No. 31176-MINAE	
El Salvador	Integrated Water Resource Management Law, 1981	Not found
	Water Quality Bylaw, Flow Control and Protected Areas, Decree No. 50, 1987	
	Irrigation and Drainage Law	
	Administration of Aqueducts and Sewers Systems Law	
	Environmental Law, Legislative Act 233, 1998	
	Special Bylaw on Residual Waters	
Guatemala	Dispersed legislation in different normative bodies. Among them:	General Water Law Project, 30 August 2004
	Civil Code, Act 1932	
	Environmental Protection and Improvement Municipal Code	
	Health Code	
Honduras	Law for the Use of National Waters, April 1927	General Water Law Project, 2004
	Law for Drinking Water and Sanitation, 2003	
Nicaragua	General Environmental Law, 1996	General Water Law Project, January 2005
	Law 440, 'Suspension of Water Use Concessions', 2003	
Panama	General Water Law, 1966	Not found
	Law 41, Panama Canal River Basin, 1998	
	Law 44, Special Administrative Regime for the Management, Protection and Conservation of Watersheds, 2002	

In Nicaragua, the proposed law was generally approved, which places it in an advanced position within the legislative procedure; however, as in Costa Rica, it is subordinate to policy priorities and to the dynamics of the local power structures. Given the serious political crisis that this country is going through, as well as being delayed while the Central America Free Trade Treaty with the USA was being discussed, it is very doubtful that the project will be voted for by Congress in the next few months.

In the case of Costa Rica, by February 2004 the Water Department of the Ministry of Environment and Energy had granted 899 concessions throughout the country. Of these concessions, 0.9% corresponds to groundwater extraction and 29.3% is for irrigation (Table 6.12). This is due to the lack of existing control on the part of the Ministry, thereby facilitating the illegal extraction of water.

In other countries of Central America, such as Nicaragua and Guatemala, regulations are lacking, resulting in the uncontrolled extraction of water. In Guatemala, groundwater is managed by a private company. In Honduras, there are no controls regarding the exploitation of aquifers; therefore, it is possible that safe extraction limits are being exceeded.

Conclusions

Central America appears on the surface to be a water-abundant region. However, as described elsewhere (e.g. Shah, Chapter 2, this volume), population and water supply do not overlap. Further, supply is not consistent throughout the year, and there are often problems with quality due to sedimentation and pollution. These issues, combined with growing populations, have already brought out general challenges for supplying water for agricultural, domestic and industrial purposes throughout the region.

However, unlike the other regions covered in this book, there is currently little groundwater used in agriculture in any of Central America's countries.

Table 6.12. Costa Rica: concessions granted by the Water Department of the Environmental Ministry by type of source and use, February 2004. (From Water Department, Environmental Ministry of Costa Rica.)

Type of use	Surface water	Wells (underground)	Total
Agricultural/fishing	28,793.6	162.3	28,955.9
Agro-industrial	6,608.9	1,693.8	8302.7
Irrigation	121,118.0	2,495.5	123,613.5
Human consumption	3,958.8	1,556.5	5,515.2
Commercial	83	98.2	181.1
Industry	4,218.9	1,956	6,175
Hydraulic energy	722,965.6		722,965.6
Tourism	2,974.3	568.2	3,542.5
Total	890,721.1	8,530.5	899,251.5

Figures collected for this chapter show a total of less than 50,000 ha under groundwater irrigation, with most of that figure in Guatamala. At least three factors help to explain this outcome. First, there is relatively little irrigation in the region in general. Second, when irrigation does exist it is (in rural areas) usually supplied through relatively abundant surface water, which is generally of lower extraction cost than groundwater. Third, the main aquifers in Central America are generally located under metropolitan areas or, in other words, metropolitan areas have tended to grow over the main aquifers.

To date, the main use of groundwater in Central America is for household consumption, followed by industry and tourism activity. Still, the experience of Central America, both in terms of its overall groundwater situation and with reference to its urban use, highlights many of the stories, issues and challenges brought up elsewhere in this volume.

First, groundwater use in Central America seems to be following the development paradigms described by Shah and Kemper (respectively, Chapter 2 and Chapter 7, this volume). Agricultural groundwater use is generally still in stage I of their typologies, the stage before significant problems have emerged. While movement down the agricultural groundwater development path will likely vary from country to country and within countries, urban and industrial use surrounding metropolitan areas is already in stages II or III, and aquifers are showing clear signs of stress.

Second, as in most of the other regions described in this book, Central America has a great vacuum regarding information on groundwater. This applies to information on the resource itself as well as on its use. Failure to create, centralize and share information means there is little basis for management and decision-making in the already stressed urban areas, as use in some areas could lead to scarcity in the future. Moreover, there is only limited information on the potential for additional agricultural development, as the information that is available is not always consistent across countries, making it difficult to establish a regional information and lessons-sharing system.

Third, as in most of the other regions described in this book, the majority of Central American countries have either no water laws or only obsolete laws with little practical application. For this reason, the groundwater governance problems already occurring in South Asia and China are also occurring in Central American cities and may impact agriculture in the future. Connected to the governance problem is the growth of cities, particularly capital cities, over the highest potential aquifers. While use of groundwater in urban areas has clear benefits, the absence of land regulation and planning has meant that many cities have expanded into recharge areas, threatening the water that helped the cities' existence in the first place. Lack of control over industrial and human contaminants is increasingly threatening water quality. Uncontrolled agricultural chemical use is high throughout the region with concomitant risks to groundwater quality through percolation. To date, measurable concentrations do not generally exceed permissible limits, but pollutants are being detected, meaning that concentrations are increasing.

Clearly there has been no 'agricultural groundwater revolution' in Central America; nor is there likely to be one in the future simply because of climatic conditions. None the less, there are critical connections between agriculture and groundwater in Central America, though the importance of these connections is not the same as in regions where direct agricultural use is much higher. It is the study of this contrast that can help us to understand how broad the connections between agriculture and groundwater can be.

People Interviewed

- Ricardo Tompson, Ministry of Agriculture, Belize.
- Antonio Gaitán, Coordinator of DIAPRYD, PLAMAR-Ministry of Agriculture and Livestock (MAGA), Guatemala.
- Alejandro Flores Bonilla, Director of the Division of Irrigation and Drainage of the Ministry of Agriculture and Livestock, El Salvador.
- Oscar Cosenza, Director, General Department of the Secretariat of Agriculture and Livestock (SAG), Honduras.
- Rigoberto Reyes, Irrigation and Drainage Unit, Ministry of Agriculture and Forestry (MAG-FOR), Nicaragua.
- Marvin Coto, Director of Operations, National Irrigation and Drainage System (SENARA), Costa Rica.
- Héctor Elías Pérez, Director of the National Department of Rural Engineering and Irrigation, Ministry of Agricultural Development, Panama.

Notes

1 Based on information provided by Ricardo Tompson, Ministry of Agriculture in Belize.
2 Based on information provided by Antonio Gaitán, Coordinator of DIAPRYD, PLAMAR, Ministry of Agriculture and Livestock (MAGA).
3 Based on information provided by Alejandro Flores Bonilla, Director of the Department of Irrigation and Drainage of the Ministry of Agriculture and Livestock in El Salvador, September 2005.
4 Based on information provided by Oscar Cosenza, Director of the General Department of Irrigation and Drainage, Secretariat of Agriculture and Livestock (SAG), Honduras.
5 Based on information provided by Rigoberto Reyes, Irrigation and Drainage Unit, Ministry of Agriculture and Forestry (MAG-FOR), Nicaragua.
6 Based on information provided by Marvin Coto, Director of Operations, National Irrigation and Drainage Service (SENARA), Costa Rica.
7 Based on information provided by Héctor Elías Pérez, Director of the National Department for Rural Engineering and Irrigation, Ministry of Agricultural Development, Panama, September 2005.
8 The number of seasons will depend on the kind of crop, if there are periods of rotation of one, two or more times a year.
9 'The Central American Environment and Development Commission (CCAD) has tried to interject the environmental variable into the regional integration process so that it would be taken into consideration in the economic, social or any other kind of decision'. Madrigal, P. (1977) Aplicación y Cumpliminto de la Legislación Ambiental en Centroamérica. *Revista Parlamentaria* 5(3) 152.

References

Aguilar Grethel (2005) Estado del Marco Normativo y Legal del Agua en Centroamérica. Documento borrador. GWP-BID.

Aquastat (2001) Land and Water Agriculture. FAO Info System on Water and Agriculture.

CEPAL (2003) Anuario Estadístico de América Latina y el Caribe. Santiago de Chile, Chile.

Choza (2002) Prevención y Control de la Contaminación de Aguas Subterráneas en Acuíferos Urbanos. Programa Asociado de la Global Water Partnership de Centroamérica.

CIEUA (1998) Evaluación de recursos de agua de la República de El Salvador, Distrito de Mobile y Centro de Ingeniería Topográfica, El Salvador.

CRRH-SICA, GWP-CATAC-UICN (2002) Diálogo Centroamericano sobre Agua y Clima. San José, Costa Rica.

GWP-Global Water Partnership – Central America (2005) Situación de los Recursos Hídricos en Centroamérica: Hacia una Gestión Integrada. San José, Costa Rica.

Hetch, G. (1989) Calidad de las aguas del acuífero regional Las Sierras. Internal report. INETER, Dpto. de Hidrogeología.

Leonard, J. (1987) Recursos Naturales y Desarrollo Económico en América Central. Un Perfil Ambiental Regional. International Environment and Development Institute (Instituto Internacional para el Ambiente y el Desarrollo).

Losilla, M., Rodríguez, H., Schosinsky, G., Stimson, J. and Bethune, D. (2001) Los Acuíferos Volcánicos y el Desarrollo Sostenible en América Central. Editorial de la Universidad de Costa Rica. Primera Edición.

Programa de Reconversión Productiva de Agricultura y Alimentación (PARPA). Información compilada con base en información suministrada por PLAMAR-MAGA 2003–2004.

Reynolds, J. (1992) Contaminación por nitratos en las aguas subterráneas de la Cuenca del Río Virilla. En: Primer Simposio Nacional sobre Plaguicidas: Problemática y Soluciones. San José, Costa Rica, octubre 1992.

SENARA/BGS (Servicio Nacional de Aguas Subterráneas y Avenamiento) (1989) Continuación de la investigación hidrogeológica en la zona norte y este del Valle Central, Costa Rica. Informe final 1984–87. SENARA, Informe Técnico No.165. San José, Costa Rica.

SENARA/BGS (1998) Continuación de las Investigaciones Hidrogeológicas en el Valle Central de Costa Rica. Servicio Nacional de Aguas Subterráneas, Riego y Avenamiento/ British Geological Survey; Final report. SENARA, San José, Costa Rica.

SENARA (2005) Area de operaciones: proyectos en diseño y en funcionamiento.

SG-SICA (2001) Plan Centroamericano para el Manejo Integrado y la Conservación de los Recursos del Agua.

Silvel, E., G. Gellert, E. Pape and E. Reyes (1997) Evaluación de la sostenibilidad. El caso deGuatemala, FLACSO/WWF, Foro Mundial para la Vida Silvestre.

TAHAL (1990) Plan Maestro de Abastecimiento de Agua Potable de la Gran Área Metropolitana. Tomo III, Aspectos Hidrológicos e Hidrogeológicos. Instituto Costarricense de Acueductos y Alcantarillados. San José, Costa Rica.

World Bank (2005) The Little Green Data Book.: From the World Development Indicators. Available at http://www.oas.org/USDE/publications/Unit/oea34s/begin.htm#Contents

II Current Management Paradigms

7 Community Management of Groundwater

Edella Schlager

University of Arizona, School of Public Administration and Policy, McClelland Hall, Room 405, Tucson, AZ 85721, USA

Introduction

Over the last 20 years, scholars have devoted considerable attention to the ability of farmers, fishermen, pastoralists and other types of resource users to organize, adopt, monitor and enforce institutional arrangements that govern their use of common pool resources (CPRs) in a sustainable manner (Ostrom *et al.*, 2002). During this period, progress has been made in carefully identifying and defining key theoretical concepts, developing typologies that organize diverse types of problems and institutional arrangements, identifying factors that help explain the circumstances under which resources users are likely to engage in collective action to develop governing arrangements, identifying design principles that account for durability of self-governing arrangements, and developing an impressive body of empirical work devoted to theory development and hypothesis testing. According to Stern *et al.* (2002, p. 445) the study of institutions for managing CPRs is sufficiently developed to be recognized as a field within the social sciences. Surface irrigation systems have been a focal resource in the development of this field. Much attention and effort has been devoted to explaining the conditions that contribute to the emergence and persistence of farmer-managed irrigation systems. Comparative analyses of farmer-managed and government-managed systems have also been conducted.

This chapter extends the work of scholars on self-governance of CPRs to groundwater in irrigation settings. While work has been conducted by such scholars on groundwater basins in the USA, little focused attention has been paid to groundwater and irrigation. The first section of the chapter covers conceptual tools and theory from the field of CPR governance. The second section applies the conceptual tools and theoretical concepts to groundwater irrigation. The arguments are illustrated in two ways: first, by a comparative analysis of surface irrigation systems and groundwater irrigation; and second, through the use of several case studies. The final section explores promising

 The Agricultural Groundwater Revolution: Opportunities and Threats to Development (M. Giordano and K.G. Villholth)

types of linkages between communities of groundwater users and higher-level governments. Local-level governance is a key component of sustainably managing groundwater basins.[1] How higher-level governments can encourage and support local management efforts is an important topic.

A Theory of Common Pool Resources

Foundational concepts

CPRs are defined as natural or man-made structures characterized by costly exclusion and subtractability of units (Ostrom and Ostrom, 1977). Examples include surface irrigation systems, groundwater basins, fisheries, forests and grazing lands. Both exclusion and subtractability present challenges for governing CPRs sustainably. Exclusion involves defining who may enter a resource and who may not – making such a determination is rarely a straightforward process. Ideally, exclusion should occur in a manner that limits access to the number of users whose use will not threaten the resource. Physical, institutional and social issues often confound such efforts (Ostrom *et al.*, 1994). The sheer size of some resources makes enforcing access limitations in any meaningful or cost-effective manner virtually impossible. In other instances, national or state constitutions forbid denying citizens access to natural resources. In other settings, there may be political or economic reasons for avoiding strict access controls. For instance, a number of surface irrigation systems have been described as long and lean – the goal being to provide at least some water to as much land as is possible. Rationales range from equity concerns, i.e. assisting many people, to cost–benefit analysis issues, i.e. the more land included in a scheme, the better the cost/benefit ratios. In either case, too much land can be included within a system with some farmers experiencing chronic water shortages.

Exclusion is critical for sustainability, but also for governance. Resource users are much less likely to undertake costly and time-consuming efforts to manage CPRs if they cannot capture many of the benefits resulting from good management. Why design a water allocation scheme that conserves water if the additional water supplies may be captured and used by someone else? Why invest in groundwater recharge projects if others can pump the recharged water? Inadequate exclusion promotes free riding, and free riding discourages collective action (Dietz *et al.*, 2002).

Even if exclusion is adequately addressed in relation to a CPR, sustainability is not ensured because of substractability. Subtractability means that each 'unit' harvested from a CPR is not available for other users to harvest. The groundwater that a well owner pumps and uses to water his crops is not available for other well owners to pump. Since each resource user gains the value of each unit harvested but imposes some of the costs of harvesting on all resource users, resource users are likely to harvest more than is economically or ecologically desirable (Gordon, 1954; Scott, 1955; Dietz *et al.*, 2002). Or, as Ostrom *et al.* (1994, p. 10) explain: '[I]ncreased water withdrawal by one pumper reduces the water other pumpers obtain from a given level of investment in pumping inputs'. The problem of exclusion may be adequately addressed but the CPR may still be overused because of the harvesting actions of the resource users. Consequently, if CPRs are to be governed sustainably, the challenges posed by difficult and costly exclusion and substractability must together be addressed.

Considerable attention has been devoted to the problem of overuse. The earliest formal models of resource use, such as those developed for fisheries (Gordon, 1954; Scott, 1955), focused on it, and many models since then have followed suit (e.g. Hardin, 1968; Clark, 1980; Norman, 1984). While overuse is problematic, resource users are likely to confront a host of CPR dilemmas (Ostrom *et al.*, 1994). Ostrom *et al.* (1994) define CPR dilemmas as suboptimal outcomes produced by the actions of resource users and the existence of feasible institutional alternatives, which, if adopted, would lead to better outcomes (Ostrom *et al.*, 1994, p. 16).

In addition to overuse, resource users may engage in a variety of actions that produce suboptimal outcomes in their use of a CPR. For instance, well owners may place their wells too close together, interfering with one another's pumping; a farmer may install a deep tube well near another farmer's shallow well, drying it up; farmers may fail to maintain a tank that would otherwise serve to capture rainwater and recharge it into the underground aquifer.

Ostrom *et al.* (1994) relax the implicit assumptions underlying formal models focused on overuse to develop a typology of CPR dilemmas. Most models assume a uniformly distributed resource. By relaxing that assumption and allowing resources to be patchy, so that some areas of a resource are more productive than others, assignment problems may emerge. Assignment problems involve resource users competing over productive areas and interfering with one another's harvesting (Ostrom *et al.*, 1994, p. 11). Furthermore, most formal models assume identical harvesting technologies among resource users. By relaxing that assumption and allowing diverse technology utilization, technological externalities may emerge among resource users. Technologies used by harvesters interfere with one another causing conflicts among resource users. For instance, a high-capacity well may dry up a shallow tube well (Ostrom *et al.*, 1994, p. 12). Thus, in addition to overuse, or what Ostrom *et al.* (1994) term appropriation externalities, resource users may experience assignment problems and technological externalities.

As Ostrom *et al.* (1994) note, appropriation problems stemming from when, where, how and how much to harvest are not the only problems resource users are likely to experience. Another class of dilemmas – provision problems – is also likely to emerge in many CPR settings. Provision problems relate to developing, maintaining and/or enhancing the productive capacity of the CPR. For instance, adequately functioning surface irrigation systems require that diversion structures, headworks, canals and outlets be regularly repaired and maintained. The productivity of an aquifer may be enhanced by capturing water during wet seasons and directing that water underground to be used during dry seasons. Provision problems are distinctly different from appropriation problems. Appropriation problems require resource users to coordinate their harvesting activities; provision problems require resource users to cooperate and contribute to the production of public goods.

Self-governing institutional arrangements

Ostrom (2001) argues that resource users are more likely to invest in designing and adopting rules to address CPR dilemmas if they perceive that (i) the benefits produced by the new sets of rules outweigh the costs of devising, monitoring and enforcement; and (ii) they will enjoy those benefits. Whether these two conditions hold depends on characteristics of the resource and characteristics of the resource users. For Ostrom (2001) four resource characteristics are crucial:

1. *Feasible improvement*: Resource conditions are not at a point of deterioration such that it is useless to organize or so underutilized that little advantage results from organizing.
2. *Indicators*: Reliable and valid indicators of the condition of the resource system are frequently available at a relatively low cost.
3. *Predictability*: The flow of resource units is relatively predictable.
4. *Spatial extent*: The resource system is sufficiently small, given the transportation and communication technology in use, that appropriators can develop accurate knowledge of external boundaries and internal microenvironments (Ostrom, 2001, p. 40).

There must be a sense among resource users that governance attempts will make a difference (attribute 1). If a resource is so degraded that users believe there is little they can do to positively affect the situation, they are unlikely to make the attempt. Conversely, appropriators may find little benefit in investing in governing arrangements if a resource is relatively abundant and of adequate quality. Whether resource users believe that feasible improvement in the productivity of the resource is possible depends on the information that they have and their ability to exercise some control over the resource. Information about a resource depends on availability of reliable and valid indicators of resource conditions, the spatial extent of the resource and the predictability of resource units (attributes 2–4). Indicators vary from resource to resource and may be as 'simple' as paying attention to wool or milk production of grazing animals or as complex as monitoring wells. The spatial extent of a resource affects both the ability of users to develop information and to assess their relative ability to capture the benefits of organization. Resource systems or subsystems that are more closely matched with the ability of resource users to monitor encourage investment in rules. Finally, predictability should be interpreted broadly to include volume and temporal and spatial patterns. Predictability provides resource users the opportunity not only to learn about the resource but also to govern harvesting activities in meaningful ways.[2]

In addition to characteristics of resources, qualities of the resource users themselves affect the benefits and costs of cooperation to devise governing arrangements. Ostrom (2001) posits the following attributes of resource users:

1. *Salience*: Appropriators are dependent on the resource system for a major portion of their livelihood or other important activity.
2. *Common understanding*: Appropriators have a shared image of how the resource system operates, and how their actions affect each other and the resource system.

3. *Low discount rate*: Appropriators use a sufficiently low discount rate in relation to future benefits to be achieved from the resource.
4. *Trust and reciprocity*: Appropriators trust one another to keep promises and relate to one another with reciprocity.
5. *Autonomy*: Appropriators are able to determine access and harvesting rules without external authorities countermanding them.
6. *Prior organizational experience and local leadership*: Appropriators have learned at least minimal skills of organization and leadership through participation in other local associations or studying ways that neighbouring groups have organized (Ostrom, 2001, p. 40).

These characteristics ease the costs of organizing, developing and adopting a common set of rules. Attributes 1 and 3 measure how appropriators value the resource. If resource users are heavily dependent on the resource for their livelihood and if they anticipate continued reliance on it well into the future, they are more likely to invest in new sets of rules. If appropriators share a common understanding of the resource and the effects of their actions on the resource and on each other, they are more likely to share a common understanding of the problems that they face and are more likely to agree upon a set of rules to address those problems. Trust and reciprocity and leadership provide resource users with 'social capital' that they can draw upon to ease bargaining and negotiation costs. Autonomy provides appropriators with the 'space' needed to engage in rule making and confidence that they will be able to capture the benefits of their institutional investments. While Ostrom (2001) separates the two sets of attributes for the sake of clarity, the attributes interact to support or discourage collective action. Resource users may have a relatively complete and accurate understanding of the resource; however, they may still be unwilling to invest in new sets of rules if the resource is of low salience to them.

Comparing surface water and groundwater irrigation

The emerging theory of CPR governance provides a consistent set of concepts and analytical tools to diagnose problems, provide a deeper understanding of the conditions under which local governance of CPRs is likely to occur, identify promising policy alternatives and shed light on the shape and form of productive relations between local-level governance arrangements and regional and national governments. One valuable use of the theory is to systematically compare surface water irrigation with groundwater irrigation. In so doing, the very real, but very different, challenges facing both types of irrigators are clarified. Local-level self-governance is possible in both settings, but it will probably exhibit different structural features and require different types of linkages with higher levels of government because of the diverse challenges presented by two contrasting physical settings: surface irrigation systems that are human-constructed CPRs, and groundwater basins that are naturally occurring CPRs.

Surface water irrigation

The governance challenges groundwater irrigators commonly face differ considerably from those faced by surface water irrigators. The differences result from the distinct physical structures of surface irrigation systems compared with groundwater basins and, consequently, the different water development paths that unfold between the two types of water systems.

To construct and operate a surface irrigation system requires considerable upfront production and transactions costs. Using the terminology discussed above, irrigators immediately confront provision problems. At a minimum, production costs entail building a diversion structure, a distribution system and field outlets and channels. A single person or family generally cannot meet such production costs; rather a collective effort is necessary, involving many people, their resources and their participation. The transaction costs of organizing people, developing information about the physical setting, negotiating over the location and design of the irrigation system, organizing labour as well as monitoring and enforcing agreements concerning contributions and work are significant. Providing an irrigation system requires upfront organization and collective action.[3]

Marshalling the participation and resources needed to provide an irrigation system is closely tied to anticipating and addressing the inevitable appropriation dilemmas that will emerge. In many instances, water will be insufficient to meet all irrigators' needs all the time. Water allocation rules must be established, at least in a rudimentary form before the system is built, to provide assurance to farmers about the benefits they will likely receive from participating in the collective undertaking. Once the system is built, it must be maintained, requiring the creation of rules governing irrigators' contributions to system upkeep (Tang, 1992, 1994). Farmers will be more likely to abide by their maintenance requirements if the water allocation rules are functioning well. In turn, water allocation rules are likely to be more productive if the system is well maintained. In other words, provision and appropriation dilemmas are closely tied together in surface irrigation systems. Adequately addressing one set of dilemmas often requires adequately addressing the other. If that is accomplished, positive feedback between the two processes acts to support and sustain the system. Of course, the opposite is true as well. If appropriation dilemmas are not adequately addressed, provision is likely to falter, which will further exacerbate the appropriation dilemmas (Tang, 1994; Lam, 1998).

Rose (2002) notes the unique character of surface irrigation systems that may make them particularly amenable to farmer-based governance. She argues, echoing Ostrom's attributes of the resource, that unlike many other CPRs,

> [r]esource-related activities involved in irrigating – taking water from ditches, laboring on infrastructure development and upkeep – are especially open to mutual monitoring. Not only can one farmer observe another farmer along the same ditch, but upstream and downstream communities can observe what other communities are doing with respect to water use and infrastructure maintenance.
> (Rose, 2002, p. 239)

Farmers can more readily determine and define the boundaries of their irrigation systems. They can monitor water flows and the variation in volume over time. They can experiment with different water allocation rules and determine which allocation methods better fit their particular physical setting. They can also readily monitor and observe one another's behaviour and determine whether water allocation or labour contribution rules are generally being followed.

While the physical setting of surface irrigation systems is more conducive to the emergence and persistence of farmer-based management compared with groundwater irrigation settings, as will be discussed below, scholars have noted the challenges that farmers face in maintaining their governing systems over time. In particular, scholars have begun to explore the effects of heterogeneity on the performance of farmer-governed surface irrigation systems (Bardhan and Dayton-Johnson, 2002; Ruttan, 2004). Bardhan and Dayton-Johnson (2002) reviewed the findings of several large *n* studies of farmer-managed irrigation systems that devoted attention to heterogeneity. Across all of the studies, the effects of heterogeneity were consistently negative (Bardhan and Dayton-Johnson, 2002, pp. 104–105). Income inequality and asymmetries between head-enders and tail-enders was associated with rule breaking, poor system maintenance and poor water delivery performance. Landholding inequalities were associated with poor canal maintenance. Differential earning opportunities among irrigators were associated with lower rule conformance and system maintenance.

Ruttan (2004) carefully reanalysed the data collected by Tang (1989, 1992) to explicitly examine the effects of different forms of heterogeneity on the performance of irrigation systems. In addition to the findings reported by Bardhan and Dayton-Johnson (2002), she found that variation in income had a negative effect on the likelihood that sanctions for rule breaking would be applied (Ruttan, 2004, p. 28). Ruttan (2004, p. 35) also found that sociocultural heterogeneity had a negative effect on rule conformance and system maintenance.

Causal mechanisms have not been identified. However, a number of attributes of appropriators could be at work. For instance, differential earning opportunities could affect the salience (appropriator attribute 1) of irrigation systems for farmers. If irrigated agriculture becomes a secondary income source for some farmers, they may be unwilling to devote resources to the irrigation system. Sociocultural heterogeneity could impact trust and reciprocity (appropriator attribute 4). If irrigators speak different languages, or if they come from different ethnic traditions, communicating and developing cooperative norms may be very difficult.

Groundwater irrigation

One of the most striking aspects of groundwater development is how rapidly it unfolds once a minimum level of technology and energy becomes widely available. Entry to groundwater basins is minimally restricted, with land ownership or leasing the only requirement for access. Depending on the setting, such as water table levels, even relatively poor farmers may access groundwater through inexpensive technologies. Even if farmers do not invest in their own wells, either because they do not have the necessary capital or their landholdings are too fragmented to justify a well, they may gain access to groundwater through markets (Shah, 1993; Dubash, 2002).

Groundwater is widely adopted because of its high value. For some farmers it may be the only source of irrigation water, either because they do not have access to surface water irrigation, or even if they are within the command area of a canal system they may not receive water. For many farmers, groundwater is more reliable, timely and adequate than the water they receive from canal systems. For other farmers, groundwater may be more 'convenient' than canal water, even if canal water is reliable and timely. A farmer who owns a well that provides enough water for irrigation needs may opt out of a communal system and its various requirements and responsibilities, such as contributing labour and materials for canal maintenance.

Compared to surface irrigation, developing groundwater entails substantially lower upfront production and transaction costs. Nature has provided a reservoir that is, at least initially, and in many cases, very easily accessed through a well. Consequently, production costs may be borne by a single individual or family. Transaction costs are also low. Farmers need not organize, bargain and negotiate over the development of an irrigation system and system design, or monitor and enforce commitments. Some farmers may form partnerships to raise the capital necessary to build a well; however, the transaction costs they face are substantially lower than those faced by farmers attempting to develop and build a surface irrigation system.

The physical setting of groundwater basins acts as a two-edged sword. Groundwater basins are a source of relatively inexpensive, reliable irrigation water that may be developed by individual families, once technology and energy are readily accessible. However, at the same time, the physical setting presents extraordinarily difficult challenges that may confound irrigators' attempts to address appropriation and provision problems. Unlike surface irrigation systems in which, through experience, observation and experimentation, the boundaries, capacity and variability of the system may be determined by irrigators, groundwater pumpers may never grasp the boundaries, structure or capacity of the 'invisible resource' they tap into without considerable assistance from engineers and hydrologists. Furthermore, unlike surface irrigation systems, in which irrigators may readily observe one another as they go about their daily farming activities, groundwater pumpers cannot easily determine the number of other pumpers, the capacities of their wells, how much water they are taking, the effects of their pumping on the overall productivity of the groundwater basin, etc. Thus, surface irrigators are more likely to develop norms of CPR management because of the information-rich environment within which they interact. Groundwater irrigators face an information-poor environment that makes it more difficult to develop self-governance norms (Rose, 2002).

Easy accessibility and limited information about the CPR combine to create significant barriers to the emergence of local-level governance of groundwater basins. Easy accessibility allows hundreds, if not thousands, of farmers across a groundwater basin to farm more intensively and to raise more high-valued crops. Only after farmers have invested heavily in wells and in productive activities and have come to appreciate and enjoy improved living standards do appropriation and provision problems emerge. As Bastasch (1998, p. 102) notes concerning groundwater development in the state of Oregon, located in northwest USA:

> Judgments about general groundwater availability, whether or not water tables are declining, impacts of new uses on nearby wells or streams and ultimately the public welfare itself, all hinge on good data. . . . When data are sufficient

to trigger groundwater controls, the damage has usually already been done and communities are heavily invested in the customary level of (over-)use.

Tackling appropriation and provision problems is not easy (i) because of information problems; (ii) because solutions often require farmers to limit well building and to adopt limits on the amount of water they may pump, which they may perceive as threatening their livelihoods; and (iii) because monitoring the use of an easily accessed, but invisible, resource is costly and difficult. Assuring thousands of farmers that their conservation actions will benefit them and will not be siphoned off by others is not likely to be easy.

The water development path in groundwater irrigation is very different than that in surface water irrigation. Farmers using groundwater do not have to organize, build and manage an irrigation system. They individually invest in wells that are used for irrigation. Farmers using groundwater are not confronted by appropriation or provision problems until long after they have become accustomed to the benefits of irrigation. When they do face dilemmas, they are more likely to face appropriation problems initially and provision problems later. Recalling Ostrom's resource and appropriator attributes, the following subsection argues that farmers are much more likely to organize to address appropriation problems than they are to address provision problems.[4] Farmers are likely to address provision problems only with considerable assistance from higher-level governments.

Appropriation problems

Appropriation problems are highly local compared to provision problems. They stem from actions and choices of appropriators whose effects become apparent within a short time frame, such as during an irrigation season.[5] Assignment problems, for instance, occur because people compete to use the most productive patches of a CPR and in the process they interfere with one another's harvesting activities. People may place wells too closely together, reducing the productive capacity of each of the wells. Technological externalities occur because the different harvesting techniques that people use interfere with one another. A high-capacity well may create a cone of depression that dries up surrounding shallow tube wells (Dubash, 2002).

Effectively and equitably addressing assignment problems and technological externalities requires considerable time and place information. Working knowledge of the types of technologies used, location of wells, uses made of the water, landholding patterns, actions causing the harvesting conflicts and so forth are necessary if rules that match a specific setting are to be devised. Such local knowledge resides with water users and not regulators. Shah (1993, pp. 129–132) notes the numerous difficulties regulators external to local communities have in devising effective rules. A common approach to address assignment problems is to impose well spacing rules. The rules only apply to more modern technologies, such as electric and diesel pumps, thus failing to afford any protection for more traditional technologies. Also, well spacing rules are enforced through banks that will not provide capital for the purchase of pumps unless well spacing rules are followed.[6] Farmers who can raise sufficient capital without relying on a bank can avoid well spacing rules.

Groundwater users can determine the causes and effects of spacing wells too close together or of allowing high-capacity wells to be situated among traditional water-lifting devices. Well owners and others who are dependent on those wells for water face incentives to problem-solve in order to protect their water sources. Depending on the social ties among groundwater users and experiences that they have had in engaging in other collective efforts, they may pursue strategies or undertake collective efforts to address assignment problems and technological externalities. For instance, Shah cites several examples of groundwater users effectively addressing such problems among themselves:

> The owners of grape orchards in Karnataka and Andhra Pradesh, for instance, are known to buy up neighbouring lots at premium prices to solve the problem of interference . . . in many parts of Gujarat, where localized water markets have assumed highly sophisticated forms, it is common for a well owner to lay underground pipelines through neighbours' fields at his own cost, and dissuade them from establishing their own wells by informal long-term contracts for the supply of water at mutually agreed prices.
>
> (Shah, 1993, p. 7)

Appropriation externalities result from overuse of CPRs in the short term. Appropriation externalities in groundwater may often be spatially and temporally confined, allowing closely situated groundwater users to learn about the effects of pumping on water tables and on one another's pumping activities. That learning can form the basis for developing locally devised solutions to appropriation externalities.[7] For instance, Sadeque (2000) examines the development of water access and allocation rules to address appropriation externalities that emerge during the dry season in Bangladesh. Domestic water uses are provided for through shallow hand pumps. During the dry season, when groundwater demand is quite high, especially to irrigate the winter rice crop, the hand pumps dry up, leaving many households without a reliable and convenient source of water. As Sadeque (2000, p. 277) notes: 'In the competition for groundwater, simple, low-cost technologies like hand tube wells, used mostly for drinking and other domestic users, lose out. The perception of affected people of the low water table areas as victims of water deprivation is becoming marked, with acrimony towards irrigation'.

In a study of two villages in northwestern Bangladesh, Sadeque (2000) found conflict between domestic users and irrigators to be widespread during the dry season. However, he also found instances of cooperation and coordination emerging to address such conflicts. For instance, a series of shallow wells installed by an international non-profit development agency for domestic water uses are carefully governed by the households who participated in their development and who are responsible for their maintenance. During the dry season, the households impose restrictions on water use to tide families over. These restrictions also affect households who did not participate in the well project. While during the wet season non-participating families are not restricted in their access to the wells, during the dry season their access and use is strictly limited. They are allowed water after the households who govern the wells have their needs met. In addition, cooperation is emerging between villagers and owners of irrigation wells. Irrigation well owners allow villagers to take water from wells to meet basic consumption and cooking needs. Also, some well owners operate wells during early morning hours for the express use of villagers' domestic water needs. Sadeque (2000, p. 286) argues that such cooperation has emerged as a means of avoiding government regulation: 'People realized that negotiation was better than having controls imposed by central and distant authorities which might not be in the interest of either party. Additionally, regulations would result in bureaucratic control and therefore encourage corruption'.

In general, appropriation problems tend to be local in nature. Furthermore, the specific types of problems that emerge and their causes tend to be highly dependent on configurations of factors unique to each situation. Consequently, workable solutions are usually those grounded in specific time and place information – information that is readily available to groundwater users, but not to regulators. In addition, groundwater users often face incentives to invest in collaborative attempts to resolve such problems. Coordination may yield substantial benefits. Thus, compared with provision problems, which will be discussed later, communities are more likely to address appropriation problems.

It is not uncommon in the emerging literature on groundwater and irrigation to find instances of groundwater users addressing appropriation problems or having the capacity to address such problems. For instance, Shah (1993) describes a village in Junagadh district, Gujarat, in which numerous irrigation wells dry up during the dry season. Shah (1993) notes that farmers have a good understanding of how their wells function, and pursue a variety of strategies to ensure water availability throughout the dry period, but with mixed success. Some farmers are more innovative than others and appear to have developed approaches that are relatively successful. Shah (1993, pp. 164–165) argues that with a little assistance, primarily in the form of information, such as location and productivity of wells over time and various successful strategies that some farmers pursue, farmers could develop collective strategies to address appropriation problems and thereby increase agriculture productivity.

Appropriation problems that emerge in groundwater aquifers may be more manageable for irrigators because they exhibit some of the resource attributes identified by Ostrom (2001). Owners of closely situated wells, for instance, may readily realize the effect that their pumping has on one another as water levels in their wells decline under heavy pumping and begin to recover as they reduce their abstractions (attribute 2 – indicators; attribute 3 – predictability). In other words, it is possible through experience and careful observation to determine the onset and causes of appropriation problems.

Provision problems

Provision problems center on maintaining, recovering or enhancing the productive capacity of a CPR. Provision problems are the undesirable effects of intensive groundwater use (Llamas and Custido, 2003). As increasing volumes of water are pumped and water tables decline, a host of problems may emerge – pumping costs may increase, and wells may need to be replaced. Soil compaction and subsidence occur as water is withdrawn and the sand and gravel that compose the basin compact. If a groundwater basin is hydrologically connected to surface streams and rivers, surface water sources may dry up as water tables decline. As surface water sources are depleted, aquatic life, riparian vegetation and the birds and animals dependent on it die off (Blomquist, 1992).

Provision problems also include water quality. Basins may be polluted by industrial and municipal wastes, agricultural runoff and inadequate or improper disposal and treatment of human and animal waste. Declining water tables and water quality problems combine in the form of salt water intrusion. Coastal basins are highly susceptible to salt water intrusion. As water tables decline, the hydrologic pressure that the fresh water of the basin exerts against the salt water declines and salt water invades the fresh water. Although it is possible to halt the spread of salt water, it is very difficult and costly to reclaim portions of basins that have been polluted by salt water (Blomquist, 1992).

Provision problems do not only centre on undesirable effects of intensive resource use; they may also include the failure to take advantage of opportunities to enhance the productive capacities of CPRs. In the case of groundwater basins this typically takes the form of failing to use their full storage capacity. The unfilled storage space may be taken advantage of and surface water may be captured and placed underground for use at a later time. Of course, enhancement, if not carefully managed or attended to, can result in degradation of surface soils in the form of waterlogging, a common problem among some canal irrigation systems.

Provision problems are especially challenging to address, both for local communities of resource users and regional and national governments. Provision problems tend to be extensive – they are caused by, and affect, many groundwater users across an entire basin. It may take well owners years to detect longer-term declines of water tables, as water tables may vary from year to year. Even if well owners suspect long-term declines, their magnitudes and causes may be difficult to determine without considerable effort and investment in hydrogeologic studies. Such studies may take years to complete as the boundaries and structure of the basin must be determined, storage capacity identified, rates of natural recharge and pumping volumes computed, and identification of different water uses and their consumptive use of water measured (Kendy, 2003). No single well owner, or community of well owners, is likely to have the expertise or sufficient resources to invest in such studies.

Even if a community undertook such a study, and developed information about a basin, it is unlikely the community, acting alone, could resolve the problem of mining. Mining affects the multiple communities or clusters of groundwater users scattered across a basin and would require widespread participation to resolve. A similar argument may be made for the other types of provision problems. Developing reliable information about groundwater basins requires considerable time and investment in technical studies; it is not information that water users can develop by monitoring their wells and speaking with their neighbours.

Even if adequate models and data have been developed for a groundwater basin, sufficient uncertainty and a weak legal system may provide groundwater pumpers the opportunity to avoid making difficult choices. For instance, the Umatilla River basin, located in northeastern Oregon, has experienced water conflicts and controversies for several decades (Oregon Water Resources Department, 2003). The Umatilla River, a tributary of the Columbia River, is hydrologically connected to alluvial and hard-rock (basalt) aquifers. The basin also includes a number of closed, or contained, deep hard-rock aquifers. Most surface and groundwater diversions are devoted to agricultural enterprises – irrigated crops and dairies. As surface water supplies became fully appropriated and rights in surface water difficult to obtain, farmers turned to groundwater, which was not heavily regulated. By the 1960s, however, a variety of groundwater problems began to emerge in different parts of the basin such as sharp water table declines, unstable water levels and interference among water appropriators. Under Oregon law, the Oregon Water Resources Commission can impose various types of control measures to address groundwater problems. Since the mid-1970s, the Commission has created four critical groundwater areas and one classified groundwater area within portions of the basin (Oregon Water Resources Department, 2003). The primary effect of designating critical and classified areas is to stop or substantially reduce the number of new well permits issued. In other words, new water rights cannot be developed in critical groundwater areas. If an individual or business wants to obtain additional water supplies, they have to acquire existing water rights.

Currently, groundwater problems persist and in some instances are becoming more acute in the Umatilla basin. In some critical groundwater areas, water levels have stabilized; in many, the rate of water level declines has slowed; and in others, declines continue unabated (Oregon Department of Water Resources, 2003). Outside of the designated critical groundwater areas, groundwater problems are emerging. These results are not surprising. Restricting or closing areas experiencing groundwater problems to new groundwater development may work to slow the intensity of groundwater use. Adequately addressing groundwater problems will likely require careful management of existing uses as well.

The Oregon Water Resources Commission finds itself in a difficult spot. Through its ongoing groundwater monitoring program in the Umatilla basin, and through a variety of hydrogeologic studies that it has carried out, it has developed a working understanding of the basin and the location as well as likely causes of groundwater problems. However, it cannot readily act on that knowledge. Designating critical groundwater areas is very unpopular among water users and is actively resisted. For instance, it took the Commission almost 14 years to designate the Butter Creek critical groundwater area in the Umatilla basin and impose pumping controls, in part because groundwater users repeatedly challenged the Commission's actions in court (Bastasch, 1998). Administratively imposed controls are unlikely to lead to the sustainable use of the Umatilla basin. Currently, Umatilla county, which is home to the four critical groundwater areas in the Umatilla basin, is attempting to create a collaborative effort involving a wide variety of stakeholders to develop and implement alternative management actions (Umatilla County Groundwater Solutions Taskforce, 2005).

As the Umatilla River basin case illustrates, addressing provision problems requires that users limit their pumping of groundwater, forego some of the income and other valued activities that pumping made possible and switch to economic activities in which the consumptive uses of water are lower (Kendy, 2003). In addition to limiting groundwater pumping, groundwater users may also have to invest in public goods to recover or maintain the groundwater basin, such as recharge projects to increase the amount of water stored in the basin, or different sources of surface water to supplement groundwater. Given the very difficult physical, social and economic challenges surrounding provision problems and their solutions, groundwater users and governments, in general, will not be able to address such problems without assistance from each other.

Shah (1993) describes the situation of a coastal village of Mangrol *taluka*, Gujarat (a taluka is an administrative division in India below a district). The wells closest to the sea are saline and unfit for irrigation and the fields watered from those wells are barely productive. A middle belt of fields and wells are just beginning to experience salinity; however, it is expected that they too will succumb to the migrating sea water within a few years. A belt of fields and wells further inland have not yet experienced salinity. While the farmers know what is happening, they are reluctant to address the problem. For those whose fields have been rendered unproductive, limiting pumping is unlikely to be effective unless it is matched with active recharge programmes. They view their situation as hopeless; the resource has been so degraded that there is little that they can do that would make a difference. For those who are just beginning to experience salinity, they are unwilling to limit their pumping. They believe that limiting pumping would not protect them from salinity, unless everyone limited pumping. That would only occur if additional sources of water were developed, so that no one would have to cut back on water use. Those further inland are not experiencing problems and are not interested in developing solutions (Shah, 1993, pp. 168–169).[8]

Relations between irrigators and governments

Surface and groundwater irrigators need the assistance of higher levels of government if they are to adequately address provision problems. The form of that assistance is not entirely clear; however, accumulated evidence suggests the form such assistance should not take. The empirical evidence from studies of surface water irrigation systems is clear and consistent. Farmer-managed irrigation systems perform better than government-managed irrigation systems. Tang (1989, 1992, 1994) studied 47 irrigation systems located around the world and Lam (1998) studied more than 100 irrigation systems in Nepal. Both studies included farmer-managed and government-managed systems. In both studies, farmer-managed irrigation systems performed significantly better than did government-managed irrigation systems. Compared with government-managed systems, irrigators in farmer-managed systems paid close attention to boundaries and to exclusion, attempting to more closely match water supply with demand. Furthermore, irrigators in farmer-managed systems devised more rich and complex sets of rules to govern access, water allocation and contributions to maintenance that better matched the physical and social settings. In addition, irrigators in farmer-managed systems had a better understanding of their systems, and were more likely to engage in attempts to revise the rules. Also, irrigators in farmer-managed systems have devised active monitoring systems, and were therefore more likely to be sanctioned if caught violating the rules. In general, the work of Tang and Lam suggests that farmer-managed systems outperform government-managed systems in terms of system maintenance, adequacy of water supply and rule-following behaviour.

Evidence from groundwater irrigation is suggestive, but few systematic comparative institutional studies of different forms of well-governing arrangements have been conducted. Shah (1993) cites a study conducted by Lowdermilk *et al.* (1978) in Pakistan of crop yields under different levels of control of water sources. Among groundwater users, crop yields were highest among farmers who owned their own wells and lowest among farmers who depended on public tube wells (Shah, 1993, p. 29). Shah (1993, p. 29) states that a number of studies have been conducted in India that suggest that farmers prefer water from privately owned tube wells over publicly owned tube wells. This is so, Shah (1993, p. 29) argues, because water service from state tube wells is inferior to that of private tube wells. State tube wells suffer from poor maintenance, long shutdown periods, erratic power supplies and so forth. The root of the problem lies in management. Shah (1993, p. 30) concludes: 'A state tubewell operator is in reality accountable to no one, for he can neither be punished nor rewarded by the community he is meant to serve'.

A case study, developed by Singh (1991), of the construction, operation and maintenance of a public tube well used for irrigation in Uttar Pradesh, India, clearly illustrates Shah's arguments. The well and its associated infrastructure were designed and built by the government irrigation department. The department is supposed to operate and maintain the well. Water allocation and distribution was turned over to farmers' committees formed by the government irrigation department. The well and its infrastructure are not well matched to the patterns of landownership. According to Singh (1991), government officials face few incentives to operate and maintain the well appropriately, water service is erratic and the farmers' organizations have slowly fallen apart.[9] The evidence from studies of well ownership and operation appears to coincide with the evidence from canal irrigation systems. Government-operated canal systems and wells perform poorly relative to farmer-operated canal systems and wells. What is not well understood in relation to wells, and consequently needs more study, is the relative performance of different types of farmer-based ownership and management structures.[10]

If governments perform poorly in the direct production and management of surface irrigation systems and wells, what should the roles of governments be? As Stern *et al.* (2002) note, one of the most understudied areas in the field of CPR governance is the linkages and relations among local communities and higher-level governments and organizations. Young (2002) argues that developing productive, complementary relations is challenging because local communities and regional and national governments often have conflicting and competing interests in how CPRs should be governed and used. For instance, national governments tend to view CPRs as valuable for producing national revenues, either through granting concessions to multinational corporations to harvest timber, or to encouraging farmers to raise multiple cash crops. Local resource users tend to view CPRs as the foundation for their livelihood and are not as interested in generating foreign exchange, or other revenue generating activities for their national governments or government officials. Young (2002) urges giving greater weight to local interests and greater decision-making authority to local resource users rather than external government officials. Local resource users are more likely to attempt to address their most pressing needs, which are directly related to the productivity of CPRs.

Generally, productive relations among different levels of government tap into the strengths of local communities and higher-level organizations and match them to the particular CPR dilemma facing the resource users. Appropriation problems, as argued earlier, tend to be localized, with both causes and solutions hinging on time and place information. Since resource users have ready access to, and familiarity with, time and place information, they are likely to be in a better position to address such problems. Consequently, government roles should be more limited, such as assisting resource users in developing information about activities and practices contributing to appropriation problems, providing users with access to conflict resolution mechanisms, and recognizing as legitimate the rules that resource users devise. Supportive roles for governments may also involve redesigning or repealing rules that adversely affect the ability of resource users to address appropriation problems. As Shah (1993; see also Shah, Chapter 2, this volume) so forcefully argued, electric board pricing policies have a powerful effect on the actions of owners of electric wells. Pricing policies may need to be redesigned to provide more appropriate incentives for well owners to address appropriation problems.

Provision problems call for the development of different types of productive and complementary relations. The causes of provision problems tend to extend across a basin, affecting many communities of groundwater users and not single communities, as appropriation problems do. Solutions, too, will often require the active participation of many of the groundwater users scattered across the basin. Consequently, communities of groundwater users will likely need the active assistance of higher levels of government in order to adequately address provision problems.

As Moench (2004) has convincingly argued, one of the most critical roles for governments to play in addressing provision problems is developing appropriate and reliable sources of groundwater information. For instance, many national governments develop and rely on 'crudely estimated extraction and recharge balances'. Such estimates are often based on outdated information and educated guesses about well numbers and extraction rates. Furthermore, water balance estimates are made at too general a level to be useful to support local management actions. Moench (2004) suggests providing direct measures of groundwater conditions, such as trends in water table levels, which groundwater users are most interested in and most affected by. Kendy (2003) too rejects the widespread practice of estimating and using water balance estimates. Instead, governments should focus on measuring the amounts of water consumed, not extracted. Water consumption is a more accurate and useful measure of water use.

Given the public goods nature of solutions to provision problems, such as developing accurate and timely data about groundwater basins and groundwater use, or developing alternative sources of water, the temptation may be to assign primary responsibility for provision problems to governments. That, however, would be a mistake, if for no other reason than the solutions and the information on which the solutions will be based, to be workable, require the active participation of groundwater users. For instance, effective solutions that slow or eliminate declines in water tables, or that stop the intrusion of salt water into a basin, require that groundwater users accept limits on wells and pumping, and explore and adopt activities that reduce water consumption. Furthermore, developing alternative sources of water will only have the desired effect of reducing or eliminating the undesirable effects of intensive groundwater use if groundwater users switch to the alternative sources and reduce their groundwater pumping.

Provision problems are difficult to resolve. In many instances in western USA, states and water users have, at best, managed to slow the progression of provision problems (Schlager, 2005). In many fewer instances, states and water users have managed to resolve provision problems and restore groundwater basins to a very productive level of functioning. Blomquist (1992) details several case studies of groundwater basins in southern California in a handful of which groundwater users, city and county governments as well as the state of California were able to arrest groundwater mining and salt water intrusion. For instance, West Basin underlies much of the coastal portion of Los Angeles county. West Basin is relatively vulnerable. It adjoins the Pacific Ocean on one side and, because the basin is covered with impermeable clays, recharge occurs almost entirely through water discharges from Central basin, the groundwater basin directly upstream of it (Blomquist, 1992, p. 33). West Basin began to experience degradation problems in 1912. By the end of the 1950s, 'with water levels down 200 feet in some places, an accumulated over-draft of more than 800,000 acre-feet, and a half-million acre-feet of salt-water underlying thousands of acres of land and advancing on two fronts, the groundwater supply in West Basin was threatened with destruction' (Blomquist, 1992, p. 102). Over the course of 50 years, groundwater users, local and regional governments, California courts and the legislature were able to craft a series of solutions that arrested groundwater mining and halted salt water intrusion. The solutions involved limiting pumping, although not to the level of natural recharge; building surface water projects to import water from other areas of the state; building and operating recharge basins in the Central basin; and investing in a series of injection wells in which a barrier of fresh water was built to halt the spread of salt water. Through a combination of pumping limits, which were developed by groundwater users bargaining with one another in the shadow of a state court, and the development of a series of public goods that required the close coordination of state and local agencies and groundwater users, some southern California basins have been protected.

Work by Lopez-Gunn (2003) illustrates, however, the delicate relations between groundwater users and regional and national government agencies. Lopez-Gunn explored three adjoining groundwater basins in Spain, only one of which is actively governed in such a way that in the long term it is likely to halt water table declines, even though pumpers and government officials in all three basins have access to similar types of groundwater management tools (Lopez-Gunn, 2003, p. 370).

The three groundwater basins are located in the interior region known as Castilla La Mancha. The region is home to three of the largest aquifers in Spain, two of which have been declared overused under the Spanish Water Act (Lopez-Gunn, 2003, p. 369). A declaration of overuse triggers a variety of actions, including the mandatory formation of a water user association and the adoption and implementation of strict pumping regulations. Thus, the relations between well owners and government agencies differ. In eastern La Mancha, one basin that has not been declared over used, well owners voluntarily formed their own water user association, which was recognized by the state regional water authority. The water user association includes all well owners in the basin and has developed its own set of pumping regulations, which have been recognized by the state. Furthermore, the water user association and the regional water authority are working together to define and allocate water rights, and they work together to actively monitor and sanction use to rule violators (Lopez-Gunn, 2003, p. 372).

Relations between well owners and government agencies are markedly different in the other two basins – western Mancha and Campo de Montiel. Water user associations were imposed in both basins and their membership does not encompass all well owners. Strict management plans were also imposed. In the case of western Mancha, rules are regularly violated and thousands of unsanctioned wells have been built. Monitoring and sanctioning are exercised entirely by the regional water authority, with the water user association turning a blind eye to rule violations (Lopez-Gunn, 2003, pp. 371–372). Surprisingly, however, water tables have stabilized in both basins. Lopez-Gunn (2003, p. 377) attributes this to a rich subsidy programme that pays farmers to limit pumping. Once the subsidy programme ends, Lopez-Gunn (2003, p. 377) expects water tables to decline once again.

The form that productive and complementary relations among communities of groundwater users and higher levels of government are likely to take will vary depending on the nature of the CPR dilemma to be addressed. In many instances, appropriation problems can be addressed by groundwater users with more limited support from governments. In general, governments can be most helpful by encouraging resource users to solve their appropriation problems and by reducing any regulatory or legal barriers standing in the way of self-governing solutions. Provision problems are much more difficult and costly to address and require close coordination between resource users and governments. Effective solutions require the expertise, resources and authority of higher-level governments to supply public goods, and the expertise,resources and authority of resource users to change how and how much they use groundwater and to help shape the type, form and location of public goods provided by governments.

Conclusion

A growing body of groundwater case studies demonstrates that groundwater users are capable of devising solutions to CPR dilemmas that are local in nature. More complex and extensive CPR dilemmas, however, often require more collaborative efforts between resource users and regional and national governments.

The shape and form of productive and complementary relations among resource users and different organizations and governments is not well understood and requires substantial investigation. Groundwater basins and large-scale canal irrigation systems present challenging governance issues that are often avoided, ignored or made to disappear within the black box of integrated management (Chambers, 1988; Ostrom, 1992; Blomquist and Schlager, 2005). Even if a workable set of arrangements are devised that adequately address appropriation and provision problems, governance challenges do not end. As Ostrom (1992, p. 63) argues:

> It is necessary to stress the ongoing nature of the process of crafting institutions, since it is so frequently described (if discussed at all) as a one-shot effort to organize farmers. . . . Without the continuing capacity to match new rules to new circumstances, successful irrigation systems face considerable difficulties in coping with the diverse environmental and strategic threats that arise in dynamic systems.

Notes

1 See Kendy (2003) for a discussion of confusion surrounding the concept of sustainability in relation to groundwater aquifers.

2 The attributes are an initial effort to identify proximate factors that directly affect self-organizing efforts among resource users. The factors require greater conceptual development and empirical testing before they may be strictly relied upon (Ostrom, 2001; Agrawal, 2002). Conceptually, the physical characteristics implicitly assume that appropriation externalities, or specific forms of provision problems, are the central problem to be addressed. For instance, feasibility of improvement centres on degradation of the resource, and predictability centres on resource flows. However, the attributes of the physical system may be interpreted more broadly to include the components and structure of resource systems and not just flows. This would allow for a wider range of problems to be captured by the characteristics.

3 Among the many criticisms of government-built and -operated surface irrigation systems is that little attention is paid to provision or appropriation dilemmas and their linkages. Once a system is built, few resources are devoted to maintaining it, and in many systems irrigators are not asked to contribute to upkeep. Also, appropriation dilemmas often emerge as the system is being built. Those at the head of the command area are often allowed to take as much water as they please, as the rest of the

system is being built. Later, they are reluctant to limit their water use. A vicious circle readily emerges: as appropriation dilemmas intensify, farmers face few incentives to contribute to system maintenance; as the system continues to decay, farmers face few incentives to take water in an orderly manner.

4 As one reviewer noted, irrigators are more likely to develop rules that address appropriation problems in alluvial aquifer settings and not hard-rock aquifer settings. In alluvial aquifers, pumpers can more readily identify the effects of their pumping on others and on the aquifer. I am grateful for the reviewer's insight.

5 As Shah (1993, p. 135) explains: 'Externalities associated with private development and exploitation of groundwater resources – and the environmental ill effects they normally produce – are generally considered and analysed from a macro perspective. The source of the problem, however, is micro and can be traced to characteristic behavioural patterns of farmers as economic agents'.

6 As a reviewer noted, well spacing rules may also be enforced through limiting electricity connections.

7 Findings from studies of CPRs such as fisheries suggest that resource users find appropriation externalities more challenging to address than assignment problems and technological externalities. In the case of fisheries, fish populations fluctuate unpredictably and fishermen find it difficult to relate their harvesting activities with fish abundance or scarcity (Schlager, 1990, 1994). The 'noise' of fish population dynamics drowns out the effects of harvesting on fish stocks. While local fishing communities do a relatively good job of addressing assignment problems and technological externalities, they rarely attempt to directly address production externalities (Schlager, 1994). Groundwater users may find appropriation externalities less challenging to address than fishermen because the interaction between pumping and water tables is more direct and observable than is the interaction between fishing and fish populations.

8 The exception to the claim that in general communities will not organize to address provision problems appears to provide support for it. Sakthivadivel (Chapter 10, this volume) notes the emergence of a people's groundwater recharge movement in India. Communities in a few states are actively investing in small-scale recharge facilities, or they are using existing canal irrigation infrastructure, such as canals, tanks and reservoirs, to percolate water underground. The purpose of such activities is to maintain the productivity of shallow wells. The water from the wells is used to ensure a reliable source of drinking water or to ensure irrigation water over the course of a season. The communities are able to capture most of the water that they recharge for their own uses. They are not engaged in attempts to restore, maintain or enhance the productivity of the groundwater aquifer as a whole. Rather, they are engaged in annual storage projects.

9 A number of other studies have noted the poor performance of government-owned tube wells (see e.g. Johnson, 1986; Meinzen-Dick, 2000).

10 Dubash (2002) provides a careful comparative institutional analysis of varying and changing groundwater exchange relations across two villages.

References

Agrawal, A. (2002) Common resources and institutional stability. In: Ostrom, E., Dietz, T., Dolsak, N., Stern, P.C., Sonich, S. and Weber, E.U. (eds) *The Drama of the Commons*. National Academy of Sciences, Washington, DC, pp. 41–86.

Bardhan, P. and Dayton-Johnson, J. (2002) Unequal irrigators: heterogeneity and commons management in large-scale multivariate research. In: Ostrom, E., Dietz, T., Dolsak, N., Stern, P.C., Sonich, S. and Weber, E.U. (eds) *The Drama of the Commons*. National Academy of Sciences, Washington, DC, pp. 87–112.

Bastasch, R. (1998) *Waters of Oregon*. Oregon State University Press, Corvallis, Oregon.

Blomquist, W. (1992) *Dividing the Waters: Governing Groundwater in Southern California*. ICS Press, San Francisco, California.

Blomquist, W. and Schlager, E. (2005) Political pitfalls of integrated watershed management. *Society and Natural Resources* 18, 101–117.

Chambers, R. (1988) *Managing Canal Irrigation: Practical Analysis from South Asia*. Cambridge University Press, Cambridge.

Clark, C.W. (1980) Restricted access to common property fishery resources: a game theoretic analysis. In: Liu, P.T. (ed.) *Dynamic Optimization and Mathmatical Economics*. Plenum Press, New York, pp. 117–132.

Dietz, T., Dolsak, N., Ostrom, E. and Stern, P.C. (2002) The drama of the commons. In: Ostrom, E., Dietz, T., Dolsak, N., Stern, P.C., Sonich, S. and Weber, E.U. (eds) *The Drama of the Commons*. National Academy of Sciences, Washington, DC, pp. 3–35.

Dubash, N.K. (2002) *Tubewell Capitalism*. Oxford University Press, New Delhi, India.

Gordon, H.S. (1954) The economic theory of a common property resource: the fishery. *Journal of Political Economy* 62, 124–142.

Hardin, G. (1968) The tragedy of the commons. *Science* 162, 1243–1248.

Johnson S.H. III. (1986) Social and economic impacts of investments in ground water: lessons from Pakistan and Bangladesh. In: Nobe, K.C. and Sampath, R.K. (eds) *Irrigation Management in Developing Countries: Current Issues and Approaches*. Westview Press, Boulder, Colorado, pp. 179–216.

Kendy, E. (2003) The false promise of sustainable pumping rates. *Ground Water* 41, 2–5.

Lam, W.F. (1998) *Governing Irrigation Systems in Nepal: Institutions, Infrastructure and Collective Action*. ICS Press, San Francisco, California.

Llamas, M.R. and Custodio, E. (2003) Intensive use of groundwater: a new situation which demands proactive action. In: Llamas, M.R. and Custodio, E. (eds) *Intensive Use of Groundwater*. A.A. Balkema Publishers, Lisse, The Netherlands, pp. 13–31.

Lopez-Gunn, E. (2003) The role of collective action in water governance: a comparative study of groundwater user associations in La Mancha Aquifers in Spain. *Water International* 28, 367–378.

Lowdermilk M.K., Early, A.C. and Freeman, D.M. (1978) Farm irrigation constraints and farmers' responses: comprehensive field survey in Pakistan. Water Management Research Project Technical Report 48, Colorado State University, September.

Meinzen-Dick, R.S. (2000) Public, private, and shared water: groundwater markets and access in Pakistan. In: Bruns, B.R. and Meinzen-Dick, R.S. (eds) *Negotiating Water Rights*. ITDG Publishing, London, pp. 245–268.

Moench, M. (2004) Groundwater: the challenge of monitoring and management. In: Gleick, P. *The World's Water 2004–2005*. Island Press, Washington, DC, pp. 79–100.

Norman, C. (1984) No panacea for the firewood crisis. *Science* 226, 676.

Oregon Water Resources Department (2003) *Ground Water Supplies in the Umatilla Basin*. Pendleton, Oregon.

Ostrom, E. (1992) *Crafting Institutions for Self-Governing Irrigation Systems*. ICS Press, San Francisco, California.

Ostrom, E. (2001) Reformulating the commons. In: Burger, J., Ostrom, E., Norgaard, R.B., Policansky, D. and Goldstein, B.D. (eds) *Protecting the Commons: A Framework for Resource Management in the Americas*. Island Press, Washington, DC, pp. 17–41.

Ostrom, V. and Ostrom, E. (1977) Public goods and public choices. In: Savas, E.S. (ed.) *Alternatives for Delivering Public Services: Toward Improved Performance*. Westview Press, Boulder, Colorado, pp. 7–49.

Ostrom, E., Gardner, R. and Walker, J. (1994) *Rules, Games, & Common-Pool Resources*. University of Michigan Press, Ann Arbor, Michigan.

Ostrom, E., *et al.* (2002) *The Drama of the Commons*. National Academy Press, Washington, DC.

Rose, C. (2002) Common property, regulatory property, and environmental protection: comparing community-based management to tradable environmental allowances. In: Ostrom, E., *et al.* (eds) *The Drama of the Commons*. National Academy Press, Washington, DC, pp. 233–257.

Ruttan, L. (2004) The effect of heterogeneity on institutional success and conservation outcomes. Paper prepared for the bi-annual meeting of the Internationl Association for the Study of Common Property, Oaxaca City, Oaxaca, Mexico, August 9–13.

Sadeque, S.Z. (2000) Competition and consensus over groundwater use in Bangladesh. In: Bryan, R.B. and Meinzen-Dick, R.S. *Negotiating Water Rights*. ITDG Publishing, London, pp. 269–291.

Schlager, E. (1990) *Model Specification and Policy Analysis: The Governance of Coastal Fisheries*. PhD dissertation, Indiana University.

Schlager, E. (1994) Fishers' institutional responses to common-pool resource dilemmas. In: Ostrom, E., Gardner, R. and Walker, J. (eds) *Rules, Games and Common-pool Resources*. University of Michigan Press, Ann Arbor, Michigan, pp. 247–266.

Schlager, E. (2005) Challenges of Governing Groundwater Among U.S. Western States. Paper prepared for the Midwestern Political Science Association Meeting, Chicago, IL, April 11–13.

Scott, A. (1955) The fishery: the objectives of sole ownership. *Journal of Political Economy* 63, 116–124.

Shah, T. (1993) Groundwater markets and irrigation development. *Political Economy and Practical Policy*. Oxford University Press, Bombay, India.

Singh, K.K. (1991) Administered people's participation in irrigation management: public tubewells. In: Singh, K.K. (ed.) *Farmers in the Management of Irrigation Systems*. Sterling Publishers, New Delhi, India, pp. 147–169.

Stern, P., *et al.* (2002) Knowledge and questions after 15 years of research. In: Ostrom, E., Dietz, T., Dolsak, N., Stern, P.C., Sonich, S. and Weber, E.U. (eds) *The Drama of the Commons*. National Academy of Sciences, Washington, DC, pp. 445–490.

Tang, Y.S. (1989) Institutions and collective action in irrigation systems. PhD dissertation, Indiana University.

Tang, Y.S. (1992) *Institutions and Collective Action: Self-Governance in Irrigation*. ICS Press, San Francisco, California.

Tang, Y.S. (1994) Institutions and performance in irrigation systems. In: Ostrom, E., Gardner, R. and Walker, J. (eds) *Rules, Games and Common-pool Resources*. University of Michigan Press, Ann Arbor, Michigan, pp. 225–246.

Umatilla County Critical Groundwater Solutions Taskforce (2005) Task Force Handout. Available at http://www.co.umatilla.or.us/Groundwater.htm

Young, O.R. (2002) Institutional interplay: the environmental consequences of cross-scale interactions. In: Ostrom, E., Dietz, T., Dolsak, N., Stern, P.C., Sonich, S. and Weber, E.U. (eds) *The Drama of the Commons*. National Academy of Sciences, Washington, DC, pp. 263–291.

8 Instruments and Institutions for Groundwater Management

Karin Erika Kemper

World Bank (South Asia Sustainable Development Department), 1818 H Street NW, Washington, DC 20433, USA

Groundwater is one of the key resources enabling agricultural development, providing farmers from Argentina and India to China with access and flexibility in water application that usually cannot be matched by surface water resources unless a farmer lives in close proximity to a perennial river or lake. No wonder, therefore, that groundwater is so popular in agriculture, as already highlighted in the regional chapters in this volume. In fact, groundwater irrigation now surpasses surface water as the main source of irrigation water in many regions. Because of the growth in groundwater irrigation, agriculture now accounts for an estimated 70% of total groundwater use with only 20% and 10% going to industry and residential uses, respectively (Brown *et al.*, 1999). However, the large-scale expansion in agricultural groundwater use is leading to the resource being overexploited in an increasing number of countries. Intensive exploitation of groundwater for agricultural uses in India, China, North Africa and the Arabian peninsula exceeds natural replenishment by at least 160 billion cubic metres per year (www.wateryear2003.org).

While published cases of agricultural groundwater use and overuse are impressive, it is important to note that groundwater in some hydrogeological settings is not used alone, but in conjunction with surface water, for instance, as a supplement when irrigation schemes are undermanaged and farmers seek reliability and flexibility provided by their own wells. This, added to the fact that rural groundwater use is generally unmonitored, means that worldwide use in agriculture is probably underestimated – as highlighted in the work on South Asia, China and sub-Saharan Africa in this volume – because often only 'pure groundwater irrigation areas' are counted.

The development of drilling technology allowed the spreading of intensive groundwater abstraction in agriculture since the 1970s. This was not accompanied simultaneously by the evolution of institutional arrangements and investments in management agencies. In most countries, groundwater has therefore traditionally been dealt with in a laissez-faire mode, i.e. farmers, be it in Brazil or Pakistan, have used groundwater to irrigate their crops, typically without attention to the sustainability of the resource. The effect has been twofold. On the one hand, this unregulated groundwater use has permitted spectacular expansion of agricultural growth and lifted millions of people out of poverty (World Bank, 2005). On the other hand, many aquifers worldwide are now under severe stress and groundwater cannot wholly sustain the production that has been initiated. In these overexploited areas, it has also become clear that introducing aquifer management is a time-consuming and politically challenging endeavour. As this chapter shows, there are a number of countries worldwide that have started to proactively manage their groundwater resources. However, there are as yet few well-established examples of good practices and effective groundwater management in developing countries. Even more than in regard to surface water management, groundwater institutions are in an evolutionary phase and no simple blueprints for management success are appropriate. The reasons for this state of affairs which relate primarily to the nature of the groundwater resource itself will be amply discussed below.

The objective of this chapter is to (i) discuss the special nature of groundwater and the resulting challenges for its effective management in agriculture; (ii) provide an overview of the institutional arrangements and instruments available for groundwater management in a variety of settings worldwide; and (iii) highlight some key issues regarding the way forward in groundwater management for the future.

The paper focuses primarily on the quantity dimension of groundwater overabstraction and briefly touches on pollution management issues, which are even more complex.

What Do We Mean by Instruments and Institutions for Groundwater Management?

Institutional arrangements, here for short called 'institutions', are described as the 'rules of the game' (North, 1990) within which stakeholders act. They include formal laws and regulations, informal norms and organizations. In the context of groundwater management, we can imagine national or state water laws dealing with groundwater, irrigation laws, their regulations and decrees, as well as norms developed and applied in communities or irrigation command areas regarding groundwater development and use (well construction and spacing norms, water abstraction rules, etc.). These latter norms may be written or informal.

Such institutional arrangements, whether devised at national, state, provincial or community levels, and whether formal or informal, define and affect *instruments* devised to manage groundwater. Typical instruments include groundwater use rights, abstraction permits or concessions, groundwater tariffs, subsidies and, to a certain extent, groundwater markets. These instruments are called *direct* instruments, given that they are designed to directly affect groundwater management decisions by stakeholders. Importantly, however, there is also a range of indirect instruments that stem from other sectors, but that have an impact on groundwater use, such as energy pricing, agricultural produce pricing and trade policies (Kemper, 2003).

A further important ingredient in the institutional framework is the organizational form for groundwater management. For instance, in most countries, groundwater is formally managed by government agencies, often at the central

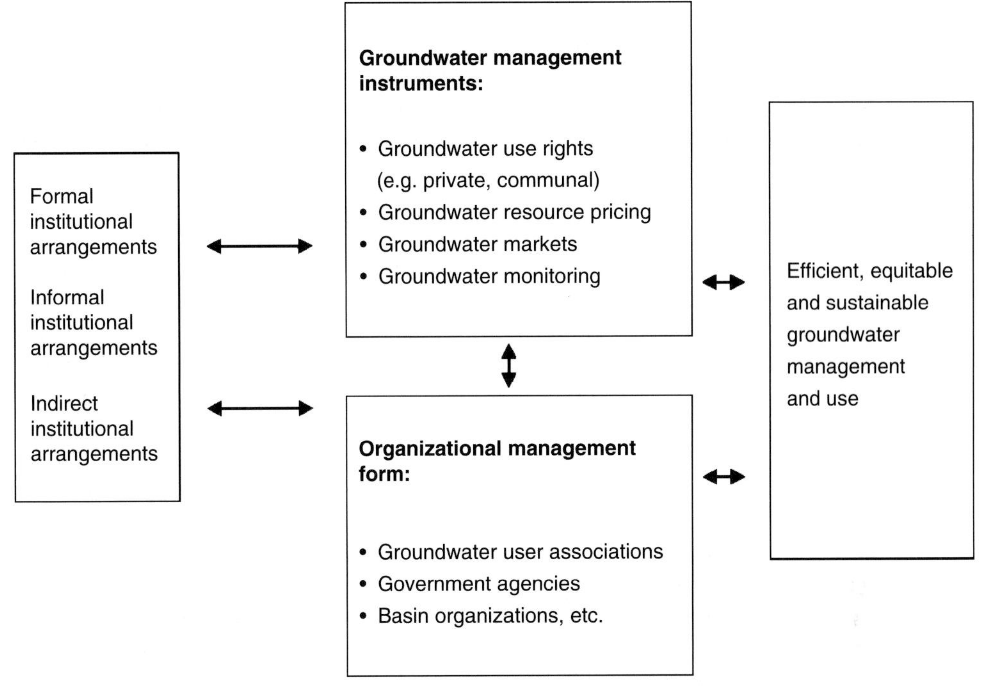

Fig. 8.1. Schematic of the institutional framework for groundwater management.

and sometimes at a lower administrative level. With increasing groundwater scarcity problems, however, aquifer management organizations, which consist of local stakeholders, have started to develop. This tends to coincide with changes in the laws governing groundwater management, but can also happen spontaneously.

Figure 8.1 illustrates schematically how all of the above constitute the institutional framework that conditions groundwater management, with the different institutional arrangements, instruments and organizational forms influencing each other. This chapter provides an overview of the 'menu of institutional ingredients' that can be combined in a variety of ways in order to achieve improved groundwater management, depending on the specific characteristics of an aquifer, a country or a region. The chapter also highlights the importance of the organizational management form with regard to the expectations that one would have concerning the performance of an institutional framework for groundwater management.

In Which Way Are the Challenges for Groundwater Management Different from Those for Surface Water?

The decentralized nature of groundwater use

In addition to being invisible, groundwater is a 'horizontal' resource (in spite of the verticality of wells that abstract groundwater from aquifers), i.e. farmers located above an aquifer can sink wells independently of each other over a significant areal extension depending on the size of the aquifer. For example, in Mexico some aquifers have an area of only a few square kilometres, whereas the Guarani aquifer system in South America has an area of 1.2 million square kilometres, i.e. the size of England, France and Spain combined (World Bank, 2003).

Therefore, groundwater as a resource – in a situation of abundance – is distributed in an equitable manner to those above a given aquifer. With the less-pronounced upstream–downstream dimension, which is so defining in surface water management, and where upstream users literally have the upper hand over downstream users, the groundwater management challenge is a radically different one. The key issue is to manage a pool resource, which any user who can afford a deep enough well has access to and which therefore can provide benefits to many, but with the focus to make it last for as many users as possible for as long as possible. Groundwater management therefore implies dealing with decentralized stakeholders who will make their decisions based on private utility, weighing their costs (sinking the well, variable abstraction costs, etc.) and their benefits (well yields, type of use, benefit derived from it, etc.). Compared to surface water management, there is no 'tap' in the form of a reservoir release or an irrigation gate intake that can control water access.

The management challenges vary, of course, from country to country and between regions within countries. The manageability of groundwater will depend on the size of the countries and of aquifers, aquifer yields, storage capacity, population density and abstraction for agriculture (since agriculture is usually the primary purpose with the largest number of users, it will have the most impact on management challenges) (Table 8.1).

The categories shown in Table 8.1 only serve as abstracts and in practice assessments will differ. Aquifers vary not only in their spatial dimensions, but also in their yields and recharge profiles. Just so do groundwater users differ, and sociopolitical settings, which influence institutional options for aquifer management, will diverge as much as aquifer characteristics. Aquifer management strategies will therefore have to be developed accordingly. The key point is, however, that the more the actors need to be involved and monitored and the more the abstraction is compared to yield, the higher will be the transaction costs to devise and implement institutional arrangements for aquifer management, and therefore the bigger the challenge to manage the aquifer in a sustainable manner.

The need for groundwater management instruments changes over time. As illustrated in Fig. 8.2, there is a logical progression to groundwater management needs (also compare with Fig. 2.5 by Shah, Chapter 2, this volume).

The figure depicts a typical curve for aquifer management needs, ranging from the baseline situation where groundwater is abundant compared to abstraction to a high-stress situation where abstraction has turned excessive and is leading to irreversible aquifer deterioration. While many will agree that groundwater management is needed in the high-stress situation in order to return to the more

Table 8.1. Management implications for some types of aquifer–groundwater user relationships.

	Low density of agricultural groundwater users and low abstraction rate compared to recharge	**High density of agricultural groundwater users and high abstraction rate compared to recharge**
Small/medium aquifer	Low transaction costs in developing and enforcing institutional arrangements for groundwater management; few instruments (e.g. monitoring network) needed Example: many aquifers in sub-Saharan Africa	Medium to high transaction costs to institute groundwater management, but probably manageable due to small areal extent of intervention needed; however, need for groundwater management in order to ensure sustainability Example: some Mexican aquifers
Large/extensive aquifer	Possibly higher transaction costs in developing and enforcing institutional arrangements for groundwater management due to spatial distribution; but few instruments needed while abstraction remains low Example: Guarani aquifer system	If extensive, major aquifer: Very high transaction costs to institute effective groundwater management, both to achieve agreement on the institutional framework and to enforce and monitor Example: North China Plain If extensive, but low-permeability aquifer: High transaction costs due to high density of users; but low transaction costs because aquifer could be managed as local units Example: Indian basement

stable development situation, we clearly face a paradox here. As can be seen in the figure, groundwater management instruments would ideally be employed at any stage of aquifer use. Even in the baseline situation, registration of abstraction wells and springs as well as source mapping are highly recommended, given that transaction costs for doing so are much lower in a situation of few users and sustainable abstraction than in a later stage when stress has set in. A simple network with a number of monitoring points would also provide important information. For instance, the state of Maharashtra, India, has been monitoring groundwater for 30 years. While the groundwater situation 30 years ago probably would not have triggered major concerns, the long-term investment in the monitoring network and data collection is now paying off because the data series provides important information, even if not sufficient to resolve the serious overabstraction problems

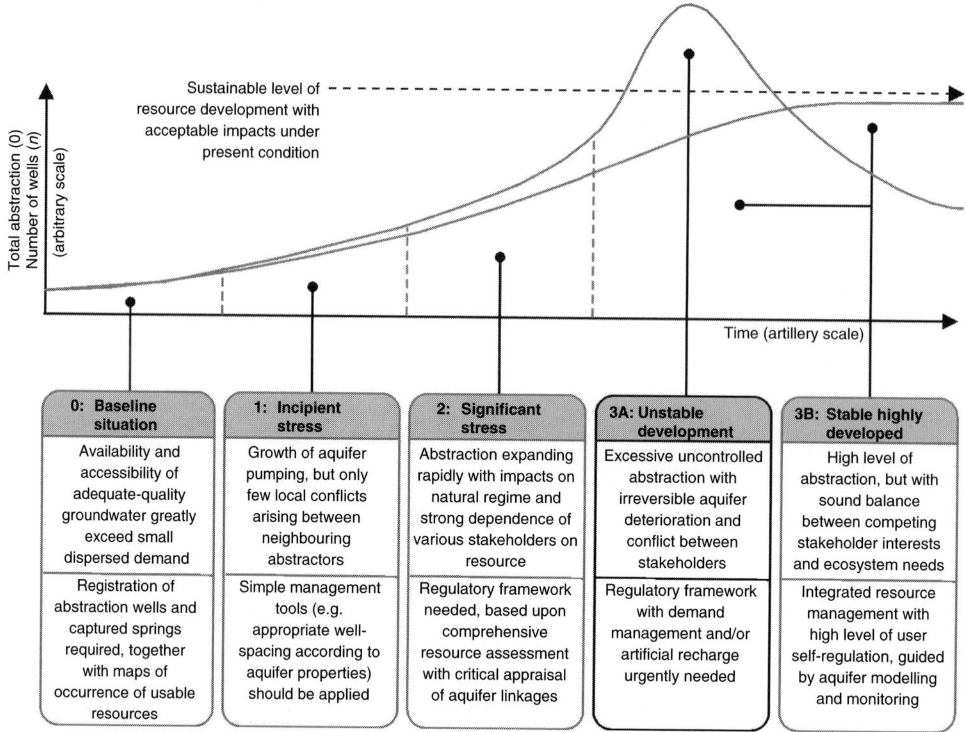

Fig. 8.2. Stages of groundwater resource development in a major aquifer and their corresponding management needs. (From World Bank, 2002–2004.)

managers now face. Regrettably, such basic steps to start building a future aquifer management system are usually not taken. One region where the option still exists is sub-Saharan Africa where groundwater is still abundant in many areas but where development requires scarce financial and human resources to meet many other needs. If action is not taken, once significant stress begins to show and crucial information about users and aquifer yields is required, it may not be available. One possible solution is to improve capture of information already collected. The groundwater development projects in place in many countries in the region generate substantive insight among the drillers and diggers, and the information they create could very well be captured to provide baseline information about aquifers and aquifer users.

Many countries that have not invested in collecting and systematizing such information start only after major aquifer stress appears with the basics, such as well and user registration, measurement or estimation of groundwater abstraction and definition of an entitlement regime, rather than being able to focus on management and fine-tuning of instruments. This way, much valuable time is lost and in many cases it is already too late.

While this chapter primarily focuses on the challenges posed by overabstraction, it is recognized that there are a number of regions in the world where groundwater still constitutes a resource to be further developed. Differing examples are presented in discussions on South Asia (Shah *et al.*, Chapter 11, this volume) and Central America (Ballestero *et al.*, Chapter 6, this volume). As pointed out by Shah *et al.* (2000): '[C]entral to appreciating the global groundwater situation . . . is the coexistence of regions with undeveloped resources and those with overdeveloped resources, and the socioeconomic dynamic that has relentlessly impelled the former to shrink and the latter to expand.'

Equity considerations

Especially in areas with relatively shallow aquifers, groundwater is a very equitable resource. For instance, the expansion of the treadle pump in India and Bangladesh in recent years shows that low-income farmers can benefit from easy *access* to groundwater and increase their productivity and economic well-being (Shah *et al.*, 2000). In large areas of Africa and South Asia, people dig their own shallow wells or are able to invest in relatively shallow boreholes.

However, when many people do the same in a given area, the many incremental uses can eventually lead to the negative impacts of overabstraction mentioned earlier and reverse the equity effect that originally existed. The first groundwater users who have to abandon their wells as groundwater levels sink can be expected to be the poor who do not have the financial resources to afford pumping water from increasing depths, or to invest in new wells. They are also the first to be hit if their wells turn saline or when their domestic well runs dry or gets polluted. As a consequence, they may have to abandon farming, turn to the city to seek their livelihood or – if they are urban dwellers – start purchasing more expensive water from private vendors (given that the public water supply system frequently does not reach the poor). Even the somewhat richer farmers may experience serious indebtedness if they overinvest to chase a falling water table in shallow aquifers – since their wells are unlikely to generate sufficient income to meet the interest on their loans.

While development efforts and the literature have focused on the access to groundwater and the potential benefits of its use as an equity issue (Kahnert and Levine, 1993), an increasing number of overexploitation and pollution scenarios are now entering the global groundwater agenda. Unfortunately, up to now very few studies have been carried out with regard to the equity impacts of groundwater overexploitation. Such research should provide clues on the costs and benefits of groundwater management actions from a societal point of view. With the prevailing attitude among many groundwater developers that groundwater is a freely exploitable resource, it is always more complex to put simple management measures in place once problems have already arisen. By then vested interests have already developed among users (e.g. relating to amounts of water used and perceived as entitlements, or provision of access to privately developed wells for monitoring purposes) that may make it difficult to develop a clear picture of an aquifer's characteristics and to put in place measures such as monitoring and agreements for more efficient use of the resource. At the same time, groundwater management does entail costs to society and to the users, so a balance needs to be found between the cost of management investments compared to the benefits of long-term sustainability of groundwater use (Kemper, 2003).

The 'Menu': What Instruments Can Be Used for Groundwater Management in Agriculture and What Are the Requisite Institutional Arrangements to Make Them Work?

In this section, the different instruments comprised in the institutional framework – groundwater use rights, monitoring and pricing – will be discussed.

Groundwater use rights, permits, concessions and licensing

In this chapter, we use the term 'groundwater use rights' as the umbrella expression for any instrument that defines the right of a user to abstract groundwater according to certain parameters, such as volume and duration. Different countries have given these rights different names, such as permits, concessions, licenses and entitlements. All these instruments confer a certain right in a defined way, and what is called a permit in one country and context may be called a concession in another.

Groundwater use rights are often ambiguous and difficult to define. This is due to the previously mentioned difficulty assessing the magnitude and availability of the resource itself. Groundwater modelling is intricate and expensive, and if no good models are available that provide information about available yield over time, the basis for giving any type of water rights, be it concessions or tradable rights, is very weak. Users and water developers' knowledge can be useful to some extent. For example, in Mexico users strongly overstated their water use. Partially as a result, the country is now considering the buy-back of water rights since effectively too many rights were given at the time of initial allocation.

Once groundwater use reaches a certain point with respect to availability, i.e. once the resource becomes scarce, well-defined groundwater use rights can become a key method to control overabstraction, and countries such as the USA[1] and Mexico have taken the step to implement groundwater rights systems. Well-defined groundwater use rights entitle individual users or user groups to an abstraction allocation at a certain point in time or during a specified time period. Without a clear definition of who the users are and how much water they are entitled to, the users themselves have no incentive to use the water efficiently, because they have no guarantee that if they save water today, the aquifer's yield will permit them to abstract what they need tomorrow. In addition, if water allocations are to be shifted to different users, without defined groundwater use rights, there is no information about how much can be reallocated, who would win and who would lose and how compensation might be structured.

It is important to note that to achieve better groundwater management, groundwater use rights need not be tradable. Obviously, tradability would introduce an increased option for efficiency, but often the first, most important step is to register the users and get a better estimate of the types and magnitude of abstraction. This information can then be compared to information about aquifer recharge and thus long-term water use sustainability (Kemper, 2001; World Bank, 2002–2005).

Proponents of water trading consider tradable rights a very powerful instrument because they induce the right holder to apply a long-term perspective. The holders will consider not only what the water can directly produce for them (e.g. tonnes of rice), but also the opportunity cost of the water (e.g. the value added by using the water in car manufacturing, which is the payment that could be expected if the groundwater right were traded). Thus, the highest value of water use is taken into account and provides an incentive for more efficient use and reallocation of surplus water to a higher-valued use.

Often even without codified rights systems, both formal and informal groundwater markets have developed in water-scarce areas. However, these markets typically do not provide incentives for long-term use perspectives, because use rights are unclear. For instance, in the informal groundwater markets in Gujarat, India, water is sold without consideration of the limits of the resource, and while the allocation of the resource may be more efficient than if the markets did not exist, the groundwater level is nevertheless being drawn down. The ability to sell whatever water is pumped may even be an added incentive to overabstraction. This serves to remind that water markets are complex institutional set-ups in themselves and need substantial regulation if they are to fulfill sustainability and equity objectives.[2]

Further, the establishment of rights and markets does not mean they will actually be used to increase use efficiency. For example, Mexico has long had a formal groundwater market, but the market has not been very active, in part due to the transaction costs built into the system (World Bank, 2006). By contrast, the groundwater market in New Mexico, USA (which is also driven by conjunctive use regulations), is very active (DuMars and Minier, 2004). Reallocation by trading means getting compensated.

Water use rights are thus rules that need to be designed, changed and adapted to different situations. They are advocated here as a tool to provide a long-term horizon to water users. As mentioned earlier it is important to note that tradable groundwater use rights per se will not resolve overexploitation of an aquifer unless a certain percentage of the aquifer volume is reserved to achieve a certain stabilization. Theoretically, this could take place, for instance, in the same way as air pollution rights trading, where each year a certain, decreasing amount of water is designated as tradable, effectively decreasing the consumptive use on a yearly basis. This implies, however, that groundwater users forego a certain amount of water every year and thereby lose income opportunities or that they have, in the meantime, implemented more efficient technologies and therefore can accept this restriction for the good of all. It will depend on the locality-specific circumstances if groundwater users will easily come together and agree on such restrictions. For instance, in Mexico, groundwater management user groups, the so-called COTAS, have now existed for about 10 years, but while they have been able to promote awareness-raising activities and also, to some extent, water-saving investments, there are very few COTAS that have as yet decided to restrict total water use of the aquifer or take active steps towards its stabilization. Also in water-scarce Yemen, where a World Bank–financed project supported the introduction of more efficient irrigation water use, the Project Implementation Completion Report pointed out that while farmers readily accepted the new technologies, they tended to use the saved water to expand their planted areas, thus leading to improved livelihood for farmers in the short run, but not leading to improvement of the aquifer conditions, which will have long-term implications for the farmers (World Bank, 2001). In Arizona, USA, on the other hand, farmers have to reapply for groundwater use permits periodically, and each time the total permitted abstraction volume is adjusted downwards, based on assumed changes in technology (Jacobs and Holway, 2004). Also on the North China Plain, Foster *et al.* (2004a) report that agricultural water-saving measures, such as improved irrigation water distribution through low pressure pipes and drip and micro-sprinkler technology, improved irrigation forecasting, and deep ploughing, straw and plastic mulching, etc. have reduced non-beneficial evapotranspiration and led to real water savings in the order of 35–40 mm/year in various pilot areas. At the same time, farmers' incomes have increased to above the national average. Clearly, these are encouraging examples, which show that the institutional arrangements need to include not only water user participation and awareness raising, but also enforcement and sanctioning mechanisms.

In the absence of prior well metering, a pragmatic first step to assign water use rights is an initial assessment by groundwater management agencies (e.g. North China Plain) or a self-assessment by groundwater users of their historical use (e.g. Mexico). These assessments may overestimate the historical use, such as in the Mexican and Chinese cases, but groundwater administrators do accept them as a starting point. The challenge then is to eventually reduce the overall volume of rights in order to arrive at the actual amount of groundwater withdrawn. Only in this next step will groundwater use actually be decreased. This gradual decrease can only take place if the institutional framework is sufficiently developed to permit follow-up actions (e.g. re-registering of wells and permits, and use of licensed drillers). Quite clearly, this is a long-term process that requires considerable resources and, perhaps more importantly, social and political will. This latter factor can be especially problematic since costs will be more immediate than benefits.

Another important aspect in the allocation of groundwater rights is the distinction between open access and common property resources (see Schlager, Chapter 7, this volume for detail). Aquifers are a typical example of a common property resource and are often also an open access resource, when neither private nor collective groundwater use rights exist. The introduction of water use rights can remediate this situation by offering an incentive towards a long-term perspective by individuals and an interest in controlling fellow users. As pointed out previously, however, high transaction costs can be expected in the introduction of groundwater use rights due to existing vested interests by current users. They can be especially high if an aquifer is already overexploited and decisions for curbing groundwater use have to be taken. For this reason, it is recommendable to start groundwater management in situations that require less sacrifice, i.e. lower costs for stakeholders, and not to wait until situations become critical.

It should also be pointed out that groundwater use rights could be accorded to groups on a collective basis. The reasoning regarding incentives remains the same, i.e. if the group has a water use right, there is an interest to preserve or stabilize the aquifer on behalf of the group. Naturally, intragroup enforcement of agreed actions is also essential in this case.

Groundwater use rights: enforcement, monitoring and sanctions

The implementation and effectiveness of a groundwater use right crucially depends on enforcement capacity, sanctioning systems, water reallocation mechanisms and the need for the generation of information and its management. There is also an important linkage to pricing mechanisms (see the following section).

As mentioned earlier a key issue in groundwater management is the size of the groundwater user community. Groundwater aquifers can be very small, with only tens or hundreds of users, such as is the case for some aquifers in Mexico, California and South Africa. It is very well conceivable that users would be able to arrive at a joint management framework, even without individual property rights. As pointed out by Shah *et al.* (2000), many aquifers, especially in Asia, have thousands of users. In that case, it is far more difficult to envision one integrated framework at the 'community level', and obviously transaction costs for both introducing and maintaining any groundwater framework increase significantly (see also Table 8.1). In such cases, submanagement structures around subaquifer units are required. The many groundwater recharge movements in India show that even if recharge and water savings do not take place across an entire large aquifer, the local impacts can be beneficial.[3]

For groundwater use rights to function as management instruments, the following need to be in place:

- initial allocation;
- registration mechanism and maintained registry system;
- functioning monitoring system;
- enforcement of the limits set by the individual or communal use rights;
- credible sanctioning system.

All of the above, i.e. the individual design and the implementation, depends on the aquifer and on local or national institutional capacity. Sandoval (2004) and Jacobs and Holway (2004) describe how the administrative systems are organized in the states of Guanajuato (Mexico) and Arizona (USA), respectively, and how these states have designed their groundwater management systems around existing capacity. In the case of Arizona, the state groundwater management agency is far stronger than the one in Guanajuato. Accordingly, in

Guanajuato an approach has been taken that strongly relies on local groundwater user groups in order to complement and enforce the groundwater permit administrative system. These examples show the local nature of designing systems to suit local conditions.

In summary, groundwater use rights are essential to provide incentives for better groundwater management, but perhaps even more than with surface water, they need to be designed in a flexible and locally adapted manner to allow for local needs and circumstances. For this, the characteristics of the aquifer, individual or common property right cultures, different lengths of validity of the rights, formality and informality as well as transferability need to be taken into account.

Groundwater pricing

When dealing with the need for more efficient groundwater use and allocation, a prime recommendation is usually the introduction of a groundwater tariff or fee. The rationale is that groundwater users have an incentive to use water efficiently when it has a price. If it is free, they will use more than they would otherwise, unnecessarily reducing the availability of water for everyone and increasing scarcity of, and thus competition for, the resource. If 'the price is right', users will have incentives to use less water and introduce water-saving technologies, thus freeing water for other uses.

In groundwater, pricing issues are distinct from surface water, given that abstraction of the groundwater resources usually takes place on private land and with private equipment. Therefore, there are actually two options for pricing: pricing the resource itself or pricing the other inputs needed in order to pump groundwater such as the pump, borehole and, most importantly, energy.

Energy pricing

The cost of energy is usually seen as the most important incentive to reduce overabstraction. Figure 8.3 depicts the Mexican situation and we can see that there was a noticeable decrease of electricity consumption in 1990, when an increase in the special rural energy tariff took place. One can infer from the results that the elasticity with regard to energy pricing in Mexico is significant, i.e. water users clearly respond to price changes that affect their energy bills. Usually, however, this type of action is not easy to apply due to political reasons – as was also the case in Mexico when the government responded to pressures and decreased the tariffs again. This is reflected in the downturn in the price curve in Fig. 8.3, and a corresponding increase in pumping from 1992 and onwards.

The Mexican situation is not unique. Many countries subsidize agricultural inputs and, among them, rural energy (e.g. a number of states in India, Brazil, etc.). Once this has happened, it is politically very difficult to return to, or start implementing, energy prices that actually reflect the cost of energy to the state. The effect is not only a clear incentive for groundwater overabstraction, but also important fiscal implications for the state. Depending on the cal-

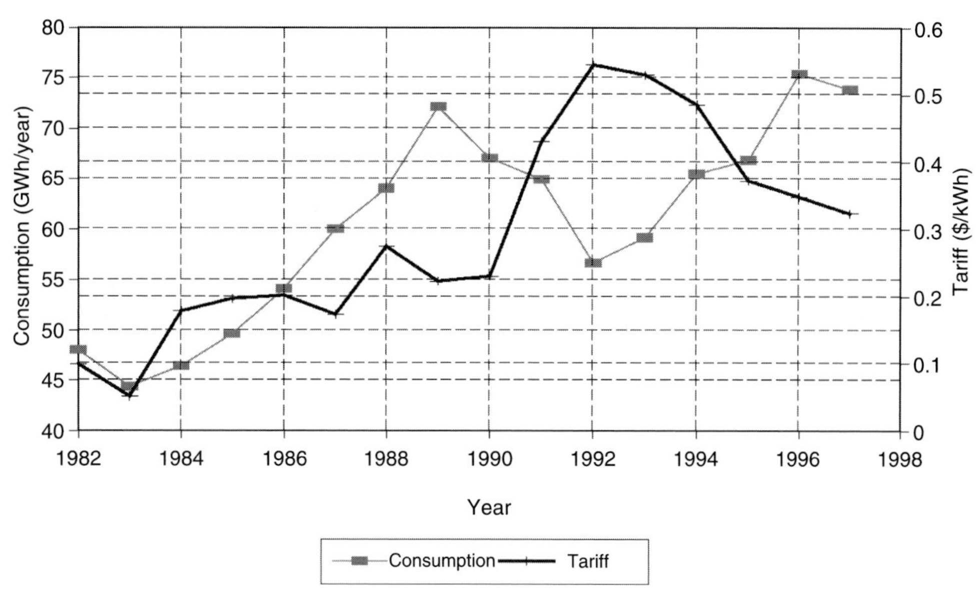

Fig. 8.3. Relationship between electricity tariff and consumption in Mexico, 1982–1997.

culation method, energy subsidies to agriculture in India amount to between $1.9 billion and $6.5 billion per year (Bhatia, 2005).[4] At the all-India level, electricity subsidies to agriculture are estimated at 26% of gross fiscal deficit. They may vary from 80% in Madhya Pradesh and Haryana to 50% in Andhra Pradesh, Gujarat and Karnataka, and to about 40% in Rajasthan, Punjab and Tamil Nadu (Bhatia, 2005).[5]

Even here innovative ways need to be sought. While energy pricing is seen by many politicians as an effective means to subsidize rural producers – and therefore a number of countries even apply zero tariffs (e.g. the states of Tamil Nadu and Andhra Pradesh in India reverted to zero tariffs after the elections in May 2004 (Bhatia, 2005)) – the detrimental effect on groundwater aquifers needs to be taken into account.[6] The well-intentioned 'pro-poor' policy may eventually turn into an 'anti-poor' policy when the aquifers become overexploited and only the rich can afford to continue pumping. That is why other types of subsidies should be contemplated. An option could, for instance, be lump sum payments to small farmers that would permit them either to pay the full electricity bill or to reduce their pumping, pay a lower bill and use the 'gain' for something else. In this way the energy tariff would not distort the true price of groundwater, and at the same time not hurt the poor (World Bank, 2006).

Pricing the groundwater resource

Another way to provide an incentive to use water more efficiently is to price the resource itself, i.e. users pay for the abstraction of the groundwater resource itself. For the maximum impact, this should be based on volumetric metering, thus providing an incentive to use less water. Many times, however, metering equipment is not installed on wells or it is not effectively monitored by the (ground)water management agency. Therefore, few countries practise direct groundwater pricing, especially in agriculture where there tend to be large numbers of users, and transaction costs for monitoring are disproportionately high. In some countries, e.g. Mexico and France, industrial and municipal users pay, but because agricultural users are exempt and they use the largest share of the water, the impact on the groundwater resource is little.

Due to the cost of monitoring individual wells – and also due to the possibilities of corruption in meter reading or tampering – there are now efforts to develop remote-sensing tools, which can help calculate groundwater use based on the observed crop cover. The advantage of these tools is their visual power and the fact that water users themselves can learn to interpret them. This affords the possibility for aquifer self-management rather than reliance on well-by-well monitoring, thus increasing transparency in aquifer management and reducing strategic transaction costs. By using remote-sensing information, users can monitor each other's groundwater use, for example, by comparing neighbours' type of crops and area under cultivation, enabling peer pressure to enforce abstraction agreements (including use efficiency) and reducing possibilities of shirking.

In spite of some caveats (e.g. how to accurately model and calculate evapotranspiration), remote sensing can develop into an important and increasingly affordable tool for groundwater management. Attempts at its use are taking place, for instance, in Idaho, USA, and in South Africa.

Another option is self-declaration as practised in New Mexico and in Arizona, USA. In these states groundwater users declare once a year what their actual abstraction has been. In Arizona, every time a permit expires, it is reconsidered from a technical point of view and the new permit will be issued taking into account the potential water savings that the user could make by installing more efficient irrigation technology. This way, total abstraction from the state's aquifers is brought down over time (Jacobs and Holway, 2004). South Africa also uses self-declaration.

Subsidies for technological improvements

As mentioned earlier, a further instrument to improve groundwater management in agriculture consists of subsidies to improve irrigation efficiency by farmers' investments in better technology. This may imply support to make investments in closed conveyance pipes instead of earth canals that are subject to evaporation, shifting from flood irrigation to drip irrigation and investments in soil levelling, mulching, etc. There are a number of examples worldwide showing that these approaches work from the technological point of view, as in China, Mexico and Yemen. However, such measures will only be effective if farmers do not at the same time expand their fields or increase their cropping cycles. The incentive to do so and to improve one's individual livelihood is significant; therefore, understanding of the reasons for these subsidies and enforcement plays an important role.

Groundwater Management Organizations – Participation, Information and Awareness Raising

As has become clear from the earlier discussions, a number of instruments exist to introduce more efficient groundwater use and allocation. These range from effective monitoring to defining groundwater use rights and to pricing the resource.

At the same time, the effectiveness of any of the instruments employed in a given situation will depend on the organizational set-up for groundwater management. Groundwater is distinct from surface water in that many different users are involved in abstracting the resource, and monitoring their individual behaviour is very costly. Users, of course, are very well aware of this fact, and therefore their incentive to comply with metering regulations and with prohibitions against sale/lease of water or tariff payments is typically very low.

Experience from many countries has shown that actively involving stakeholders, and providing them with information and with a say in the management of their resource, is essential to create incentives for compliance, be it in regard to groundwater or to surface water management. As previously mentioned, the COTAS in Mexico have had a very important role in raising awareness and providing information to groundwater users. As pointed out by Foster *et al.* (2004b):

[The] fundamental goal of the COTAS (as conceived) is to provide the social foundation to promote measures to slow down, and eventually eliminate, aquifer depletion. It is clear from the experience to date that the COTAS cannot achieve this goal alone – but neither could the 'water administration' achieve it without the COTAS.

The Government of Jordan came to the same conclusion when well abstraction limits were not followed by users, and it started implementing a promising, stakeholder-based approach (Chebaane *et al.*, 2004). The experience of river basin organizations worldwide (although not focused on groundwater) has shown the power of information and of stakeholder involvement in achieving better water resource management performance (Dinar *et al.*, 2005).

The reasoning is simple: (ground)water users who do not know what the conditions of their resource are will be less willing to sacrifice their current income than those who are aware that overexploitation is going to hurt them in the foreseeable future. For this, they need comprehensible and reliable information and a voice in shaping the institutional framework.

Blomquist (1992) provides a comprehensive description and analysis of the development of local management structures in eight Californian groundwater basins. Interestingly, each development started with (i) the recognition that the groundwater resource was under increasing stress (as noticed by sinking water levels and sometimes saltwater intrusion) and (ii) the collection of data about the aquifer, its recharge and potential safe yield. Once the data were obtained and confirmed on the ongoing overdraft, water users were able to forecast the potential consequences of non-action and started to organize for more sustainable use and management of their aquifers.

These examples illustrate that groundwater users need to be recognized as true stakeholders who are entitled to information about the resource they are so dependent on. For many water agencies, this implies a significant shift, not only in technology, from being centralized agencies that keep the information about water availability to themselves and take decisions without the participation of other stakeholders. Obviously, the trend towards definition and official allocation of (ground)water use rights (such as in Brazil, Chile, Mexico and South Africa) contributes to a move towards transparency. Information is essential for decision making among all levels of stakeholders to determine what planning horizon to consider, which savings measures to propose and accept, what investments to make as well as what service to require from water agencies and government authorities. With a better-defined basis of groundwater use rights – and responsibilities – information becomes more valuable and more crucial to the different stakeholders.

A number of countries, including the USA, Mexico and India are thus moving towards the management of aquifers by groundwater user associations of various types, in an attempt to involve users in decision making and increase compliance with decisions that have been taken collectively. In those cases, these developments are accompanied by a range of other demand management instruments discussed in this chapter. In the USA, this shift has been taking place over the last five decades and is showing good results (Blomquist, 1992; Jacobs and Holway, 2004; Sandoval, 2004). This topic is presented in depth by Schlager (Chapter 7, this volume) on community participation and communal approaches.

Conjunctive Use of Surface and Groundwaters

Groundwater use within surface water irrigation projects

As outlined at the beginning of this chapter, groundwater management is often treated as if it took place in areas distinct from surface irrigation schemes. However, when looking at many such schemes, ranging from India to Pakistan and Mexico, farmers actually use surface and groundwaters in conjunction. This implies that groundwater use is probably even more widespread than it seems. Many times farmers use groundwater because surface water schemes are not functioning, not delivering water on time or not timely enough to grow sensitive (and often high-value) crops. If groundwater were managed better – and surface water more effectively – significant benefits could be achieved.

One of the key disadvantages of unmanaged conjunctive use is that without control, groundwater use is usually concentrated at the tail ends or around the margins of surface water irrigation areas. This is suboptimal because excessive groundwater abstraction here often aggravates natural salinity problems, and meanwhile excessive groundwater laminae in the main riparian areas can cause rising water tables and water logging. Planned conjunctive use would optimize the situation by spreading both uses.

Obstacles to managed conjunctive use include distortions between surface and groundwater abstraction costs. Why would farmers upstream – where they receive abundant surface water through their irrigation canals – want to irrigate with groundwater, which would be far more expensive than the highly subsidized surface scheme only to benefit the tail-end farmers? Often there are also legal impediments to doing so. Therefore, the incentive structure needs to be examined in order to move towards more sustainable management of the physical system.

Agricultural water use in highly populated regions

There is evidence from the North China Plain, where the growth of small towns that are reliant on groundwater for their populations and industries is significant, that the impact on the – rapidly diminishing – groundwater source is large (Foster *et al.*, 2004a). In such situations a groundwater management strategy needs to take into account both agricultural and urban uses. Similarly, interaction between surface water and groundwater needs to be taken into account (e.g. the Rio Grande in New Mexico) in providing new permits for surface water abstractions since there will be impacts on groundwater abstractors.

Nevertheless, an important issue in this regard is the recurring assertion that since surface and groundwaters are hydrologically connected, aquifers cannot be managed in isolation. This argument is relatively weak, however, given that in many cases surface waters are managed – if at all – without ever taking into account the connected groundwater resources. Thus, while the principle to apply a conjunctive management approach is desirable, nowadays many aquifers are under such pressure that pragmatism would dictate tackling them directly, without neglecting basic principles of integrated groundwater management as identified in the course of time (Kemper and Alvarado, 2001; Foster *et al.*, 2004b). Thus, in cases in which the hydrological connection to surface water resources is very significant, conjunctive use could and should appropriately be taken into account – such as in New Mexico, USA (DuMars and Minier, 2004) – but the pros and cons of doing so need to be carefully assessed.

Groundwater Quality Management

Management of groundwater quality in an agricultural context has several dimensions: the pollution caused by agriculture (e.g. salinization due to fertilizer use, contamination of groundwater by pesticides, overpumping of coastal aquifers and sea water intrusion, overabstraction of aquifers with underlying saline water); and the pollution caused by other actors, but with a negative impact on water quality also for irrigators.

In terms of the management instruments to be used in the first case, these range from educating farmers about the appropriate amounts of fertilizers and options for integrated pest management to avoid contamination of the groundwater to phasing out certain products, to increasing prices of harmful products in order to discourage their use. Unfortunately, non-point source pollution is very difficult to manage and there are not many successful examples.

With regard to salinization due to overabstraction, the same approaches as discussed earlier apply: groundwater abstractors need to be made aware of the problem, solutions need to be developed and a number of instruments are available – ranging from peer pressure to introduction of groundwater use rights and pricing instruments – to curb demand. Unfortunately, salinization is reversible only at enormously high costs and should therefore be avoided rather than mitigated.

The pollution by growing urban centres and industries is not the topic of this chapter; therefore suffice it to say that even here integrated approaches are needed and that with growing populations, especially in Asia, the interface between urban and agricultural water quality management is becoming more pronounced.

Conclusion – Is Groundwater Manageable?

Groundwater management was neglected for a long time due to the apparent abundance of the resource. With population and economic growth and the technological options to abstract groundwater at reasonable prices from ever-greater depths, the need to actively manage the resource has become clear. This is especially the case in developing countries where the poorer segments of rural society do not have other livelihood options available, should they lose access to their safe water source, both regarding production and drinking water supply.

This chapter argues that institutional frameworks for groundwater management need to comprise a range of instruments to manage the resource. Contrary to a mechanistic belief, however, the need to fully integrate the human dimension is highlighted. Thus, the creation of incentives through the introduction of groundwater use rights, direct and indirect pricing, or water trading is an important step. However, the horizontal dimension of groundwater use makes it hard to fully control the application of such instruments unless a given aquifer has very few users and the responsible authority, a very clear mandate and sufficient capacity. In most cases, the users themselves are the most important stakeholders in devising groundwater management schemes as well as in devising and choosing the most applicable instruments.

As countries move towards actively managing their groundwater resources, their approaches are taking this interplay into account. Some countries rely more strongly on formal institutional arrangements such as regulations and official monitoring and sanctioning mechanisms; others try to combine both formal arrangements and informal water user agreements; and still others focus primarily on water users in order to deal with their specific groundwater management challenges. The choice of these approaches is related not only to the institutional strengths in the individual countries but also to the type of hydrogeological regime and population and economic profile they have to deal with.

While there are very few success stories as yet – and these are essentially in developed countries – increasing groundwater scarcity and pollution are providing an impulse for central and local governments worldwide to introduce groundwater management frameworks and instruments, adapted to their needs. The toolbox for groundwater management already exists. Now the political will needs to be developed in order to bring about – or intensify – change. This will imply reviews of existing groundwater management structures, the costs that current institutional arrangements have for specific groups in the medium and long terms as well as the costs to society at large. This information needs to be made available to decision makers to provide an impetus for the use and further development of existing groundwater management tools.

Acknowledgements

The author would like to thank Stephen Foster and Barbara van Koppen for their very useful review comments and the Groundwater Management Advisory Team of the World Bank/GWP for the many valuable discussions over the years that have influenced the content of this chapter.

Notes

1 Each state in the USA has a different system.
2 See Mariño and Kemper (1999) for an in-depth analysis of water markets and the needed institutional arrangements to make them function.
3 Although recently concerns have been voiced that groundwater recharge in an upstream area, due to water-harvesting structures, may impede flow to downstream areas, effectively leading to a reallocation of the water resource. This issue needs further study.
4 That is, Rs 80 billion and Rs 281.2 billion, respectively (Rs/$ exchange rate used 43:1).
5 According to Bhatia (2005), these estimates may be on the higher side, given that State Electricity Boards tend to lump transmission losses into agricultural subsidies. Nevertheless, the subsidies do constitute a large part of the states' deficits, illustrating that not only the groundwater situation, but the entire states' finances are affected.
6 Politically, an important issue relates to the fact that groundwater users point out that surface water users are usually highly subsidized because frequently neither the capital nor the operation and maintenance costs of surface water irrigation systems are recovered. Accordingly, this leads to a political dilemma, with groundwater users questioning why they should be paying higher prices for water than surface water users do.

References

Bhatia, R. (2005) Water and energy, Background Paper for the report *India's water economy: bracing for a turbulent future*. World Bank, 2006.

Blomquist, W. (1992) *Dividing the Waters: Governing Groundwater in Southern California*. ICS Press, San Francisco, California.

Brown, L., Gardner, G. and Halweil, B. (1999) *Beyond Malthus*. The Worldwatch Environmental Alert Series, Worldwatch Institute, Washington, DC.

Chebaane, M., *et al.* (2004) Participatory groundwater management in Jordan: development and analysis of options. In: Kemper, K.E. (ed.) *Groundwater – From Development to Management*. Hydrogeology Journal Theme Issue, Springer, New York/The World Bank,Washington, DC.

Dinar *et al.* (2005) *Decentralization of River Basin Management: A Global Analysis*. Policy Research Working Paper No. 3637. The World Bank, Washington, DC.

DuMars, C.T. and Minier, J.D. (2004) The evolution of groundwater rights and groundwater management in New Mexico and the Western United States. In: Kemper, K.E. (ed.) *Groundwater – From Development to Management*. Hydrogeology Journal Theme Issue, Springer, New York/The World Bank, Washington, DC.

Foster, S.S.D., *et al.* (2004a) Quaternary aquifer of the North China Plain – assessing and achieving groundwater resource sustainability. In: Kemper, K.E. (ed.) *Groundwater – From Development to Management*. Hydrogeology Journal Theme Issue, Springer, New York/The World Bank, Washington, DC.

Foster, S., Garduño, H. and Kemper, K. (2004b) *The 'COTAS' – Progress with Stakeholder Participation in Groundwater Management in Guanajuato-Mexico*. GWMATE Case Profile Collection, No. 10. World Bank, Washington, DC. Available at: www.worldbank.org/gwmate

Jacobs, K. and Holway, J.M. (2004) Managing for sustainability in an arid climate: lessons learned from 20 years of groundwater management in arizona, USA. In: Kemper, K.E. (ed.) *Groundwater – From Development to Management*. Hydrogeology Journal Theme Issue, Springer, New York/The World Bank, Washington, DC.

Kahnert, F. and Levine, G. (1993) *Groundwater Irrigation and the Rural Poor: Options for Development in the Gangetic Basin*. The World Bank, Washington, DC.

Kemper, K.E. (2001) *Markets for Tradable Water Rights*. Overcoming Water Scarcity and Quality Constraints. 2020 Focus 9, Brief 11. IFPRI, Washington, DC.

Kemper, K.E. (2003) Rethinking groundwater management. In: Figuères, C., Rockström, J. and Tortajada, C. (eds) *Rethinking Water Management: Innovative Approaches to Contemporary Issues*. Earthscan, London.

Kemper, K.E. and Alvarado, O. (2001) 'Water,' in Mexico: a comprehensive development agenda for the new era. In: Giugale, M., Lafourcade, O. and Nguyen, V. (eds) *Policy Notes*. The World Bank, Washington, DC.

Mariño, M. and Kemper K.E. (eds) (1999) *Institutional Frameworks in Successful Water Markets – Brazil, Spain and Colorado/USA*. The World Bank, Washington, DC.

North, D.C. (1990) Institutions, *Institutional Changes and Economic Performance*. Cambridge University Press, Cambridge.

Sandoval, R. (2004) A participatory approach to integrated aquifer management: the case of Guanajuato state, Mexico. In: Kemper, K.E. (ed.) *Groundwater – From Development to Management*. Hydrogeology Journal Theme Issue, Springer, New York/The World Bank, Washington, DC.

Shah, T., *et al.* (2000) *The Global Groundwater Situation: Overview of Opportunities and Challenges*. International Water Management Institute (IWMI). Colombo, Sri Lanka.

World Bank (2001) *Yemen – Land and Water Conservation Project*, Vol. 1. Project Completion Report. The World Bank, Washington, DC.

World Bank (2002–2005) *Sustainable Groundwater Management – Concepts and Tools*. Groundwater Management Advisory Team (GWMATE) Briefing Note Series, Note 1. Washington, DC.

World Bank (2003) *Environmental Protection and Sustainable Development of the Guarani Aquifer System Project*. Appraisal Document. The World Bank, Washington, DC.

World Bank (2005) *Pakistan Country Water Resources Assistance Strategy. Pakistan's Water Economy: Running Dry*. Washington, DC.

World Bank (2006) *Analisis comparativo de politicas relacionadas con el sector agua: exploracion sobre los impactos en la productividade del agua*. Working Paper No. 2. Serie de Agua de Mexico. Report No. 36854.

Available at: www.wateryear2003.org

9 When the Well Runs Dry but Livelihood Continues: Adaptive Responses to Groundwater Depletion and Strategies for Mitigating the Associated Impacts

Marcus Moench

Institute for Social and Environmental Transition (ISET), 948 North Street, Suite 7, Boulder, CO 80304, USA

Core Arguments

Groundwater level decline, pollution and quality degradation are widely recognized as major emerging problems in many parts of the world. This recognition has not, however, translated into equally wide management responses. The reverse has, in fact, often proved true. In parts of India, groundwater over-extraction and quality decline have been recognized since the 1970s (United Nations Development Program, 1976; Bandara, 1977). With a few possible exceptions, little has been done to regulate groundwater extraction or control degradation of the resource base. This is also the case across Latin America and Africa and in countries as diverse as China, Spain and the western USA (Ballester *et al.*, Chapter 6; Masiyandima and Giordano, Chapter 5; Wang *et al.*, Chapter 3; Llamas and Garrido, Chapter 13; and Schlager, Chapter 7, this volume, respectively). This situation is, in fact, mirrored across much of the globe.

This chapter argues that the lack of progress in implementing conventional management responses to groundwater problems reflects a combination of technical, social, behavioural and organizational limitations that are inherent features in most contexts. Such limitations are often compounded by the growth of competing demands and social 'conflict' over access to the resource and the manner in which it is used. In some cases, such conflicts are fundamental, i.e. one set of objectives or uses cannot be satisfied unless other sets of objectives and uses are modified in fundamental ways. Recognizing the importance of an emerging problem or the 'need' for management does not change the fundamental nature of the limitations or reduce the inherent nature of some conflicts. As a result, whatever the 'need' for management, alternative or complementary approaches that are adapted to the inherent limitations present in a given context are often essential. In many cases, such adaptive approaches will involve courses of action that fall outside the limits of conventional groundwater management. Furthermore, at least in some cases, adaptive approaches may be more effective in addressing the societal impacts of groundwater problems than even the best-implemented forms of conventional 'water-focused' management.

What is an adaptive approach? Research conducted by the Institute for Social and Environmental Transition (ISET) and our partners in India and other locations (Moench, 1994; Moench *et al.*, 1999, 2003) suggests that adaptation is a continuous process and adaptive approaches need to be designed in ways that:

- Encourage evolution of strategies as conditions change over time or, to put it another way, have in-built mechanisms to respond to ongoing change processes;
- Reflect the social, political, economic and technical context in which groundwater problems are occurring and the types of response – including or excluding conventional management – that are likely viable within that context;
- Respond to inherent limitations on scientific knowledge;
- Build off the incentives and courses of action households, communities and regions are already undertaking or have a strong incentive to undertake in response to a given problem;
- Are strategic in that they focus on core objectives (livelihood and environmental values as opposed to specific groundwater parameters) and respond to the spatial and temporal factors that influence the probable effectiveness of response strategies rather than attempting to be 'comprehensive' or 'fully integrated'.

The above criteria indicate that adaptive responses do not exclude conventional water management techniques. Instead, they identify such conventional techniques as one among many avenues for responding to groundwater problems. Conventional 'water-focused' techniques are, in essence, one subset of a much larger set of techniques, each of which may be more or less effective in any given context for addressing core social objectives that are threatened by groundwater problems. Strategic 'adaptive' approaches can be viewed as including the full array of conventional water-focused management techniques while also moving beyond them to encompass a potentially very wide range of interventions designed to reduce or eliminate the negative impact of groundwater conditions on livelihoods and environmental values. This can involve fundamental shifts in livelihoods (e.g. changing from agricultural to non-farm systems) or it can involve shifts within livelihoods (e.g. crop choice or technology shifting within agriculture). Furthermore, the element of change or 'process' is central. Strategies need to recognize and be able to respond as economic, social, hydrological and other conditions change over time. The core difference between the approaches suggested here and most conventional management is the explicit focus on: (i) core livelihood and environmental objectives rather than groundwater per se; (ii) the inclusion of response strategies that do not attempt to influence groundwater resource conditions directly; (iii) the 'strategic' element – tailoring water- and non-water-related responses to a given moment and socio-ecological context rather than attempting to develop 'comprehensive' 'integrated' strategies; and (iv) the concept and role of adaptation – the manner in which strategies can continuously be shaped to reflect ongoing change processes. This last element – a tautology at present (adaptive approaches are defined as approaches that focus on adaptation) – is explored in detail later.

This chapter begins with a section that briefly outlines a series of key factors that limit the viability of conventional approaches to groundwater management in many, if not most, contexts. Following this, the conceptual foundations for alternative, more adaptive approaches to groundwater management and the mitigation of impacts from emerging groundwater problems are discussed. Illustrative examples of adaptation drawn from specific field areas in India, Mexico and western USA are presented next. A diverse selection of examples has been utilized to highlight both the similarity of many key issues and the fact that solutions appropriate in one region usually cannot be generalized to other areas. The final section outlines strategic implications for organizations seeking to catalyse effective responses to groundwater problems.

The Limits of Management

Conventional approaches to the sustainable management of groundwater supply generally consist – at least on the conceptual level – of techniques designed to balance extraction within any given aquifer to levels that do not exceed long-term recharge rates, i.e. on the 'sustained yield'. Extraction levels that exceed recharge rates over the short term – e.g. during a 3- to 4-year drought period – should ideally be balanced by other periods when high levels of precipitation or artificial recharge activities ensure that recharge exceeds extraction. This approach is often enshrined in law. According to Llamas and Garrido (Chapter 13, this volume), for example, the Spanish Water Act of 1985 'basically considers an aquifer to be overexploited when the pumpage is close or larger than the natural recharge'. This is also the case with estimation procedures in India (World Bank and Ministry of Water Resources – Government of India, 1998). While the validity of the sustained yield concept is widely debated (Llamas and Garrido, Chapter 13, this volume), in practice it generally forms the basis for most legislation and management attempts designed to regulate groundwater supply.

The above ideal is rarely met. In some cases this is an inherent characteristic of the resource: natural recharge rates can be extremely low and extraction is, in essence, an inherently unsustainable activity that involves mining a finite supply. Such situations are common in many arid parts of the world. Where they exist, the technically 'ideal' goal of groundwater supply management would consist of a planned depletion schedule along with longer-term strategies for replacing supplies or shifting demand as the aquifer is depleted. More importantly from the perspective of this chapter, however, are the much more common situations in which substantial natural and/or induced recharge *does occur* but extraction rates are well above sustainable levels. In these situations the ideal image of managing aquifers to achieve long-term sustained yields may be technically feasible but is in fact rarely achieved. Even where aquifer recharge rates are known to be extremely low, effective attempts to develop depletion schedules and manage extraction to achieve them are extremely rare.

The reasons why management rarely reflects technical ideals are important to understand. They may reflect a fundamental disjuncture between management concepts and social, economic and scientific ground realities (Moench, 1994, 2002, 2004; Burke and Moench, 2000 COMMAN, 2004). While a detailed discussion of the reasons that management concepts often fail is beyond the scope of this chapter, key elements include:

1. *Timescales and the rapid process of social and economic change*. Conventional approaches to groundwater management necessitate an inherently long-term perspective. Precipitation and recharge fluctuate greatly over periods of time that often involve decades. Groundwater systems are lagged, and the effectiveness of conventional management approaches depends on the ability to take action on a sustained and consistent basis. The factors driving demand for groundwater and, more importantly, the incentives local populations have to participate in management and the vulnerability of livelihood to groundwater conditions are generally influenced by factors that operate on much shorter timescales. In many parts of the world, rapid processes of economic, social and political change have a major impact on the nature of local livelihood systems and, through those systems, on groundwater dependence and the incentives individuals, communities and regions have to invest in longer-term groundwater management initiatives. The short-term nature of incentives conflicts with long-term management ideals.

2. *Inherent scientific limitations on the ability to quantify water availability and hydrologic dynamics within aquifers*. These include: (i) the absence of data, particularly the long-term monitoring information required to define basic hydrological and water use parameters; (ii) ongoing climatic and other change processes; and (iii) the hydrogeological complexity of aquifer systems. As a result, the dynamics of even the best-monitored systems in wealthy locations such as western USA are often poorly understood. Where monitoring systems are weak, as they are throughout much of the less industrialized world, the scientific understanding of aquifer systems necessary for conventional management is even less sound. As Llamas and Garrido (Chapter 13, this volume) note, 'uncertainty is an integral part of water management'. As a result, response strategies need to be developed in ways that incorporate, rather than attempt to eliminate, such uncertainties.

3. *Mismatches between the scale and boundaries of aquifer systems and the scale and boundaries of human institutions*. The fact that human institutions rarely match with the boundaries of hydrologic systems has been widely recognized for decades as a critical factor constraining water management and underlies the emphasis on developing watershed institutions as the critical unit for management that runs throughout the integrated water resources management (IWRM) literature. This constraint is particularly severe in the case of groundwater where boundaries may not have any surface representation and rarely, if ever, match with existing human institutions. Furthermore, in many cases the core 'human unit' determining groundwater use lies at an extremely micro-level – the individual well owner. As a result, groundwater conditions are affected by the aggregate demand coming from thousands of individual disconnected decision makers.

4. *Disjuncture between the factors that have been shown through recent research on common property as important for the formation of effective management institutions and the nature of groundwater occurrence and use.* A substantial literature has developed over the last two decades that documents the conditions common to successful management of common pool resources (see e.g. BOSTID, 1986; Ostrom, 1990, 1993; Bromley, 1998; Schlager and Blomquist, 2000; and Schlager, Chapter 7, this volume) such as groundwater. Some of the most important factors that emerge regularly in this literature include: (i) a high level of – and broadly felt – need for management; (ii) clear systems of rights or rules-in-use governing access and resource utilization; (iii) clear boundaries on the resource and the user group; (iv) mechanisms to control free riders (including ways to restrict access for non-members or those not holding resource use rights); (v) clear systems for monitoring resource conditions and use including documentation of the benefits from management; (vi) relative economic and cultural homogeneity among group members; (vii) a proportional equivalence between the costs and benefits from management; (viii) effective mechanisms for enforcement; and (ix) small primary management group size often accompanied by the nesting of institutions where some management functions need to occur at regional or system scales rather than local scales. These conditions are generally violated in groundwater management contexts.

The above issues may be conceptually clear, but the practical constraints they impose on management are rarely recognized. These constraints are illustrated below using examples from India, Mexico and western USA.

In India, debates over the need for groundwater regulation and management have been ongoing since the mid-1970s. While these have led to numerous proposals for augmentation, regulation, rights reform and the implementation of economic incentives for efficient use, with the exception of extensive efforts to harvest and recharge rainwater, relatively few reforms have actually been implemented (World Bank and Ministry of Water Resources – Government of India, 1998; Shah *et al.*, 2003a; COMMAN, 2004). Even reduction of electricity subsidies to agriculture – which are widely recognized as a major incentive encouraging groundwater overdraft – and/or reforming tariff structures to reduce such incentives has proved difficult (World Bank, 1999; Kumar and Singh, 2001; Shah *et al.*, 2003b). Why has the initiation of management been so difficult? While a very wide variety of factors play a role, key elements include:

1. The extremely large number of wells. Recent estimates suggest that the number of wells exceeds 20 million (World Bank and Ministry of Water Resources – Government of India, 1998). As a consequence, tens of thousands of individuals often use any given aquifer, and the ability of the government to register wells or move beyond this to establish and enforce volumetric rights systems is extremely limited (Dhawan, 1990; Moench, 1991, 1994; Moench *et al.*, 1997). Similarly, the large number of users complicates virtually all of the factors known to contribute to management through common property approaches.

2. The rapid pace of economic and demographic change affecting much of the country. India is undergoing a process of 'peri-urbanization' involving, in many areas, diversification in the nature of rural livelihoods to include many non-agricultural elements (Moench and Dixit, 2004). Nevertheless, at present many rural livelihoods are heavily dependent on groundwater-irrigated agriculture. This appears to have two effects. First, because people depend on agriculture to meet current needs, management activities that would require reductions in groundwater use are seen as having an immediate impact on current income. Second, despite current dependency, growing aspirations and the vision of opportunities beyond agriculture limit user concerns over the longer-term impact groundwater depletion may have. In combination, these factors reduce the incentive to manage groundwater resources in order to protect livelihood in the future while increasing the incentive to exploit the resource base to support current needs.

3. Limited scientific information on groundwater conditions. Although groundwater monitoring networks were established in parts of India during the 1970s and have been substantially expanded since then under programmes such as the 'Hydrology Project', basic scientific information on aquifer conditions is often extremely limited (World Bank and Ministry of Water Resources – Government of India, 1998). The problem is compounded because key elements in any water balance equation – such as evapotranspiration by native vegetation – are not estimated. Furthermore much of India is underlain by hard-rock systems that can make the identification and modelling of groundwater flow systems extremely complex – a factor noted more than a decade ago (Narasimhan, 1990).

Overall, conditions in India clearly illustrate many of the factors constraining conventional management approaches.

The difficulties inherent in establishing the information base required for conventional management approaches are also clearly illustrated in the case of Mexico. There, despite substantial support from the World Bank, even the precursor activities required for management have not proceeded rapidly or smoothly. Although Mexico has invested more than a decade of effort on well registration, it has so far proved impossible to develop a systematic register of operational wells (Garduno, 1999; Foster *et al.*, 2004). Registration of wells is an essential first step required to enable the state (or any other organization) to monitor any water rights or regulatory system. A recent review summarizes the situation well:

> In the 1990s major efforts were made by federal government (the CNA) to register and administer the groundwater abstraction and use rights system. However, lack

> of local operational resources and failure to mobilize user cooperation has eroded the system. Lack of consistent enforcement has meant that those who decide not to follow the rules are usually not sanctioned, thus deterring the rest of the user community to cooperate or comply with the regulation processes.
>
> Attempts to constrain groundwater exploitation in Guanajuato included three periods of nominal 'waterwell drilling bans', but the number of deep wells appears to have more or less doubled during each of these periods.
>
> (Foster *et al.*, 2004)

This note also highlights scientific uncertainties and voices concern that attempts to develop local management institutions (called COTAS) may 'flounder because of lack of action on complementary "top-down" legal procedures and policy decisions'(Foster *et al.*, 2004, p. 8). Different pieces of the information and institutional environment are moving at different rates and are, in fact, being driven by different social forces. This disjuncture, one that is often inherent in social processes operating at local and national levels, undermines the development of the consistent information and institutional framework required for groundwater management. Overall, major difficulties in developing a systematic register and the institutions for groundwater management have emerged despite the fact that the number of wells in Mexico is much smaller and typical well capacities are much larger than in India – factors that should, at least conceptually, make the task of registration and institutional development substantially more straightforward (Shah *et al.*, 2000).

The problems in India and Mexico are similar to those found in the USA and Europe. In Spain, Llamas and Garrido (Chapter 13, this volume) point out that only 2 out of 17 'groundwater user communities' that are supposed to manage groundwater in areas identified as 'overexploited' are operative. In western USA, while some systems for groundwater management that are at least partially effective have evolved in locations such as the Central Valley of California, these systems involve a very limited number of actors – in the Central Valley case between 100 and 200 large utilities, corporate and agricultural entities – that pump most of the water (Blomquist, 1992). Groundwater resources in the Central Valley remain, however, under stress. As T.N. Narasimhan, a noted groundwater expert in California, comments: 'Major regions of California such as the San Joaquin, Salinas, Owens, and Santa Clara Valleys have supported extensive groundwater use by agriculture, industry and municipalities. These resources are also presently over-developed' (Narasimhan and Kretsinger, 2003). Thus, despite some success in organizing a management system, sustainability of the groundwater resource base remains far from assured. In other regions, management is even less effective. In Arizona, for example, strong regulatory agencies were established in the 1980s to address overdraft in what were termed 'active management areas' (AMAs). This was done as part of a quid pro quo for federal investment in the Central Arizona Project, a major diversion to supply water to Arizona from the Colorado River. Despite this, groundwater levels continue to fall under many major cities in Arizona and overdraft concerns have not been resolved. The situation is of particular concern in the context of climatic variability and change where 'safe yield' policies are intended to provide a solid water-supply buffer that could reduce drought impacts. According to the US Global Change Research Program:

> A team from the University of Arizona analyzed the water budgets of several Arizona cities to determine how severe the drought impacts would be from the deepest one-year (1900), five-year (1900–19), and ten-year (1946–1955) droughts on record. Case study sites included two of the fastest growing areas in the U.S. – the Phoenix and Tuscon Active Management Areas (AMAs). In these AMAs, stringent groundwater management is mandated under the 1980 Arizona Groundwater Management Act. The study showed that, even under assumptions of continuing 'average' climate conditions, the possibility of achieving 'safe yield,' as articulated in the Act (i.e., supply and demand are in balance), remains uncertain.[1]

More to the point, in many areas information that is fundamental for effective management remains unavailable. Climatic variability and change predictions are widely recognized as involving high levels of uncertainty. As a result, defining 'safe yield' in locations such as the example from Arizona is, at best, a complex effort that will not resolve uncertainties in the information needed for management. This is, however, equally often the case even when climate change is not a central concern. Take, for example, the case of the city of Albuquerque and the Middle Rio Grande. Despite the relatively strong institutional capacity of the US Geological Survey (USGS) and the New Mexico State Engineer's office – which are responsible for groundwater monitoring – a recent report on the region states:

> Until the locations and pumping characteristics of the major supply wells in the Middle Rio Grande Basin are known with more certainty, estimates of these important parameters will introduce error into simulations and estimations of ground-water behavior. However, it may be impossible to know exactly the locations of all domestic-supply wells in the basin and the volumes of water pumped from each.
>
> (Bartolino and Cole, 2003, p. 128)

The report goes on to highlight uncertainties related to the quantity and quality of water available deeper in the basin and, probably more importantly, difficulties in measuring evapotranspiration from the *bosque* riparian vegetation a habitat that has recently been recognized as having an important environmental value. According to the report: '[E]stimates of evapotranspiration in the Middle Rio Grande Basin vary because it is a difficult parameter to measure directly. Because maintenance of the *bosque* has become a priority for esthetic and wildlife purposes, the measurement of actual evapotranspiration is of critical importance' (Bartolino and Cole, 2003). This quote illustrates two points: (i) the inherent uncertainty (despite ongoing efforts to improve estimates) in the ability to measure and monitor key parameters (domestic pumping and evapotranspiration); and (ii) the fact that new values emerge over time – historically the *bosque* was not widely recognized as having a key environmental value.

How important are such factors in the overall water balance equation and how might they affect management institutions? According to the USGS: 'The Middle Rio Grande water budget of the Action Committee of the Middle Rio

Grande Water Assembly (1999) estimated that between 75,000 and 195,000 acre-feet of water is lost annually to evapotranspiration by the *bosque* in the river reach between Otowi (north of the basin) and San Acacia' (Bartolino and Cole, 2003, p. 84). This is equivalent to 24–63% of the total groundwater withdrawals of 309,890 acre-feet reported in 1995 for the region (Bartolino and Cole, 2003, p. 61). Overall, despite intensive research on groundwater in the Middle Rio Grande that dates back at least to the 1950s, new issues are continuously emerging and fundamental scientific uncertainties regarding key parameters in the water balance equation remain high. The science is weak, and the institutions even weaker. It is, for example, extremely difficult to allocate 'secure' water rights when uncertainties regarding the quantity available account for almost 40% of current extraction.

This uncertainty has huge economic and political implications. As with many cities, Albuquerque is growing rapidly and is facing a situation in which urban and industrial demands are competing heavily with established agricultural users for access to water. As part of this general dynamic, Albuquerque has encouraged the development of new high-tech industries – including an Intel factory to produce next-generation computer chips. This new industry, the cutting edge of technological development, also happens to require substantial amounts of pure water. Steve Reynolds, who was the New Mexico state engineer for 30 years, wrote in 1980: 'Albuquerque is probably better situated with respect to water than any large city in the Southwest'.[2] With substantial water available, rapid expansion of water-intensive high-tech industries appeared to be a logical step. In 1993, following substantial investment in plant facilities, Intel requested a permit to extract 4500 acre-feet of groundwater to support its operations. This request happened to arrive at the same time that a new USGS report was released, showing that the city was pumping groundwater at nearly three times the natural replenishment rate.[3] The economic and political consequences continue to reverberate. Scientific uncertainty had, in essence, allowed (or perhaps been used) as a mechanism to enable forms of development that were ultimately unsustainable.

The above situation is common in many other parts of western USA and has a major impact on the technical ability to manage groundwater resources in an effective manner. Furthermore, limitations of technical understanding often contribute substantively to political limitations on the ability to make management decisions – it is politically difficult to argue that users should make major cutbacks in extraction unless the need is clear. When the understanding of underlying groundwater system dynamics contains huge uncertainties, defining the 'need' for management in a way that is convincing to key actors (from politicians to individual farmers) is difficult. This lag between identification of a potential problem and the gradual growth of information necessary to 'prove' makes timely responses difficult and allows use patterns to become deeply embedded. In Colorado, for example, difficulties in defining the degree to which groundwater resources are hydrologically linked to specific surface sources have been a major point of contention within many major river basins since the 1950s. Although the links between surface flows and groundwater extraction in the basins are increasingly well understood, well owners have been using groundwater for decades and it is politically difficult to reverse established use patterns. As a result, legal 'wars' between groundwater users and surface water right holders are an increasingly common part of the institutional landscape of water in Colorado. These have become 'battles of the experts' in which competing models and competing analyses seek to gain legal ground in the context of substantial scientific uncertainty. Issues may be resolved – but often only incrementally and with huge investments in legal and technical resources.

Overall, conventional approaches to groundwater management are commonly constrained by the large number of wells, the rapid pace of social and economic change as well as scientific knowledge and data limitations. It is important to emphasize that many of these limitations are *inherent* rather than situational. The challenges reflect scientific uncertainties that cannot be eliminated, political dynamics that reflect basic human nature and scale issues that emerge due to the number of users and the fundamental mismatch between patterns of human organization and patterns within the physical groundwater system. While such challenges may be narrowed by advances in knowledge or organizational systems, they are unlikely to be eliminated. As a result, the ability of society to manage groundwater resources effectively has clear limitations. Adaptation to emerging problems – rather than attempts to 'solve' them directly – thus appears essential.

Concepts of Adaptation

What does adaptation mean? In relation to natural resource issues, at least two core concepts are common. The first involves the growing field of 'adaptive' management. This term is generally used to describe management systems that focus on the resource base itself but are intended to work in a flexible manner and that respond to changes as they occur over time. The management system is 'adaptive' in that it has inbuilt mechanisms whereby the tools and objectives of management can be adjusted as new information becomes available or other conditions change. In many situations, adaptive management approaches provide for review processes at specific intervals so that such adjustments can occur.

The second concept, which I am primarily dealing with in this chapter, involves tailoring responses to the larger context in which they fit – i.e. adapting to the context rather than defining response strategies based on a relatively narrow predefined set of hydrologically focused management objectives and techniques. From this perspective, adaptation emphasizes approaches that focus on adjusting livelihood, economic and other systems in ways that mitigate the impact of groundwater problems – i.e. the core goal is to adjust society to groundwater conditions rather than attempting to 'manage' the resource base itself. This is, however, simply a matter of emphasis and does not exclude direct attempts to manage the resource base. In some contexts, direct management may be viable and adaptation in the larger sense would include both conventional and water-focused iterative 'adaptive' groundwater management strategies. Except where explicitly noted, the term 'adaptation' in the remainder of this chapter refers to the second larger concept rather than the more narrowly defined adaptive management processes.

Substantive work related to the conceptual foundations of adaptation processes has been undertaken by a number of authors loosely grouped into the Resilience Alliance.[4] Many of the concepts underlying adaptation in social systems have emerged from research on systems dynamics and the application of insights on adaptation gained from ecosystems research (Holling, 2001; Gunderson and Holling, 2002; Holling *et al.*, 2002). This research emphasizes core phases in the ongoing processes by which natural systems evolve. These phases can be seen as a loop, outlined in Fig. 9.1 from the Resilience Alliance, that consists of: 'entrepreneurial exploitation (r), organizational consolidation (K), creative destruction (Ω), and re- or destructuring (α)' (Holling, 2004). The core insight here is that systems – whether in the natural environment or, as many members of the Resilience Alliance would argue, in the social and institutional environment, progress through very clear phases. These phases start with initial expansion under conditions in which core resources (nutrients, energy, finances, etc.) are readily available and the system is relatively unstructured, followed by a phase of consolidation or restructuring.

In the case of groundwater, these two phases would be represented by the initial spread of energized irrigation when expansion led to rapid increases in productivity and has now transited into highly efficient, but much less diversified (more structured), forms of intensive agricultural production. The process of increasing structure as resources are captured leads to efficient systems that are also increasingly rigid in that the available resources tend to be evermore fully captured and utilized. Rigidity, in turn, creates the conditions for creative destruction when surprises or extreme events shake the system. Again, in the groundwater case, this might be a drought hitting when aquifers have been fully exploited. When intensification captures all recharge, the groundwater buffer that provided resilience against drought is no longer reliable and intensive agriculture cannot continue in locations where access to groundwater fails. The disruption caused by this phase breaks system rigidity and frees resources (in

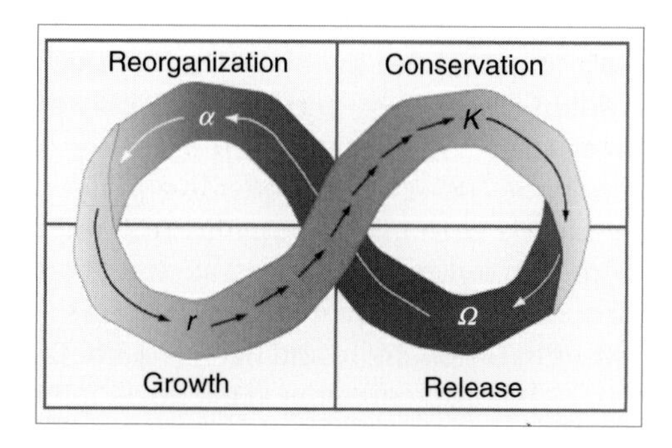

Fig. 9.1. System dynamics. (From the Resilience Alliance.)[5]

the groundwater case, an example would be the human and financial capital that had been fully invested in intensive agriculture), and leads, in turn, to a phase of reorganization.

Studies on system resilience indicate that the process of growth, conservation, release and reorganization occurs at all levels in a system and is interlinked across such levels, as the household, livelihood system, village and regional economic levels within any given agricultural economy. The studies also indicate that the nature of creative destruction is heavily dependent on the degree to which the *r–K* growth to conservation phase has resulted in the creation of a highly structured system. Basically, the more structured a system becomes, the greater is the destruction when it fails. When systems are constantly responding to small destructive events across several levels, the degree to which they become rigidly structured is limited by continuous processes of release and reorganization – that is adaptation. Such systems tend to be much more resilient, and much less subject to fundamental reorganization, when surprise events occur. However, when disruption is limited, efficiency and rigidity both grow. Again, to relate this to groundwater, by eliminating the constant need to adjust to variations in precipitation and water supply, access to groundwater has enabled the development of efficient and intensive, but also rigid, agricultural livelihood systems. When drought and groundwater overdraft in conjunction eliminate access to secure water supplies, livelihood systems based primarily on intensive irrigated agriculture have low resilience and lack capacity to adapt without first undergoing significant reorganization.

The conceptual frameworks described are closely related to concepts of risk. As the risk management literature clearly documents, exposure to, and familiarity with, risk is a key factor underlying the incentive to develop coping and avoidance strategies (Wisner *et al.*, 2004).

What does this imply for responses to emerging groundwater problems? While exploration of all the implications is well beyond the scope of this chapter, a few observations appear central:

1. It indicates that disruption, change and adaptation are inherently interlinked processes and that periods of disruption should be recognized as opportunities as well as times of crisis. Probably the most 'natural' time of change in an agricultural economy based on intensive groundwater use is during drought or similar times of 'crisis'. This is the time when creative destruction is likely to occur and livelihood, economic and political systems will be forced to adapt. In contrast, periods of stability are likely to be periods when change is resisted.

2. It implies that activities designed to buffer livelihood or agricultural systems against variability are likely to increase rigidity and reduce adaptive capacity. The shift away from rain-fed systems into systems based on groundwater irrigation that has occurred in India over the last 50 years – while it has allowed tremendous increases in productivity and been a major factor for reducing poverty (Moench, 2003) – has also encouraged the development of systems that are much more dependent on secure water supplies and much less resilient when water supply reliability declines. While it is important to avoid large-scale disruption of agricultural systems, exposure to risk and variability is essential in order to maintain resilience of the system as a whole. In the case of groundwater overdraft, actions that protect people against immediate loss also reduce the incentive to undertake structural changes that would, in a larger sense, reduce vulnerability.

3. It suggests that principles central to resilience in ecological systems (such as diversification) are also central to the development of resilient livelihood systems. Livelihood founded primarily on groundwater-dependent agricultural systems is likely to be much less resilient when the resource fails than that based on a diversified portfolio of agricultural and non-agricultural activities. Similarly, within agricultural livelihood systems resilience is likely to be enhanced when those systems involve a mix of crops, crop varieties and agriculturally related activities (e.g. animal husbandry). Diversification both between and within livelihood systems is important.

Insights from the Field

Detailed field research on what households and communities actually do to 'adapt' or respond to groundwater problems was recently undertaken as part of a larger programme on adaptation by a consortium of organizations across India and Nepal (Moench and Dixit, 2004). Aspects of this research in Gujarat, India, are discussed by Mudrakartha (Chapter 12, this volume) from the Vikram Sarabhai Centre for Development Interaction (VIKSAT), one of the partner organizations (Mudrakartha, 2005). As a result, only the broad insights generated by the project are discussed here.

The project focused on areas in which long-term groundwater overdraft conditions were compounded by drought. In these areas, people responded to both the creeping process of increasing water scarcity and the immediate impact of drought primarily by:

1. Attempting to diversify income sources away from water-dependent, agricultural forms of livelihood;

2. Attempting to increase access to water, particularly secure sources of water for domestic and livestock use, through water-harvesting activities, by drilling ever-deeper wells and by purchasing water through informal markets supplied by water tankers (commonly known as 'tanker markets');

3. When all else failed, by coping through reduced consumption.

Those who successfully 'adapted' – i.e. those who were able to maintain living standards and avoid the coping strategies that involved reductions in consumption and other indicators of living standards – were the ones who succeeded in diversifying their livelihood and obtaining secure sources of domestic water supply. The story is not, however, simply one of economic diversification and domestic water security. To be effective, these core strategies depended upon sets of linkages with higher-level economic, information and social systems. In addition, while some of the linkages occurred due to the immediate pressure of drought, many of them evolved over much longer periods of time. To be more specific, in most areas, successful diversification into non-farm activities was enabled by a combination of factors including:

1. Proactive migration and commuting: Migration and commuting were core strategies that enabled families and communities to obtain access to outside labour markets and sources of non-farm income. In some cases, this occurred over a generational basis. Families invested in efforts to find non-agricultural work for at least one key member in an urban area. In other cases, the strategy involved long-term investments in education. When drought hit, the income generated by family members working in urban areas served as a critical buffer for livelihood or as the source of capital for recovery or investment for those still living in rural areas. In yet other cases, migration involved short-term travel to work in regional labour markets – or even commuting to access specific local work opportunities. The core points here, however, are that (i) mobility was a core factor enabling diversification and (ii) in many cases it was a *proactive* strategy that occurred over long periods of time and not a short-term *reactive* response to immediate drought impacts.
2. Access to transport and markets: The ability to diversify depended heavily on the presence of transport systems and access to regional markets. In the Gujarat drought case, many farmers increased dairy production by using fodder grown in distant locations and transported it into their area. Similarly, access to secure domestic water supplies – which are in effect the single most essential requirement for people to remain in any given region – was often enabled by access to regional tanker markets for water. This was also the case for many other non-farm activities such as woodworking, diamond polishing and developing other small businesses. Transport and market access were, as a result, core prerequisites, enabling diversification both within the agricultural economy and between agriculture and non-farm activities.
3. Access to social networks: Familiarity with regional labour markets and access to key individuals already working within them were commonly mentioned as important factors enabling diversification. This was also the case for other resources such as credit.
4. Access to credit and financial institutions: Without credit, the investments required for diversification – including the costs associated with migration to distant labour markets – are often impossible to make. Similarly, without financial institutions (whether formal or informal), the ability to send remittances earned through migration and commuting for investment in local livelihood systems is greatly limited.
5. Access to education: Key skills, such as basic literacy and any higher levels of education, played an important role in the ability to diversify. In many cases, the least educated only had access to jobs as wage labourers in agriculture or the construction industry, whereas those with higher levels of education had access to a more diversified portfolio of job opportunities.
6. Presence of local institutions: Self-help groups and other community organizations along with access to higher-level non-governmental organizations (NGOs), government organizations and private businesses played a critical role. To take another example from the Gujarat drought case, cooperatives played a major role in

organizing fodder transport, enabling bulk purchase and providing credit. This was a critical input that enabled farmers to shift out of irrigated agriculture and diversify into dairy production as groundwater levels declined and the drought increased.

In most of the areas studied, the ability to adapt was not uniform across the communities. As with many other studies in the development literature, existing patterns of vulnerability created by gender, income and social position played a critical role. Women, for example, had far less access to many of the opportunities for income diversification than men, and this was often the case for other economic or socially constrained groups also. The poor were not, however, always the most vulnerable or the least adaptable. In many areas marginal farmers had much greater familiarity than larger farmers with regional labour markets as well as the established social networks required to access them. As a result, when the wells of the larger farmers failed, they lacked the experience and contacts necessary to diversify. In some cases, these larger farmers invested virtually all their capital and acquired substantial debt in unsuccessful attempts to obtain access to groundwater by deepening their wells. As a result, they suffered a far greater decline in income than smaller farmers who already had contacts and familiarity with regional labour markets. When the wells were successful, however, the larger farmers were able to maintain existing livelihood systems and experienced little, if any, decline in living standards. They, however, remained vulnerable to the next drought and the potential failure of the new wells in which most of their wealth was now invested. Instead of reducing vulnerability, the ability to maintain access to groundwater may have, in fact, reduced the incentive to adapt, and therefore increased vulnerability over the longer term.

Overall, the ability to adapt to the combination of drought and long-term groundwater overdraft was heavily influenced by a combination of location-specific factors and wider regional conditions. Where access to groundwater remained secure, the incentive to adapt to any specific drought event was low, but farmers were often aware of longer-term threats to the resource base and were taking proactive measures (e.g. investing in education and economic diversification) that would increase the ability of their family to adapt if secure water supplies were lost. Other local courses of action, such as investments in water harvesting, were a way of increasing local water security. Critical points of transition occurred when wells failed and the ability to continue intensive irrigated agriculture failed with them. At this point, the impact on livelihood depended on the degree to which diversification out of vulnerable activities had already occurred and the immediate ability of livelihood units to shift on relatively short notice. This ability depended, in turn, on social position and the ability of information, goods, services and people to flow into , and out of, affected areas.

Strategic Implications

Our recent research on responses to the combination of drought and groundwater overdraft provides a series of initial insights that raise as many questions as they answer. As a result, the implications of research on adaptation for proactive responses to emerging groundwater problems can only be identified in a preliminary manner. However, at least some implications appear to be clear.

1. Adaptation at the household livelihood level can form a central part of a proactive strategy for responding to groundwater problems. Farmers are in most cases fully aware that water level declines in their wells and changes in water quality represent a fundamental threat to agricultural livelihood. They often take proactive steps to reduce their vulnerability by investing in education and taking other long-term actions to diversify income sources. Action by governments and development agencies to support existing adaptive responses of this type are as important and may be more effective in reducing the social, livelihood and economic impacts of groundwater problems than attempts to manage the groundwater resource base directly.
2. From a hydrological point of view, it should be relatively straightforward to identify vulnerable regions in which support for adaptation in livelihood and economic systems is likely to be particularly critical. Regions where water levels (or water quality) are declining rapidly and where hydrological conditions suggest that additional groundwater resources are insufficient to sustain intensive agriculture in zones in which adaptive change is required. In such areas, interventions to assist economic transition and secure domestic water supplies are particularly critical.
3. Response strategies need to combine activities that are 'water-focused' with others that may have little direct relationship to groundwater. Particularly in areas where groundwater resources are under threat and conventional management strategies appear unfeasible, responses focused on developing the physical and social infrastructure for diversification into non-agricultural activities (roads, communications networks, financial and other institutions) are as important and need to be combined with steps to protect domestic water supplies.
4. The central role of mobility – migration and commuting – in adaptation needs to be recognized and appreciated. Globally, urban populations are growing and rural areas are taking on urban economic characteristics. This trend towards social and economic integration is a major factor contributing to the ability of populations to adapt to location-specific constraints such as groundwater overdraft. Rather than seeing migration as a signal of distress to be resisted, strategies to support it and mitigate the negative effects need to be an integral part of both long-term responses to groundwater problems and short-term responses to drought.
5. Far more attention needs to be paid to the linkages between long-term groundwater management issues and issues such as drought that are often treated as completely separate short-term crises. Droughts represent times of crisis and opportunity. They are the points of time when groundwater problems come to a head, when livelihood must shift and when the foundations for adaptation are tested. They are also times of political and social opportunity when governments, communities and households are most aware of the need for effective responses. While attempts to manage the groundwater resource base and protect domestic water supplies are inherently long-term, droughts

can represent a window of opportunity for the establishment of management systems. They are also windows of opportunity when resources (drought relief) are often available that could be used to support longer-term economic diversification and other adaptive responses. In sum, it appears particularly important to develop links between the traditionally quite separate water management, disaster relief and long-term economic development communities.

6. Explicit frameworks and approaches need to be developed that reflect the *process nature* of both conventional 'water-focused' and the wider strategies emphasized here. The adaptive management paradigm that is now emerging in a number of natural resource management fields generally incorporates explicit opportunities for revisiting decisions and strategies within management plans. This represents a starting point. The ability to adapt depends, however, on the flexibility of institutions and management concepts as much as it depends on explicit mechanisms for iteration within a management programme. Where institutions are concerned, for example, water rights that are granted in perpetuity and that relate to specific quantities (or even proportions) of the resource base are likely to be much less flexible than systems that are conditional depending on (changing) social interests, climatic conditions and a host of other factors affecting water use and availability. The same can be said for management concepts. Education and training systems that define water management in terms of 'engineering hydrology' are unlikely to produce managers capable of understanding or utilizing approaches that rely heavily on social change processes or indirect management tools. Approaches that define effective units for management based only on hydrological characteristics (aquifers and basins) are unlikely to be effective when water use within those units depends on virtual flows of water (grain, jobs, etc.) that are driven by regional or global factors. Building temporal and geographic flexibility in, and emphasizing the process nature of, management is central to the development of more adaptive approaches.

7. Although many of the issues related to the development of effective responses to emerging groundwater problems in different areas are similar, responses need to emerge from processes tailored to the specific local context. Much can be learned through comparisons between regions, and underlying response strategies (emphasizing, for example, processes and a mix of conventional, indirect and adaptive techniques) may be similar, but the details of specific management interventions and tools need to reflect location-specific conditions. In many regions, globally accepted 'best practices' are used with little evaluation of whether or not they will work. Adaptation requires learning from other areas but 'adapting' the approach to suit local conditions.

8. It is important to emphasize that adaptation and conventional approaches to groundwater management are not mutually exclusive approaches. Economic transition for populations that currently depend on intensive irrigated agriculture could be used to reduce pressure on groundwater resources and create the political space for direct management of the resource base. Strategically, approaches that focus technical resources on the protection of key groundwater resources that are of particular importance for domestic water supply would directly complement indirect actions to encourage economic transition.

Such points of synergy between conventional groundwater management and more adaptive approaches appear particularly important to explore.

Capturing benefits from the above potential synergies requires strategy. Those actors concerned with emerging groundwater problems need to move beyond the linear step-by-step attempts to develop integrated water management programmes and institutions that characterize current global water debates. As Schlager (Chapter 7, this volume) comments: 'Groundwater basins and large-scale canal irrigation systems present challenging governance issues that are often avoided, ignored or made to disappear within the black box of integrated management'. As a result, instead of attempting to formulate comprehensive approaches, more incremental 'clumsy' solutions tuned to specific times, physical contexts and institutional settings are needed. As any commercial organization interested in building a dam knows, ideas need to be developed and multiple plans formulated so that they can be put into action when a drought or other crisis creates a window of opportunity. Similarly, instead of advocating a single 'best practice' model for groundwater management, multiple models tuned to time and place are needed. It may be possible to mobilize the technical, scientific, institutional and political resources necessary to protect a key strategic aquifer supplying water to an urban area using conventional management techniques. In such cases, the aggregate political weight of urban domestic users is often articulated through a single well-organized water utility. This is far less likely to be the case in rural India or China where the social structure of resource use revolves around the livelihood of numerous individual farmers. As a result, the strategy for responding to groundwater problems in each situation needs to be different in fundamental ways.

Notes

1 Available at: http://www.usgcrp.gov/usgcrp/images/ocp2003/ocpfy2003-fig8-3.htm
2 Quote from Albuquerque Tribune in High Country News. Feature article for 26 December 1994 . Available at: http://www.hcn.org/servlets/hcn.Article?article_id=728
3 High Country News. Feature article for 26 December 1994. Available at: http://www.hcn.org/servlets/hcn.Article?article_id=728
4 Available at: www.resalliance.org

References

Bandara, C.M.M. (1977) Hydrological consequences of agrarian change. In: Farmer, B.H. (ed.) *Green Revolution? Technology and Change in Rice Growing Areas of Tamil Nadu and Sri Lanka*. Westview, Boulder, Colorado, pp. 323–339.

Bartolino, J.R. and Cole, J.C. (2003) *Ground-Water Resources of the Middel Rio Grande Basin*. US Geological Survey, Reston, Virginia.

Blomquist, W. (1992) *Dividing the Waters: Governing Groundwater in Southern California*. Institute for Contemporary Studies, San Francisco, California.

BOSTID (1986) *Proceedings of the Conference on Common Property Resource Management.*

National Research Council, National Academy Press, Washington, DC.

Bromley, D.W. (1998) Determinants of cooperation and management of local common property resources: discussion. *American Journal of Agricultural Economics* 80(3), 665–670.

Burke, J. and Moench, M. (2000) *Groundwater and Society: Resources, Tensions, Opportunities*. United Nations, New York.

COMMAN (2004) Community management of groundwater resources in rural India. British Geological Survey, Wallingford, UK.

Dhawan, B.D. (1990) *Studies in Minor Irrigation With Special Reference to Ground Water*. Commonwealth, New Delhi, India.

Foster, S., *et al.* (2004) The 'COTAS': progress with stakeholder participation in groundwater management in Guanajato-Mexico. *GW Mate Case Profile Collection*. The World Bank, Washington, DC, p. 10.

Garduno, H. (1999) Water rights administration for groundwater management in Mexico. In: Moench, M. and Burke, J. (eds) *Groundwater: The Underlying Resource: Themes in Groundwater Management for the 21st Century*. United Nations, New York, in press.

Gunderson, L.H. and Holling, C.S. (eds) (2002) *Panarchy: Understanding Transformations in Human and Natural Systems*. Island Press, Washington, DC.

Holling, C.S. (2001) Understanding the complexity of economic, ecological and social systems. *Ecosystems* 4, 390–405.

Holling, C.S. (2004) From complex regions to complex worlds. *Ecology and Society* 9(1), 11.

Holling, C.S., *et al.* (2002) Sustainability and panarchies. In: Gunderson, L.H. and Holling, C.S. (eds) *Panarchy: Understanding Transformations in Human and Natural Systems*. Island Press, Washington, DC, pp. 63–102.

Kumar, M.D. and Singh, O.P. (2001) Market instruments for demand management in the face of scarcity and overuse of water in Gujarat, Western India. *Water Policy* 3, 387–403.

Moench, M. (1991) *Sustainability, Efficiency, & Equity in Ground Water Development: Issues in the Western U.S. and India*. Pacific Institute for Studies in Environment, Development and Security, Oakland, California.

Moench, M. (1994) Approaches to groundwater management: to control or enable. *Economic & Political Weekly* (September 24), A135–A146.

Moench, M. (2002) When management fails: evolutionary perspectives and adaptive frameworks for responding to water problems. In: Moench, M. and Dixit, A. (eds) *Understanding the Mosaic (tentative)*. In press.

Moench, M. (2003) Groundwater and poverty: exploring the connections. In: Llamas, R. and Custodio, E. (eds) *Intensive Use of Groundwater: Challenges and Opportunities*. A.A. Balkema, Lisse, The Netherlands, pp. 441–456.

Moench, M. (2004) Groundwater: the challenge of monitoring and management. In: Gleick, P. (ed.) *The World's Water: 2004–2005*. Island Press, Washington, DC, pp. 79–98.

Moench, M. and Dixit, A. (eds) (2004) *Adaptive Capacity and Livelihood Resilience: Adaptive Strategies for Responding to Floods and Droughts in South Asia*. Institute for Social and Environmental Transition, Kathmandu, Nepal.

Moench, M., *et al.* (eds) (1999) *Rethinking the Mosaic: Investigations into Local Water Management*. Nepal Water Conservation Foundation and the Institute for Social and Environment Transition, Kathmandu, Nepal.

Moench, M., *et al.* (1997) *ground water regulation and management*. The World Bank, Washington, DC/Government of India, Ministry of Water Resources, New Delhi, India.

Moench, M., *et al.* (2003) *The Fluid Mosaic: Water Governance in the Context of Variability, Uncertainty and Change*. Nepal Water Conservation Foundation and the Institute for Social and Environmental Transition, Kathmandu, Nepal.

Narasimhan, T.N. (1990) Groundwater in the peninsular Indian shield: a framework for rational assessment. *Journal of the Geological Society of India* 36, 353–363.

Narasimhan, T.N. and Kretsinger, V. (2003) Developing, managing and sustaining California's groundwater resources. *A White Paper of the Groundwater Resources Association of California*. Groundwater Resources Association of California, California.

Ostrom, E. (1990) *Governing the Commons: The Evolution of Institutions for Collective Action*. Cambridge University Press, Cambridge.

Ostrom, E. (1993) Design principles in long-enduring irrigation institutions. *Water Resources Research* 29(7), 1907–1912.

Schlager, E. and Blomquist, W. (2000) *Local Communities, Policy Prescriptions, and Watershed Management in Arizona, California, and Colorado*. "Constituting the Commons: Crafting Sustainable Commons in the New Millenium", the Eighth Conference of the International Association for the Study of Common Property, Bloomington, Indiana.

Shah, T., *et al.* (2000) *The Global Groundwater Situation: Overview of Opportunities and Challenges*. International Water Management Institute, Colombo, Sri Lanka.

Shah, T., *et al.* (2003a) Sustaining Asia's groundwater boom: an overview of issues and evidence. *Natural Resources Forum* 27, 130–141.

Shah, T., *et al.* (2003b) *Energy-Irrigation Nexus in South Asia: Improving Groundwater Conservation and Power Sector Viability*. International Water Management Institute, Colombo, Sri Lanka.

United Nations Development Program (1976) Ground-water surveys in Rajasthan and Gujarat, United Nations Development Program.

Wisner, B., *et al.* (2004) *At Risk: Natural Hazards, People's Vulnerability and Disasters*. Routledge, London.

World Bank (1999) *Meeting India's Future Power Needs: Planning for Environmentally Sustainable Development*. The World Bank, Washington, DC.

World Bank and Ministry of Water Resources – Government of India (1998) *India – Water Resources Management Sector Review*, Groundwater Regulation and Management Report. The World Bank, Washington, DC/Government of India, New Delhi, India.

III Case Studies and Innovative Experience

10 The Groundwater Recharge Movement in India

RAMASWAMY SAKTHIVADIVEL

33 First East Street, Kamarajnagar, Thiruvanmiyur, Chennai 600 041, Tamil Nadu, India

Introduction

The easy accessibility of groundwater by even small-scale users, its local availability and the difficulty of coordinating and governing many users of the same aquifers across wide geographic spaces has frequently led to indiscriminate extraction of this precious natural resource for domestic, industrial and agricultural uses around the world. Groundwater exploitation, particularly in India, has increased by leaps and bounds over the last 50 years along with the expansion of shallow, mostly private, wells. The growth of groundwater abstraction structures from 1950 to 1990 clearly depicts the increasing use of groundwater utilization across sectors. As per available published statistics, the number of dug wells increased from 3.86 million (1951) to 9.49 million (1990) and of shallow tube wells from 3000 (1951) to 4.75 million (1990) (Muralidharan and Athavale, 1998). Shah (Chapter 2, this volume) highlights that these trends continue to the present.

The reasons for the increase in groundwater use in India are varied and include technological, hydrologic and policy factors. Technologically, developments over the last 50 years in the construction of deep tube wells, water abstraction devices and pumping methods have made large-scale exploitation of groundwater both possible and economic. At the same time, changes in hydrologic regimes, in particular the growing scarcity of surface water supplies as agricultural and other users have expanded, have pushed water users to seek groundwater alternatives. Finally, government policy has tended to support groundwater use. Easy availability of credit from financial institutions for sinking tube wells coupled with the provision of subsidized or free electricity (see Shah *et al.*, Chapter 11, this volume) for pumping in many states has encouraged increased extraction.

As demand for groundwater has gone up, rapid urbanization and land use changes have decreased drastically the already low infiltration rates of rainfall

 The Agricultural Groundwater Revolution: Opportunities and Threats to Development (M. Giordano and K.G. Villholth)

into the soil and have diminished the natural recharging of aquifers. Natural recharge measurements carried out in about 20 river basins across India suggest that about 15–20% of seasonal rainfall contributes to groundwater recharge in the Indo-Gangetic plains, figures that fall to only 5–10% in the peninsular hard-rock regions (Athavale *et al.*, 1992).

Increased use and limited recharge have contributed to the lowering of the water table so much that yields of many dug and tube wells have decreased substantially or even fallen to zero, particularly during the summer. The drinking water crisis that ensues in many villages in summer imposes serious health hazards to the rural masses and is responsible for the huge loss of livestock populations for want of drinking water and fodder (Shah, 1998). The general implications for agriculture are no less severe.

To respond to the growing groundwater crisis and take advantage of the high levels of runoff not captured by natural recharge, augmentation of groundwater resources through artificial recharge of aquifers has become widespread in India over the last 3–4 decades. In fact, the growth in the use of artificial recharge has expanded to such an extent that it can be called a 'groundwater recharge movement', which has behind it both secular and spiritual proponents. This recent movement builds on artificial groundwater recharge concepts that have been practised from time immemorial in the hard-rock, semi-arid regions of south- and north-western India.

In some senses, the artificial recharge movement in India can be considered as a successful example of community-based efforts to manage common property resources. However, because of the distributed nature of aquifers and their interconnectivity across space and with surface water supplies, recharge by one group or community may impact water availability for other neighbouring or downstream groups. Thus, the artificial recharge movement in India highlights both the benefits and problems of community-based approaches highlighted by Schlager (Chapter 7, this volume).

This chapter looks at the historical evolution of the groundwater recharge movement in India, how it has gathered momentum, who has been responsible for it and what it has achieved to date. Through two contrasting case studies, it then highlights both the clear local benefits of artificial recharge and the potential for negative impacts at larger scales. Together these studies show the potential gains and the governance problems that will necessarily follow the artificial recharge movement as it continues to move forward in India. The studies also provide some guidance in how artificial recharge can, and cannot, be used to solve groundwater problems.

The Artificial Recharge Movement in India

Artificial recharge, one of the oldest activities undertaken in India to conserve rainwater both above-ground and underground, is as old as the irrigated agriculture in the arid and semi-arid regions. In the olden days, the recharge movement initiated by the local communities was aided and supported by kings, chieftains, philanthropists and by those who valued water and practised conservation. There are numerous examples and stone inscriptions from as early as AD 600, citing that ancient kings and other benevolent persons considered as one of their bound duties the construction of *ooranies* (ponds) to collect rainwater and use it to recharge wells constructed within or outside *ooranies* to serve as drinking water source. Even today, thousands of such structures exist and are in use for multiple purposes in the southern coastal towns and villages of Tamil Nadu where groundwater is saline (DHAN Fondation, 2002).

Similarly, more than 500,000 tanks and ponds, big and small, are dotted all over the country, particularly in peninsular India. These tanks were constructed thousands of years ago for catering to the multiple uses of irrigated agriculture, livestock and human uses such as drinking, bathing and washing. The command area of these tanks has numerous shallow dug wells that are recharged with tank water and accessed to augment surface supplies. Many drinking water wells located within the tank bed and/or on the tank bund are artificially recharged from the tank into these wells to provide clean water supply throughout the year with natural filtering (DHAN Foundation, 2002).

In traditionally managed tank irrigation systems, when gravity-supplied water from the tank is insufficient for crop production, it is not uncommon that the village community decides to close all the tank sluices and allow the tank to act as a percolation unit to recharge the wells in the command area; the recharged water is then shared by the beneficiary farmers. This has been done to distribute the limited water to the crops without any line losses due to gravity flow. This practice is in use even today in many traditionally managed irrigation systems. However, with water supply to many tanks dwindling, converting irrigation tanks to purely percolation tanks for artificial recharge of wells in the command is increasing day by day. The trend has essentially become a movement by itself and even some state governments such as Karnataka are encouraging the practice through enactment of law enforcement (Sakthivadivel and Gomathinayagam, 2004).

Rooftop rainwater harvesting and the storage of harvested water in underground tanks is also a very common phenomenon in many Indian states experiencing acute shortage of drinking water supplies. Similarly, pumping induced recharge water from wells located near water storage structures like tanks, irrigation canals and river courses, and transporting it to a long distance through pipelines for irrigation is a common sight in many water-deficient basins. These activities can also be considered a social movement that originated spontaneously from local necessity. Further details on traditional water harvesting and recharge structures can be found in *Dying Wisdom* (Agrawal and Narain, 2001).

Progression of the Artificial Recharge Movement

The spread of the artificial recharge movement in India (ARMI) can be broadly classified under three phases: the first relates to the period before the Green Revolution when limited exploitation of groundwater was taking place, i.e. before 1960; the second is the period between 1960 and 1990 in which intense groundwater exploitation took place, leading to signs of overexploitation;

and the third is the period from 1990 to date when water scarcity became increasingly severe, and groundwater level decline became alarming in many pockets of the country.

In the first phase, which extended from early historic times until approximately 1960, traditional water-harvesting methods were given impetus through unorganized yet spontaneous movement by the local communities aided by kings and benevolent persons to meet the local requirement at times of crisis. During this period, there was very little knowledge-based input from the government or other organizations to provide assistance for understanding and systematically putting into practice artificial recharging. Instead, local communities used their intimate knowledge of terrain, topography and hydrogeology of their areas to construct and operate successful artificial recharge structures, some of which have managed to survive even today. In this phase, there was little application of science related to artificial recharging; most work was based on local knowledge and perceived wisdom. Very little understanding existed about the consequences of, and the knowledge required for, artificial recharging of underground aquifers.

The second phase, from 1960 to 1990, coincides with the period of large-scale extraction of groundwater that resulted in many aquifer systems showing signs of overexploitation, especially in arid and semi-arid regions. During this phase, curriculum relating to hydrogeology and groundwater engineering was introduced in many universities in India and the science of groundwater hydrology was better understood. Both the public and government had started realizing the importance of recharging aquifers to arrest groundwater decline and maintain groundwater levels. As a consequence, pilot studies of artificial recharge of aquifers were carried out by a number of agencies including central and state groundwater boards, water supply and drainage boards, research institutes such as National Geophysical Research Institute (NGRI), Physical Research Laboratory (PRL), National Environmental Engineering Research Institute (NEERI), agricultural and other academic institutions, and non-governmental organizations (NGOs) such as the Centre for Science and Environment (CSE).

During this period, various pilot studies were carried out and technical feasibility of artificial recharging and recovery of recharged water were established. Two important events with respect to artificial recharge also took place that are of relevance to the movement today. One is the synthesis of research and development works (Mission WatSan, 1997) carried out in India in artificial recharging by a team of experts under the Rajiv Gandhi National Drinking Water Mission (RGNDWM), constituted by the Ministry of Rural Areas and Development, Government of India, New Delhi. The second is the effort provided by the Bureau of Indian Standards (BIS) to bring out technical guidelines and specifications for artificial recharging. These have given impetus for further experimentation on artificial recharging.

In the third current phase, from 1990 to the present, water scarcity, continuous droughts in certain pockets of India and continuously declining groundwater levels in many parts of India have led both the public and government to become more aware of artificial recharge and to take it up on a war footing. Four major events that have taken place during this period are especially significant to the movement. One is the spontaneous uprising and cooperation from the public supported by religious leaders, philanthropists and committed individuals to take up artificial recharging through dug and bore wells, check dams and percolation ponds and, later – with the government joining hands with the local community – in implementing such schemes on a mass scale (Shah, 1998). The second is the action taken by state governments such as Tamil Nadu in promulgating the groundwater regulation acts pertaining to metropolitan areas and ordering the communities to implement rainwater-harvesting schemes and artificial recharging on a compulsory basis. The third event relates to awareness created among the public by NGOs such as the CSE and Tarun Bharat Sangh (TBS), and media exposure to the importance of artificial recharging.

The fourth event is the recently increasing trend of large-scale abstraction of induced recharge witnessed in many gravity irrigation systems in states like Tamil Nadu. With increasing water scarcity in many irrigation systems and availability of large-scale pumping machinery at affordable prices and subsidized power, many enterprising farmers have turned to wells near storage reservoirs, and on the sides of canals and riverine courses to create induced recharge in their own wells. The induced recharge water is transported through pipelines sometimes many kilometres away from the pumping site to irrigate non-command areas of orchards and other high-value crops, often using drip and sprinkler irrigation. This practice of pumping-induced recharge water outside the command area has had a very negative effect in managing large irrigation systems due to the siphoning of a considerable quantity of water to areas not originally included in the command. This is more so in years of inadequate water supplies to reservoirs as well as in drought years. This is a spontaneous movement, which is spreading like wildfire; if it is not controlled and regulated, many surface irrigation systems will see their death in the very near future (Neelakantan, 2003).

Revered Shri Panduranga Shastri Athavale of the 'Swadhyay Parivar' has introduced a movement in Gujarat called 'Nirmal Neer' (clean water) with an aim to provide drinking water and support irrigation through effective rainwater harvesting. Under his inspiration, schemes such as recharging of wells and tube wells, diverting rainwater into the existing ponds as well as construction and maintenance of check dams and ponds have been taken up by the villagers. During 1995, in Saurashtra region alone, people have adopted the recharging of wells scheme in 98,000 wells (Parthasarathi and Patel, 1997). The massive adoption of the scheme explicitly indicates the awareness of conservation and better utilization of rainwater.

Another interesting and innovative initiative by Rajendra singh of TBS has revolutionized the mass movement of the people of Alwar district in the semi-arid Rajasthan state and built bridges of cooperation and solidarity among them. A group of young individuals from TBS took it upon their shoulders, with people's participation and contribution, to rejuvenate defunct *johads* and construct new ones in the Aravari catchment at the foothills of the Shivalik hill ranges of Alwar district. *Johads* are ancient water-harvesting structures, constructed by the people, to store rainwater for multiple uses and to recharge groundwater.

Many *johads* have come up on the tributary streams of the Aravari catchment in the last decade, raising water level in the wells and facilitating irrigation on the cultivated lands. The dead, dry watercourses of the Aravari, which had flowing water only during rainy days in the monsoon months, came alive for the full year. Today, there are more than 200 *johads* in the catchment of Aravari. The successful water harvesting and recharging of groundwater in the upstream of the river followed by scores of *johads* along the main river had transformed the once ephemeral stream into a perennial river. These and other similar movements that are instrumental in achieving productive benefits locally have given rise to many such initiatives in other parts of India.

Artificial Recharging Methods

Definition of artificial groundwater recharge

Artificial recharge is the planned, human activity of augmenting the amount of groundwater available through works designed to increase the natural replenishment or percolation of surface waters into the groundwater aquifers, resulting in a corresponding increase in the amount of groundwater available for abstraction (http://www.unep.or.jp/ietc/publications/techpublications/techpub-8e/artificial.asp). Although artificial recharging is primarily used to preserve or enhance groundwater resources, it has also been used for many other beneficial purposes such as conservation of surface runoff and disposal of flood waters, control of salt water intrusion, storage of water to reduce pumping and piping costs, temporary regulation of groundwater abstraction, and water quality improvement through filtration of suspended solids through soils and other materials or via dilution with naturally occurring groundwater (Asano, 1985). Other areas in which artificial recharge has been used are in wastewater disposal, waste treatment, secondary oil recovery, prevention of land subsidence, storage of fresh water with saline aquifers, crop development and stream flow augmentation (Oaksford, 1985).

The various techniques used for the artificial recharge mentioned earlier have been used successfully throughout India with the notable exceptions of a few areas including Saurashtra and the Karnataka coastal zone. In those areas, the extreme porosity of the aquifer and its connection to the sea means that less water is available for harvest than is injected. In general, artificial recharge works because it is effective in minimizing water loss due to evaporation compared with similar surface storage systems. Many environmental problems arising out of surface storage are also avoided using the method. For example, there is generally no loss of agricultural or other lands by inundation as would occur behind a surface storage structure. In cases where channels are used for groundwater recharge, 'multiple use' benefits have also been achieved.

Classification of recharging methods

To artificially recharge groundwater, different methods have been developed and applied in various parts of the world. Details of these methods, as well as related topics, can be found in the literature (e.g. Todd, 1980; Huisman and Olsthoorn, 1983; Asano, 1985; CGWB, 1994). In summary, artificial recharging may be carried out by direct or indirect methods, or by a combination of methods in an integrated water resources management context.

Direct recharge

Direct surface techniques are among the simplest and most widely used methods for groundwater recharge. Using these techniques, water is moved from the land surface to the aquifer by means of simple infiltration. The infiltrated water percolates through the vadose (unsaturated) zone to reach the groundwater table. Through this process, the recharged water is filtrated and oxidized. Direct recharge methods can be grouped into three categories: (i) when the aquifer is shallow, water may be spread over fields or conveyed to basins and ditches from which it percolates; (ii) when an aquifer is situated at greater depths, recharge can be facilitated by flooding pits and dug shafts; and (iii) in cases of high overburden thickness or confining aquifer conditions, recharge can be effected by injecting surface water directly into the aquifer using boreholes or tube wells.

Water spreading is practised on an increasing scale all over the world and more so in India (Muralidharan and Athavale, 1998). Recharge through pits and shafts has limited applications as recharge capacity is low. However, abandoned stone quarries or open (dug) wells located where the water table has dropped below the excavated depths can be used as ready-made recharge pits and shafts. Recharge through wells can be applied to all hydrogeological situations; however, it might require higher capital and technological requirements. Only surface spreading methods and recharge through dug wells are focused on in this chapter as they are widely used in India.

In some cases where aquifers are under stress, irrigation tanks originally built for surface supplies are being converted into percolation tanks by closing the outlet sluices and allowing the stored tank water to recharge the aquifers. In some groundwater-only areas, surface storage structures (percolation ponds) are being constructed purely for groundwater recharge. In both cases, the percolation tanks (or ponds) are water-harvesting structures constructed across or near streams to impound rainwater and to retain it for a longer time to increase the opportunity time for infiltration. The water storage is expected to induce percolation and replenish the aquifer, which is then exploited through wells located down the gradient. Check dams, generally constructed for soil conservation, can be considered mini-or micro-percolation tanks from which water is not directly drawn for irrigation but is allowed to percolate into subsurface strata, thus augmenting the groundwater.

Most of the evaluation studies on percolation tanks at present are of qualitative nature with limited objective. These are based on the hydrogeological response of the aquifer system or the increase in crop yield. The evaluation studies in southern peninsular India indicate that the recharge efficiency varies between 30% and 60% depending upon the prevailing hydrogeological situation (Muralidharan and Athavale, 1998). The role of the percolation pond in recharging the groundwater has no doubt been realized and appreciated by the farming community in recent years, explaining the growth in application of the technique. However, an improved understanding of the performance of such systems would lead to better siting and effectiveness. Studies on water balances and the interaction between surface and subsurface reservoirs, a critical issue highlighted later, would also help greatly in understanding the value of the practice.

Indirect recharge

Indirect methods of artificial recharge include the installation of groundwater pumping facilities or infiltration galleries near hydraulically connected surface water bodies (e.g. streams or lakes) to lower groundwater levels and induce infiltration from surface water bodies. The effectiveness of induced recharge methods depends upon the number and proximity of surface water bodies, hydraulic conductivity (or transmissivity) of the aquifer, area and permeability of the streambed or lake bottom, and hydraulic gradient created by pumping. Indirect methods generally provide less control over the quantity and quality of the water than do the direct methods. When indirect methods of recharge and retrieval are practised, the water recovered consists of a small fraction of groundwater while a larger fraction of abstracted water comes from river or lake.

There are a number of other indirect recharge methods. For example, in flood irrigation, excess water percolates to the groundwater table. Similarly, seepage from lake beds, irrigation tanks, streams and canals recharges groundwater as does the use of terracing and contour bunds. While recharge from these processes can in some senses be considered accidental, each method's recharge potential can be purposefully enhanced and a combination of techniques, both direct and indirect, can be used to meet specific terrain and topography conditions and recharge needs.

Recharge through integrated water resource development

Groundwater recharge is often best accomplished as a by-product of an integrated water resources development scheme, for example, by increasing groundwater recharge by way of reservoir and canal seepage, injection and infiltration of return flow from irrigation, enhanced infiltration of rainfall as a result of levelling fields for irrigation purposes, and basin development schemes involving the construction of check dams and minor irrigation dams. The Central Groundwater Board (CGWB, 1995) states that nearly 30–40% of applied irrigation water goes as seepage from irrigation fields, a portion of which recharges groundwater. Rates for paddy are much higher than average, ranging from 55% to 88% of the applied irrigation water (Karanth and Prasad, 1979).

An experiment by the Uttar Pradesh government to develop a new and practical way to conserve and rejuvenate falling groundwater reserves through use of flood water highlights the potential for integrated artificial recharge methods. The Madhya Ganga Canal Project (MGCP) located in the lower Ganga canal commands was initiated in 1988. In 2000, the International Water Management Institute (IWMI) carried out a study (Chawla, 2000) on the Lakhaoti branch canal of the MGCP to assess the impact of diversion of surplus Ganga water, during the *kharif* season, on groundwater levels and cropping patterns. The Lakhaoti branch is spread over more than 200,000 ha and covers the districts of

Ghaziabad, Bulandsher and Aligarh in western Uttar Pradesh. It is bounded by the drainage canals of the Kali and Nim rivers.

According to the study, the canal project has helped to raise the groundwater table from 6.6 m to 12.0 m, and brought down the cost of pumping for irrigation from Rs 4500/ha to Rs 2700/ha. Previously, farmers pumped water to irrigate their crops even during the monsoons. This monsoon period pumping lowered the groundwater levels causing severe water shortages during the dry season.

Following the introduction of the MGCP, seepage from the canals and flooded paddy fields helped recharge underlying aquifers. The irrigated area in the project region increased from 1251 ha in 1988/89 to 35,798 ha in 1999/2000, and the area under paddy irrigation was increased to 14,419 ha from 83 ha. The total annual cost of pumping for paddy cultivation due to canal seepage has declined by about Rs 100/ha, resulting in a saving of Rs 180 million for the project as a whole.

Recharge for domestic supplies

The previous section illustrated how irrigation and irrigation-related storage structures can effectively be used to indirectly recharge groundwater aquifers. This section illustrates the role of groundwater recharging for meeting domestic supplies. Artificial recharging of groundwater for domestic supplies assumes significance both in the urban and rural areas of India. Projected water supply requirements for domestic and drinking water needs in 2005 were about 41,000 million cubic metres (Roy, 1993) of which 24,000 million cubic metres was for urban areas and the remaining 17,000 million cubic metres for rural areas. Currently most of the rural and part of the urban drinking water requirements are met from groundwater. In many parts of India, rural drinking water supply programmes often witness shortage of supply from bore wells because of increase in groundwater use for irrigation from boreholes in and around the drinking water bores. The example from different villages of Thumbadi watershed in Karnataka state indicates how lack of effective zoning and regulation of irrigation wells within a 250 m radius of drinking water wells allows irrigation water supply wells to come into existence, hindering the performance of public water supply wells.

The natural recharge studies carried out over different hard-rock terrain indicated that only 5–10% of the seasonal rainfall recharges the groundwater (Athavale *et al.*, 1992). The meagre annual replenishment of natural recharge to the groundwater alone with multiple uses may not be able to meet the projected demand of 17,000 million cubic metres per year by 2050. It is therefore necessary in future to have an independent groundwater source for rural drinking water supply and to protect it as a sanctuary (Muralidharan and Athavale, 1998).

Enhancement of recharge to the groundwater has therefore become mandatory in areas where groundwater is the sole source of drinking water supply. The NGRI (National Geophysical Research Institute) recommends that the methodology of artificial recharge and retrieval (ARR) developed by them can profitably be used for recharging a well during monsoon and using it for drinking water during summer months. Two or three such wells may be declared as sanctuary wells for each village and the ARR scheme may be implemented.

Alternatively, the concept of captive management practice of storing part of the runoff volume in the catchment area through a mini-percolation tank and developing a source well on the downstream side to provide adequate drinking water can be thought of (Muralidharan, 1997). Since the catchment area generally has a shallow basement and low transmissivity, the source well proposed is a large-diameter dug well. Sustainability of the well supply is achieved by constructing a subsurface barrier further downstream on the side of the source well.

Some of the requirement of urban centres in alluvial belts (river alluvium and coastal alluvium) can be met from the groundwater sources. ARR can be implemented on a macro-scale in which millions of litres of good-quality water is transferred to the aquifer every day during monsoon months through a battery of wells, and the same wells can be pumped in summer for feeding water for urban supply of potable water. The NGRI has tentatively identified five areas (Chennai, Kolkata, Mahesana and Chorwond in Gujarat and Jalgaon in Maharashtra) in the country, which have favourable hydrogeological situations for implementing such macro-ARR schemes.

Urban rainwater harvesting and groundwater recharge is catching up in many cities such as Delhi, Chennai and Ahmedabad that are facing acute water supplies, especially in summer and in deficient rainfall years. Many state governments have taken up rooftop rainwater harvesting in a big way. Although a model Groundwater Regulation Bill was circulated among the states by the Government of India as early as 1987, none of the states has adopted either this bill or its modified version to date. Only, the government of Tamil Nadu has enacted a groundwater regulation act pertaining to Chennai metropolitan area to overcome the grave situation it had faced due to severe drinking water crisis.

The Chennai Metropolitan Area Groundwater (Regulation) Act, 27 of 1987, which came into force with effect from February 1988, envisages (i) registration of existing wells; (ii) regulation of sinking new wells; (iii) issue of licence to extract groundwater for non-domestic purposes by the Revenue officials on payment of prescribed fees after getting technical clearance from Chennai Metropolitan Water Supply and Sewerage Board (CMWSSB).

It is due to the implementation of the groundwater regulation act, and other artificial recharge measures adopted, that the water table near the northern part of the city which had an average depth of 8 m before 1988 has risen to an average depth of 4 m below ground level in 2001–2002. As a result of this increase, Metro Water Board has been able to increase the withdrawal from 55 million to 100 million litres per day of water from these well fields during 2003–2004, a year of drought and water crisis in Chennai.

The potential for rooftop rainwater harvesting, based on estimates of rainfall in different parts of the country and the available area of rooftops, has been estimated by the Water Management Forum (WMF, 2003) to be roughly 1 km^3/year. Although this quantity may look small from the overall requirement of the country, this water is critical for drinking water requirements at times of crisis in drought-prone areas.

Costs of Artificial Recharge

For wider adoption of artificial recharging and use of a particular method, the cost of recharge and recovery of various artificial recharge methods is an important parameter that needs to be determined. Full-scale artificial recharge operations in India are limited and, as a consequence, cost information from such operations is incomplete.

The cost of recharge schemes, in general, depends upon the degree of treatment of the source water, the distance over which the source water must be transported and stability of recharge structures and resistance to siltation and/or clogging. In general, the costs of construction and of operation of the recharge structures, except in the case of injection wells in alluvial areas, are reasonable; the comparative costs of recharged water per 1000 m^3 in such cases works out to \$1–3. On the other hand, the cost of using recharged groundwater for domestic water supply purposes, varying from \$0.05 to \$0.15/person/year, is very reasonable, especially in areas where there is shortage of water (CGWB, 1984). The initial investment and operating costs are many times less than those required for supplying potable water using tankers; combining technologies can also result in cost savings. For example, in Maharashtra, the capital cost of combining connector well and tank into a hybrid scheme was about \$900 (the cost of a borehole) compared to the cost of a comparable percolation tank system needed to achieve a similar degree of recharge (estimated to be about \$120,000). Table 10.1 summarizes the estimated costs of various artificial recharge methods.

Contrasting Local and Basin Perspectives On Artifical Recharge

The existence of more than 250,000 tanks and ponds in hard-rock-covered areas of peninsular India itself shows the importance accorded by agriculturalists and rulers for managing the surface water sources locally. However, most of the tanks are old and their storage capacity has reduced due to siltation, and recharge volume of

Table 10.1. Economics of various artificial recharge methods. (From UNEP International Environment Centre, 2004.)

Artificial recharge structure type	Capital cost/1000 m^3 of recharge structure (\$)	Operational cost/ 1000 m^3/year (\$)
Injection well (alluvial area)	551	21
Injection well (hard-rock area)	2	5
Spreading channel (alluvial area)	8	20
Recharge pit (alluvial area)	515	2
Recharge pond or percolation pond (alluvial area)	1	1
Percolation tank (hard-rock area)	5	1
Check dam	1	1

water through the tanks has been considerably reduced. At the same time, the tank command areas have increasingly been put to multiseason cropping use with higher cropping intensities than they were originally designed to meet. As a result, farmers have turned in increasing numbers to the utilization of groundwater through dug and bore wells. The increase in the extraction of the limited renewable groundwater resources has led to a decline in water tables, especially in areas where density of wells is high and rainfall is moderate to low. This in turn has provided the impetus for the groundwater recharge movement.

As briefly discussed, there have been many studies on artificial groundwater recharge that have shown its technical effectiveness. In Maharashtra, it was shown that when tank bottoms were maintained by removing accumulated sediment and debris prior to the annual monsoon, the average recharge volume was 50% of the capacity of the tank (Muralidharan and Athavale, 1998). In Tamil Nadu and Kerala, studies carried out by CGWB on nine percolation tanks in the semi-arid regions of the Noyyal, Ponani and Vattamalai river basins showed that percolation rates were as high as 163 mm/day at the beginning of the rainy season, but diminished thereafter mainly due to the accumulation of silt at the bottom of the tanks (Raju, 1998). In Punjab, studies of artificial recharge using injection wells were carried out in the Ghagger River basin, where using canal water as the primary surface water source showed that the recharge rate from pressure injection was ten times that of gravity systems and that maintenance was required to preserve efficiency (Muralidharan and Athavale, 1998). In Gujarat, studies of artificial recharge were carried out that showed a recharge rate of 260 m^3/day with an infiltration rate of 17 cm/h (Phadtare *et al.*, 1982).

Local level benefits of groundwater recharge in Gujarat

Why artificial recharge is growing in popularity can be seen from an example from India's arid western region. The year 2000 was an unprecedented drought year in Gujarat. The water crisis that year had created an intense awakening among the people of the Saurashtra and Kutch regions about the importance of water. Social workers and NGOs undertook numerous water-harvesting projects to recharge groundwater for domestic and agricultural uses. These projects were often funded by voluntary contributions from affected people. Because of the apparent success of these efforts, under the Sardar Patel Participatory Water Conservation Programme (SPPWCP), the government of Gujarat invested more than Rs 1180 ($28) million in construction of more than 10,000 check dams across Saurashtra, Kutch, Ahmedabad and Sabarkantha regions in 2000/01, which was co-financed by beneficiary contributions. Overall, 60% of the funds was supplied by the government and 40% by direct stakeholders. The responsibility for managing the quality of construction works fell to beneficiary groups and NGOs.

An independent evaluation of the check dams in Gujarat was carried out in 2002 by the Indian Institute of Management (IIM), Ahmedabad, which covered vital aspects of the project including advantages of people's participation and impacts on agricultural production, drinking water supply and availability of fodder as well as overall socio-economic cost–benefit analysis (Shingi and Asopa, 2002).

From the analysis of survey data covering over 100 check dams, personal visits by the evaluation team to a large number of other check dams and interviews with more than 500 farmers, the team concluded that:

1. Localized rainwater harvesting systems in the form of check dams in Saurashtra were an effective solution to the water crisis through their ability to channel rainfall runoff into the underground aquifer. This offered a decentralized system for decreasing the impact of drought and allowed the people's involvement in critical water management tasks with simple, local skill-based, cost-effective and environment-friendly technologies.
2. The rainwater harvesting efforts initiated with people's participation and support from SPPWCP should be relaunched and reimplemented on a larger scale.
3. The 60:40 scheme (60% by government and 40% by beneficiaries) had six major features capable of attracting donor investment: (i) ecologically sound principles behind the concept; (ii) highly participatory nature of the programme, which allowed beneficiaries to contribute their share of the investment through labour, equipment and/or money; (iii) gendered nature of the outcome in that women were the major beneficiaries of the alleviation of drinking water and livestock feed problems; (iv) the fact that the project did not replace or endanger human or wildlife habitat; (v) focus on equitably using renewable resource like rainwater; and (vi) economic and financially sound nature of the work and its short payback period.
4. The 60:40 scheme has been, and should continue to remain, a people's programme, and it is unlikely to survive otherwise. It is felt that only the people's involvement would ensure the survival of critical components like (i) quality of works; (ii) prevention of undesirable contractor's entry into partnership with government; (iii) sustainable maintenance and supervision; (iv) speed of implementation; (v) ingenuity and innovation in implementation; and (vi) cost-efficient technical guidance.

Basin-level costs of groundwater recharge in Gujarat

Some believe that local efforts to increase artificial recharge account for only a small fraction of the massive amount of rainfall on the vast area of any particular catchment or basin. As a result of this thinking, artificial recharge by scattered local communities will not have a perceptible impact on downstream flows or impact downstream surface or groundwater users. However, from a basin perspective, all water use is likely to have some impact on users elsewhere in the system. These impacts are likely to be greatest when basins are 'closed' (i.e. all available supplies have been fully allocated) and in cases with marked inter- and intra-annual variation in rainfall. These are precisely the places and conditions under which groundwater recharge is likely to have the largest local appeal. The potential problems behind this issue are brought out by the following example, which is also from Gujarat.

The watershed known as Aji1 in the Saurashtra region of Gujarat is considered water-scarce and closed, and has high variation in rainfall ranging from

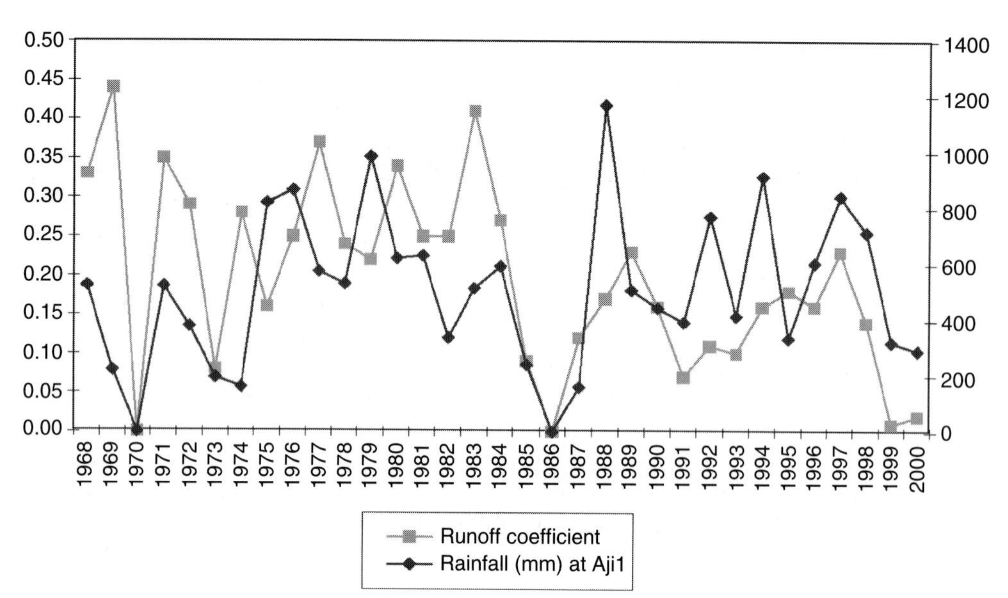

Fig. 10.1. Rainfall and runoff variations in Aji1 watershed from 1968 to 2000.

200 to 1100 mm/year. Aji1 reservoir supplies water to the city of Rajkot, located at the downstream end of the watershed. Starting around 1985, the flow to the reservoir began to decline sharply. It was hypothesized that the decline was caused by the construction of thousands of check dams and percolation ponds within the watershed, the result of a recharge movement initiated by Shri Panduranga Athavale, a religious Guru of Saurashtra, and later supported by the government of Gujarat.

In order to verify whether the decrease in flow into the downstream reservoir was related to the proliferation of upstream check dams and percolation ponds constructed to recharge the groundwater aquifer, rainfall and inflow data to the reservoir was collected for the years 1968–2000 and a simple analysis was made to compute the runoff coefficient as shown in Fig. 10.1. Although rainfall remained approximately the same throughout the period, the runoff coefficient declined markedly, especially after 1985 when the recharge movement reached its full impact. The average reduction in the runoff coefficient after 1985, almost to half of its original value, suggests the extent of the impact the upstream water-harvesting structures had on the downstream reservoir. While the impact of upstream artificial recharge on downstream users in other basins may differ from this example, what is clear is that there will be an impact (Molden and Sakthivadivel, 1999).

Conclusion

The number of groundwater wells in India has increased from less than 100,000 in 1960 to nearly 12 million in 2006. With clear signs of aquifer depletion and continued erratic rainfall, local communities as well as governments are turning to local water-harvesting and recharge structures on a massive scale. The primary objectives of this groundwater recharge movement are to increase groundwater availability for improved security of domestic supplies and to drought-proof and protect rural livelihood. This situation calls for conjunctive water management in a basin context with recharge of groundwater assuming a pivotal role with a caveat that upstream and downstream impact needs to be considered and accounted for.

As described in this chapter, groundwater recharge has a long history in India and there are a variety of direct, indirect and integrated methods at the disposal of farmers, community leaders, NGOs and the government to further expand the movement. The technical issues that these various methods must consider include recovery efficiency, cost-effectiveness, contamination risks due to injection of poor-quality recharge water, and clogging of aquifers. Numerous artificial recharge experiments have been carried out in India and have established the technical feasibility of various approaches and combinations of approaches in unconfined, semi-confined and confined aquifer systems as well as the economic viability.

What is less well understood and appreciated is the potential impacts of numerous local recharge efforts on basin-scale water availability and distribution. The popularity of groundwater recharge is a function of its local success. This, and the critical role of local involvement, highlights the advantages of community approaches to groundwater management described by Schlager (Chapter 7, this volume). However, as shown by the contrast in the two case studies from Gujarat, 'successful' local efforts at recharge can cause problems further downstream. The possible impacts of local action on regional outcomes highlights the key challenge of community-based groundwater governance also described by Schlager – the potential conflict as one moves from local to basin scales.

The reality is that the groundwater recharge movement in India, initiated by local elites and later aided by government and NGOs, has become a people's movement and is likely to stay long into the future. In order to maximize its possible benefits and minimize costs, it has to be nurtured and carried forward with systematic research and development programmes covering its physical, economic, environmental and – what is most lacking now – institutional aspects that can resolve problems across scales.

References

Agrawal, A. and Narain, S. (2001) *Dying Wisdom*. Centre for Science and Environment, New Delhi, India.

Asano, T. (1985) *Artificial Recharge of Groundwater*. Butterworth Publishers, Boston, Masachussetts.

Athavale, R.N., Rangarajan, R. and Muralidharan, D. (1992) Measurement of natural recharge in India. *Journal of Geological Society of India* 39(3), 235–244.

Central Ground Water Board (CGWB) (1994) *Manual on Artificial Recharge of Groundwater*,

Technical Series: Monograph No. 3. Ministry of Water Resources, Government of India.
Central Ground Water Board (CGWB) (1995) Report on Groundwater Resources of India. Government of India, New Delhi, India.
Chawla, A.S. (2000) *Ground Water Recharge Studies in Madhya Ganga Canal Project*. Consultancy Report to IWMI, Colombo, Sri Lanka.
DHAN Foundation. (2002) Revisiting Tanks in India. National Seminar on Conservation and Development of Tanks, New Delhi, India.
Huisman, L. and Olsthoorn, T.N. (1983) *Artificial Groundwater Recharge*. Pitman Publishing, Massachusetts.
Karanth, K.R. and Prasad, P.S. (1979) Some studies on hydrologic parameters of groundwater recharge in Andhra Pradesh. *Journal of Geological Society of India* 20, 404–414.
Mission WatSan (1997) Rajiv Gandhi National Drinking Water Mission. Quarterly Bulletin, No. 1.
Molden, D. and Sakthivadivel, R. (1999) Water accounting to assess use and productivity of water. *International Journal of Water Resources Development* Special Double Issue 15(1 & 2), 55–71.
Muralidharan, D. (1997) Captive sustainable drinking water sources well and wasteland development in hard rock areas. Proceedings of National Conference on Emerging Trends in the Development of Sustainable Groundwater Sources, Jawaharlal National Technical University, Hyderabad.
Muralidharan, D. and Athavale, R.N. (1998) Base paper on artificial recharge in India. National Geophysical Research Institute, CSRI, Hyderabad, India.
Neelakantan, S. (2003) *A Gossipmonger's Revisit to Chettipalayam*. Working Paper No. 142. Madras Institute of Development Studies, Chennai, India.
Oaksford, E.T. (1985) Artificial recharge: methods, hydraulics, and monitoring. In: Asano, T. (ed.) *Artificial Recharge of Ground water*. Butterworth Publishers, Boston, Massachusetts, pp. 69–127.
Parthasarathi, G.S. and Patel, A.S. (1997) *Groundwater Recharge Through People's Participation in Jamnagar Region*. Indian Water Works Association, pp. 51–56.
Phadtare, P.N., Tiwari, S.C., Bagade, S.P., Banerjee, A.K., Srivastava, N.K. and Manocha, O.P. (1982) Interim Report on Concept, Methodology and Status of work on Pilot Project for Artificial Recharge, CGWB, Ahmedabad, India.
Raju, K.C.B. (1998) Importance of recharging depleted aquifers: state of the art of artificial recharge in India, *Journal of Geological Society of India* 51, 429–454.
Roy, B.K. (1993) *Population and Reginalisation: The Inevitable Billion Plus*. Editor, Vasant Gowariker. pp. 215–224.
Sakthivadivel, R. and Gomathinayagam, P. (2004) Case studies of locally managed tank systems in Karnataka, Andhra Pradesh, Gujarat, Madhya Pradesh, Gujarat, Orissa, and Maharashtra. Report submitted to IWMI-Tata Policy Programme, Anand, India.
Shah, T. (1998) *The Deepening Divide: Diverse Responses to the Challenge of Ground water Depletion in Gujarat*. Policy School, Anand, India.
Shingi, P.M. and Asopa, V.N. (2002) *Independent Evaluation of Check Dams in Gujarat: Strategies and Impacts*. Executive Summary. Centre for Management in Agriculture, Indian Institute of Management, Ahmedabad, India.
Todd, D.K. (1980) *Groundwater Hydrology*, 2nd edn. Wiley , New York.
WMF (2003) Theme Paper on Inter-Basin Transfer of Water in India–Prospects and Problems, Water Management Forum. The Institution of Engineers (India), New Delhi, India.

11 Energy–Irrigation Nexus in South Asia: Improving Groundwater Conservation and Power Sector Viability

TUSHAAR SHAH[1], CHRISTOPHER SCOTT[2], AVINASH KISHORE[3] AND ABHISHEK SHARMA[4]

[1]*IWMI-TATA Water Policy Program, Anand Field Office, Elecon Premises, Anand-Sojitra Road, Vallabh Vidyanagar, 388 129, Anand, Gujarat, India;* [2]*2601 Spencer Road, Chevy Chase, MD 20815, USA;* [3]*Woodrow Wilson School of Public and International Affairs, Princeton University, Robertson Hall, Princeton, NJ 08544-1013, USA;* [4]*PricewaterhouseCoopers Pvt. Ltd., PwC Centre Saidulajab, Opposite D-Block Saket, Mehrauli Badarpur Road, New Delhi–110 030, India*

Introduction

In the populous South Asian region, power utilities have been at loggerheads with the region's groundwater economy for more than 15 years. As groundwater irrigation has come to be the mainstay of irrigated agriculture in much of India, Pakistan Punjab and Sind provinces, Nepal Terai and Bangladesh, the energy sector's stakes in agriculture have risen sharply. Way back in the 1950s, when raising energy consumption was considered synonymous with economic progress, government-owned state power utilities aggressively persuaded unwilling farmers to install electric tube wells. In states like Punjab and Uttar Pradesh, chief ministers gave steep targets to district-level officials to sell electricity connections to farmers. All manner of loans and concessions were made available to popularize tube well irrigation. During the 1960s and 1970s, the World Bank supported huge investments in rural electrification infrastructure to stimulate groundwater irrigation and agricultural growth. These policies were vindicated when the Green Revolution was found to follow the tube well revolution with a lag of 3–5 years; and researchers like Robert Repetto (1994) asserted that 'the Green Revolution is more tubewell revolution than wheat revolution'. By the 1970s, the energy–irrigation nexus had already become a prominent feature of the region's agrarian boom; even in canal commands, such as in India Punjab and Pakistan Punjab, groundwater irrigation had grown rapidly.

 The Agricultural Groundwater Revolution: Opportunities and Threats to Development (M. Giordano and K.G. Villholth)

However, the enthusiasm of state electricity boards (SEBs) towards their agricultural customers had begun to gradually wane. All of them were charging tube well owners based on metered consumption; however, as the number of tube wells increased, SEBs found it costly and difficult to manage metering and billing. The cost of meters and their maintenance was the least of the worry; but the transaction costs of farm power supply – in terms of containing rampant tampering of meters, underbilling, corruption at the level of meter readers, the cost of maintaining an army of meter readers, increasing pilferage of power – were far higher and more difficult to control. Introduction of flat tariff in state after state during the 1970s and 1980s was a response to this high and rising transaction costs of metered power supply. Flat tariff linked to the horsepower rating of the pump eliminated the hassle and cost of metering in one go; it still afforded scope for malpractices, such as underreporting the horsepower rating, but controlling this was easier than controlling pilferage under metered tariff. Flat tariffs, however, became 'sticky'. As power supply to agriculture emerged as a major driver of irrigated agriculture, chief ministers found its pricing a powerful weapon in the populist vote bank politics. Unable to raise flat tariff for years on end and under pressure to supply abundant farm power, power utilities began to find their balance sheets turning red; and the industry as well as its protagonists and multilateral donors veered around to the view that reverting to metered tariff for farm power supply is a precondition to restoring its viability. This view based on neoclassical economic theory considered only the 'transformation cost' of generating and distributing power, but overlooked the 'transaction costs' of volumetric pricing of power supply to farmers.

In this chapter, our objective is to re-evaluate the entire debate by putting it in the perspective of the New Institutional Economics, which shows how some activities we all know have high pay-offs in terms of productivity fail to get undertaken because of the presence of transaction costs, which neoclassical economics ignores (North, 1997). We begin with the premise that electricity pricing and supply policies in South Asia are closely linked with the policy goals of managing groundwater irrigation for efficiency, equity and sustainability. Analysing the energy and groundwater economies as a nexus could evolve joint strategies to help South Asia conserve its groundwater while at the same time improving the viability of its power industry.

Energy–Irrigation Nexus

Energy–irrigation nexus focuses on a class of issues that are unique to the South Asian region as well as the North China Plain. Many countries – the USA, Iran, Mexico – make intensive use of groundwater in their agriculture sectors. However, in these countries, groundwater irrigation affects a small proportion of their people; energy use by agriculture is a small proportion of their total energy use and the cost of energy use in farming is a small proportion of the total value added in farming.

India, Pakistan, Bangladesh and Nepal (but not Bhutan, Myanmar, Sri Lanka and Maldives) are the biggest groundwater users in the world. Between them, they pump around 210 km^3 of groundwater every year (Fig. 11.1). In doing so, they use approximately 21–23 million pump sets, of which about 13–14 million

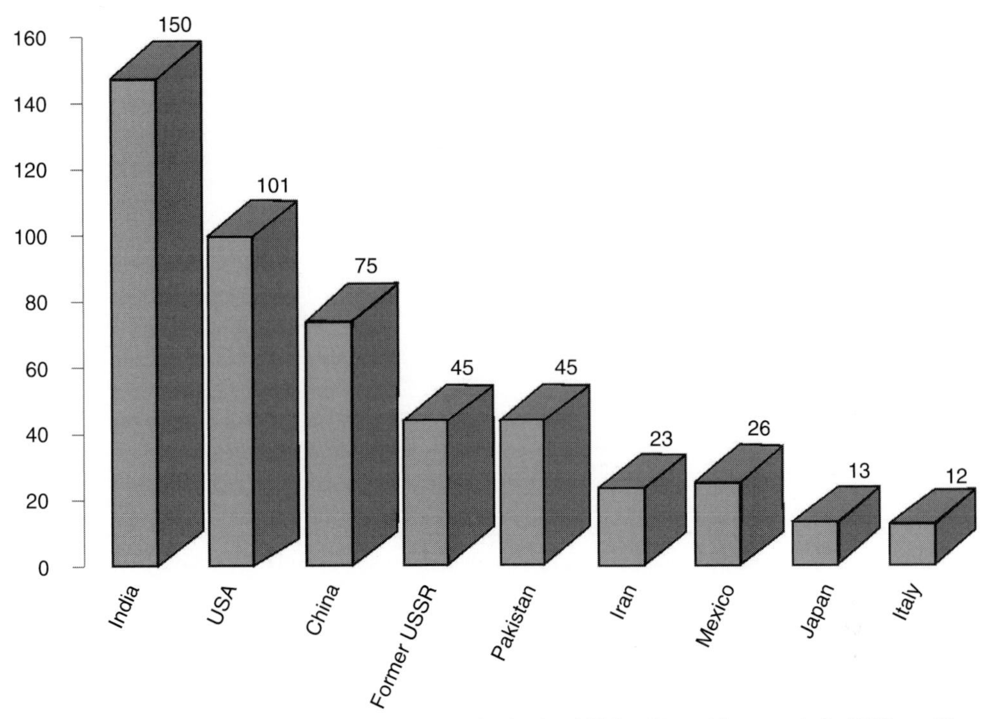

Fig. 11.1. Groundwater use in selected countries in the 1980s. (From Llamas *et al.*, 1992, p. 4.)

are electric and around 8–9 million are powered by diesel engines (NSSO, 1999). If we assume that an average electric tube well (with pumping efficiency of say 25%) lifts water to an average head of 30 m, the total electricity equivalent of energy used in these countries for lifting 210 km^3 of groundwater is around 69.6 billion kilowatt-hour per year.[1] At an alternative cost of Rs 2.5 ($0.05)/kWh, supplying this energy costs the region's energy industry Rs 174 ($3.78) billion[2]; the market value of the irrigation produced is around Rs 450–550[3] ($9.8–12) billion and its contribution to agricultural output is Rs 1350–1650 ($29.3–35.9) billion.[4] In these emerging low-income economies, pump irrigation is a serious business with economy-wide impacts, positive and negative.

Unlike in other groundwater-using countries, the pump irrigation economy in South Asia also affects vast numbers of low-income households and large proportions of people. This growth in groundwater irrigation in the region is relatively recent (Fig. 11.2). In India, gravity systems dominated irrigated agriculture until the 1970s; but by the early 1990s, groundwater irrigation had far surpassed surface irrigation in terms of area served as well as proportion of agricultural output supported (Debroy and Shah, 2003; Shah *et al.*, 2003). According to Government of India estimates, 60% of India's irrigated lands are served by groundwater wells (GOI Ministry of Water Resources, 1999); however, independent surveys suggest the proportion may be more like 75% (Shah *et al.*, 2004a; NSS 54th round).

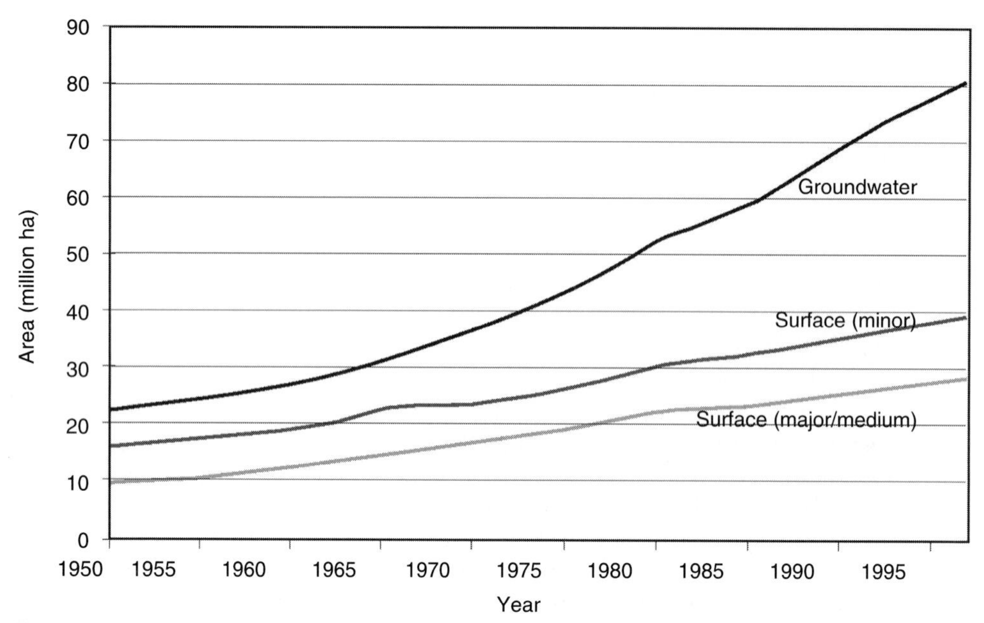

Fig. 11.2. India, irrigated area by source. (From GOI Ministry of Water Resources, 1999.)

In 1999/2000, India's 81 million landowning families (http://labourbureau.nic.in/) had more than 20 million tube wells and pump sets among them; on average, roughly every fourth landowning household has a pump set and a well; and a large proportion of non-owners depend on pump set owners for supplying pump irrigation to them through local, fragmented groundwater markets (Shah, 1993). According to a World Bank estimate, groundwater irrigation contributes around 10% of India's gross domestic product (GDP) (World Bank and Government of India, 1998), but this is made possible because groundwater irrigation uses up around 15–20% of total electricity used in the country.

Large number of small pumpers is a peculiarly South Asian aspect. In countries like the USA, Mexico and Iran, which have large groundwater irrigation economies, tube wells are fewer and larger, typically irrigating 10–500 times larger areas compared to groundwater users in India, Bangladesh and Nepal. In Mexico's Guanajuato province, the heartland of its intensive groundwater-irrigated agriculture, a typical tube well is run by a 100–150 hp pump and operates for more than 4000 h/year (Scott *et al.*, 2002). In India, Bangladesh and Nepal, the modal pump size is 6.5 hp and average operation is around 400–500 h/year (Shah, 1993). In Iran, only 365,000 tube wells are pumped to produce 45 km^3 of groundwater (Hekmat, 2002); India uses 60 times more wells compared to Iran to extract three times more groundwater.

From the viewpoint of managing groundwater as well as of transaction costs of energy supply to irrigation, these differences prove crucial. In Iran, when groundwater overdraft in the hinterland threatened water supply to cities in the plains, the Ministry of Power (which also manages water resources) was able to enforce a complete ban (provided under its Water Law) on new groundwater structures coming up in two-fifths of its plains (Hekmat, 2002). In Mexico, the Commission National de Aqua (CNA) has struggled to establish and enforce a system of water rights in the form of concessions and initiate a programme to create groundwater user organizations to promote sustainable resource management; however, while this has helped register most of its 90,000 tube well owners, Mexico is finding it impossible to limit pumping to quotas assigned to them (Scott *et al.*, 2002). Among the many factors that help Mexico make such direct management work, a very important one is that groundwater administrations in these countries have to deal with a relatively small number of fairly large irrigators.

A related aspect is the relation between groundwater irrigation, food security and livelihood. In countries with shrinking agriculture, the proportion of people dependent on groundwater-irrigated agriculture tends to be small (last column in Table 11.1). This, for example, is the case in the USA, Mexico and Iran. One would have normally thought that in such situations, it would be easier for governments to adopt a tough position with irrigators, especially if serious environmental anomalies were involved. However, we find that this is not so; Mexico has been unable to remove substantial energy subsidies to agriculture or rein in groundwater depletion (Scott *et al.*, 2002); and the USA has found it possible to only restrict the rate of, but not quite stop, the mining of the great Ogallala aquifer. Even after imposing a ban, Iran is still struggling to eliminate its annual groundwater overdraft of 5 km^3 (Hekmat, 2002). In South Asia, the dependence on groundwater is far greater, and not for wealth creation as much as to support the livelihood of millions of rural poor households. In India, for instance, pump irrigation has emerged as the backbone of its agriculture and accounts for 70–80% of the value of irrigated farm output; rapid groundwater development is at the heart of the agrarian dynamism found in some areas in eastern India that remained stagnant for a long time (Sharma and Mehta, 2002). The greatest social value of groundwater irrigation is that it has helped make famines a matter of history: during 1963–1966, a small deficit in

Table 11.1. Extent of dependence of population on groundwater and average size of WEMs in different countries. (From Hekmat, 2002, for Iran; Mukherji and Shah, 2002, for India; Scott *et al.*, 2002, for Mexico; and Shah *et al.*, 2003, for China and Pakistan.)

Country	Annual groundwater use (km^3)	Number of groundwater structures (million)	Extraction per structure (m^3/year)	Population dependent on groundwater (%)
Pakistan Punjab	45	0.5	90,000	60–65
India	150	21.28	7,900	55–60
China	75	3.5	21,500	22–25
Iran	29	0.5	58,000	12–18
Mexico	29	0.07	414,285	5–6
USA	100	0.2	500,000	<1–2

rainfall left reservoirs empty and sent food production plummeting by 19%, whereas in the 1987/1988 drought, when rainfall deficit was 19%, food production fell by only 2%, thanks to widespread groundwater irrigation (Sharma and Mehta, 2002).

It is often argued that with 60 million tonnes of food stocks, India can now take a tough stand on groundwater abuse. However, this view misses an important point; groundwater contribution to farm incomes and rural livelihood is far more crucial than its contribution to food security, especially outside canal commands.[5] In South Asia, the proportion of total population that is directly or indirectly dependent on groundwater irrigation for farm-based livelihood is many times larger than in Iran and Mexico. Indeed, our surmise is that by the turn of the millennium, three-fourths of the rural population and more than half of the total populations of India, Pakistan, Bangladesh and Nepal depended for their livelihood directly or indirectly on groundwater irrigation. It is not surprising therefore that the energy–irrigation nexus has been at the centre of the vote-bank politics in the region.

Sectoral Policy Perspectives

Groundwater policymakers face conflicting challenges in managing this chaotic economy in different areas. Particularly after 1970, agrarian growth in the region has been sustained primarily by private investments in pump irrigation. However, the development of the resource has been highly uneven. In the groundwater-abundant Ganga–Brahmaputra–Meghana basin – home to 400 million of the world's rural poor in Bangladesh, Nepal Terai and eastern India – groundwater development can produce stupendous livelihood and ecological benefits (Shah, 2001). However, it is precisely here that it is slow and halting. In contrast, Pakistan Punjab as well as India Punjab, Haryana and all of peninsular India are rapidly overdeveloping their groundwater to a stage where agriculture in these parts faces serious threats from resource depletion and degradation. The priority here is to find ways of restricting groundwater use to make it socially and environmentally sustainable. In stimulating or regulating groundwater use as appropriate, the tools available to resource managers are few and inadequate. Regulating groundwater draft and protecting the resource is proving far more complex and difficult. Direct management of an economy with such a vast number of small players would be a Herculean task in most circumstances. In South Asia, it is even more so because the groundwater bureaucracies are small, ill-equipped and outmoded. For instance, India's Central Ground Water Board, which was created during the 1950s for monitoring the resource, has no field force or operational experience and capability in managing groundwater. Direct management of groundwater economy will therefore remain an impractical idea for a long time in South Asia.

This makes *indirect* management relevant and appealing; and electricity supply and pricing policies offer a potent tool kit for indirect management provided these are used as such. Regrettably, these have so far not been used with imagination and thoughtfulness. In the groundwater-abundant Ganga basin, favourable power supply environment can stimulate livelihood creation for the poor through accelerated groundwater development; but as described later in this chapter, this region has been very nearly de-electrified (Shah, 2001). Elsewhere, there is a dire need to restrict groundwater draft as abundant power supply and perverse subsidies are accelerating the depletion of the resource. All in all, power supply and pricing policies in the region have so far been an outstanding case of perverse targeting.

A major reason for this is the lack of dialogue between the two sectors and their pursuit of sectoral optima rather than managing the nexus. The groundwater economy is an anathema to the power industry in the region. Agricultural use accounts for 15–20% of total power consumption; and power pricing to agriculture is a hot political issue. In states like Tamil Nadu, power supply to farmers is free; and in all other states, the flat electricity tariff – based on horsepower rating of the pump rather than actual metered consumption – charged to farmers is heavily subsidised. Annual losses to electricity boards on account of power subsidies to agriculture are estimated at Rs 260 ($5.65) billion in India; and these are growing at a compound annual growth rate of 26% (Lim, 2001; Gulati, 2002). If these estimates are to be believed, it will not be long before power industry finances are completely in the red. These estimates have, however, been widely contested; it is found that SEBs have been showing their growing T&D losses in domestic and industrial sectors as agricultural consumption, which is unmetered and therefore unverifiable.[6] However, the fact remains that agricultural power supply under the existing regime is the prime cause of the bankruptcy of SEBs in India.

As a result, there is a growing movement now to revert to metered power supply. The power industry has been leading this movement from the front; but international agencies – particularly, the World Bank, the US Agency for International Development (USAID) and the Asian Development Bank (ADB) – have begun to insist on metered power supply to agriculture as the key condition for financing new power projects. The Central and State Electricity Regulatory Commissions have been setting deadlines for SEBs and governments to make a transition to universal metering. The Government of India has resolved (i) to provide power on demand by 2012; (ii) to meter all consumers in two phases, with phase I to cover metering of all 11 kVA (kilovolt-ampere) feeders and high tension consumers, and phase II to cover all consumers; and (iii) to install regular energy audits to assess T&D losses and eliminate all power thefts as soon as possible (Godbole, 2002). This is an ambitious agenda indeed. However, all moves towards metered power consumption have met with farmer opposition on unprecedented scale in Andhra Pradesh, Gujarat, Kerala and in other states of India. All new tube well connections now come with metered tariff; and most states have been offering major inducements to tube well owners to opt for metered connections. Until it announced free power to farmers in June 2004, Andhra Pradesh charged metered tube wells at only Rs 0.20–0.35 ($0.4–0.7)/kWh, and Gujarat and several other states charged up to only Rs 2180.50–0.70/kWh against the supply cost of Rs 2.50–3.80 ($5–8)/kWh. In a recent move, the Gujarat government has offered a drip irrigation system free to any farmer who opts for metering.

Yet, there are few takers for metered connections; instead, demand for free power to agriculture has gathered momentum in many states.[7] Farmers' opposition to metered tariff has only partly to do with the subsidy contained in flat tariff; they find flat tariff more transparent and simple to understand. It also spares them the tyranny of the meter readers. Moreover, there are fears that once under metered tariff, SEBs will start loading all manner of new charges under different names. Finally, groundwater irrigators also raise the issue of equity with canal irrigators: if the latter can be provided irrigation at subsidized flat rates by public irrigation systems, they too deserve the same terms for groundwater irrigation.

Despite this opposition, power industry persists in its belief that its fortunes would not change until agriculture is put back on metered electricity tariff. Strong additional support to this is lent by those working in the groundwater sector where it is widely, and rightly, held that zero and flat power tariff produce strong perverse incentives for farmers to indulge in profligate and wasteful use of water as well as power because it reduces the marginal cost of water extraction to nearly zero. This preoccupation of water and power sector professionals in aggressively advocating reversion to metered tariff regime – and of farmers to frustrate their design – is, in our view, detracting the region from transforming a vicious energy–irrigation nexus into a virtuous one in which a booming, and better managed, groundwater-based agrarian economy can coexist with a viable electricity industry.

Making Metered Tariff Regime Work

Arguments in favour of metered tariff regime are several. First, it is considered essential for SEBs to manage their commercial losses; you cannot manage what you do not monitor, and you cannot monitor what you do not measure. Second, once farm power is metered, SEBs cannot use agricultural consumption as a carpet under which they can sweep their T&D losses in other segments. Third, metering would give farmers correct signals about the real cost of power and water, and force them to economize on their use. Fourth, for reasons that are not entirely clear, it is often suggested that compared to flat tariff regime, metered tariff would be less amenable to political manipulation and easier to raise as the cost of supplying power rises. Finally, flat tariff is widely argued to be inequitable towards small landowners and to irrigators in regions with limited availability of groundwater.

The logic in support of metered tariff is obvious and unexceptionable. The problem is how to make metered tariff work as envisaged. Two issues seem critical: (i) How to deal with the relentless opposition from farmers to metering? (ii) How will SEBs now deal with the problems that forced them to switch to flat tariff during the 1970s in the first place?

The extent of farmer resistance to metering is evident in the repeated failure of SEBs in various states to entice farmers to accept metering by offering metered power at subsidized rates ranging from Rs 0.20 to Rs 0.70 ($0.4–1.3)/kWh as against the actual cost of supply of about Rs 2.50 to Rs 3.80 ($5–8)/kWh. In late 2002, Batra and Singh (2003) interviewed 188 water extraction mechanism

(WEM) owners in Punjab, Haryana and western Uttar Pradesh to understand their WEM pumping behaviour. They noted that in Punjab and Haryana, an average electric WEM owner would spend Rs 2529.65 ($54.99) and Rs 6805.42 ($147.94)/year *less* on their total power bill if they accepted metering at prevailing rates of Rs 0.50 ($1)/kWh and Rs 0.65 ($1.4)/kWh, respectively, and yet would not accept metering. In effect, this is the price they are willing to pay to avoid the hassle and costs of metering.[8]

Besides dealing with mass farmer resistence, protagonists of metering also need to consider that the numbers of electric tube wells – and alongside, the problems associated with metering them – are now ten times larger than when flat tariff was first introduced. Before 1975, when all SEBs charged farm power on metered basis, the logistical difficulty and transaction costs of metering had become so high that flat tariff seemed the only way of containing it. A 1985 study by the Rural Electrification Corporation in Uttar Pradesh and Maharashtra had estimated that the cost of metering rural power supply was 26% and 16%, respectively, of the total revenue of the SEB from the farm sector (Shah, 1993). This estimate included only the direct costs, such as those of the meter, its maintenance, the power it consumes, its reading, billing and collecting. These costs are not insignificant[9]; however, the far bigger part of the transaction costs of metering is the cost of containing pilferage, tampering with meters, underreading and underbilling by meter readers in cohort with farmers.

All in all, the power sector's aggressive advocacy for introducing metered tariff regime in agriculture is based, in our view, on an excessively low estimation of the transaction costs of metering, meter reading, billing and collecting from several hundred thousand tube well connections scattered over a vast area[10] that each SEB serves. Most SEBs find it difficult to manage metered power supply even in industrial and domestic sectors where the transaction costs involved are bound to be lower than in the agriculture sector. Even where meters are installed, many SEBs are unable to collect based on metered consumption. In Uttar Pradesh, 40% of low tension (LT) consumers are metered but only 11% are billed on the basis of metered use; the remaining are billed based on minimum charge or an average of past months of metered use (Kishore and Sharma, 2002). In Orissa, under far-reaching power sector reforms, private distribution companies have brought all users under the metered tariff regime; however, 100% collection of amounts billed has worked only for industries, as in the domestic and farm sector – subject to a large number of scattered small users – collection as a percentage of billing declined from 90.5% in 1995/1996 to 74.6% in 1999/2000 (Panda, 2002).

In order to make metered tariff regime work reasonably well, three things are essential: (i) the metering and collection agent must have the requisite authority to deal with deviant behaviour amongst users; (ii) the agent should be subject to a tight control system so that he or she can neither behave arbitrarily with consumers[11] nor form an unholy collusion with them; and (iii) the agent must have proper incentives to enforce metered tariff regime. In agrarian conditions comparable to South Asia's, a quick assessment by Shah *et al.* (2003) suggested that all these conditions obtain in some way, and therefore metered tariff regime works reasonably well in North China (Shah *et al.*, 2004b).

The Chinese electricity supply industry operates on two principles: (i) total cost recovery in generation, transmission and distribution at each level with some minor cross-subsidization across user groups and areas; and (ii) each user pays in proportion to his or her use. Unlike in much of India where farmers pay either nothing or much less than domestic and industrial consumers do for power, agricultural electricity use in many parts of North China attracts the highest charge per unit, followed by household users and then industries. Operation and maintenance of local power infrastructure is the responsibility of local units, the village committee at the village level, the Township Electricity Bureau at the township level and the County Electricity Bureau at the county level. The responsibility of collecting electricity charges is also vested in local units in ways that ensure that the power used at each level is paid for in full. At the village level, this implies that the sum of power use recorded in the meters attached to all irrigation pumps has to tally with the power supply recorded at the transformer for any given period. The unit or person charged with the fee collection responsibility has to pay the Township Electricity Bureau for power use recorded at the transformer level. In many areas, where power supply infrastructure is old and worn out, line losses below the transformer make this difficult. To allow for normal line losses, 10% allowance is given by the Township Electricity Bureau to the village unit. However, even this made it difficult for the latter to tally the two; as a result, an Electricity Network Reform programme was undertaken by the National Government to modernize and rehabilitate rural power infrastructure. Where this was done, line losses fell sharply[12]; among the nine villages Shah visited in three counties of Hanan and Hebei provinces in early 2002, none of the village electricians he interviewed had a problem tallying power consumption recorded at the transformer level with the sum of the consumption recorded by individual users, especially with the line loss allowance of 10%.

An important reason why this institutional arrangement works is the strong local authority structures in Chinese villages: the electrician is feared because he is backed by the village committee and the powerful party leader at the village level; and the new service orientation is designed partly to project the electrician as the friend of the people. The same village committee and party leader can also keep in check flagrantly arbitrary behaviour of the electrician with the users. The hypothesis that with better quality of power and support service, farmers would be willing to pay a high price for power is best exemplified in Hanan where at 0.7 yuan ($8.75; Rs 4.03)/kWh[13] farmers pay a higher electricity rate compared to most categories of users in India and Pakistan, as also compared to the diesel price at 2.1 yuan/l.

In India, there has been some discussion about the level of incentive needed to make privatization of electricity retailing attractive at the village level. The village electrician in Hanan and Hebei is able to deliver on a reward of 200 yuan/month, which is equivalent to half the value of wheat produced on a mu (or one-thirtieth of the value of output on a hectare of land). For this rather modest wage, the village electrician undertakes to make good to the Township Electricity Bureau full amount on line and commercial losses in excess of 10% of the power consumption recorded on the transformers. If he can manage to keep losses to less than 10%, he can keep 40% of the value of power saved.

All in all, the Chinese have all along had a working solution to a problem that has befuddled South Asia for nearly two decades. Following Deng

Xiaoping who famously asserted that 'it does not matter whether the cat is black or white, as long as it catches mice', the Chinese built an incentive-compatible system that delivered quickly rather than wasting time on rural electricity cooperatives and village Vidyut Sanghas (electricity user associations) being tried in India and Bangladesh. The way the Chinese collect metered electricity charges, it is well nigh impossible for the power industry to lose money in distribution since losses there are firmly passed on downstream from one level to the one below.

If South Asia is to revert to metered tariff regime, the Chinese offer a good model. But there are two problems. First, the Chinese agricultural productivity is so much higher than most regions in South Asia that even with power charged for at real cost, the cost of tube well irrigation constitutes a relatively small proportion of the gross value of output. In South Asia, irrigation cost of this order – i.e. Rs 2100–8600 ($45.65–186.96)/ha – would make groundwater irrigation unviable in all regions except parts of Punjab and Haryana where farm productivity approaches the Chinese levels.

The second problem is that while South Asian power industry can mimic – or even outdo – the Chinese incentive system, it cannot replicate the Chinese authority system at the village level. Absence of an effective local authority that can guard the farmers from arbitrary behaviour of the metering agent or protect the latter from non-compliance by the users may create unforeseen complications in adapting the Chinese model to South Asia. India has begun experiments to find new metering solutions only recently. Indian Grameen Services, a non-governmental organization (NGO), tried an experiment to organize Transformer User Associations in Hoshangabad district of Madhya Pradesh; the idea was that the SEB would set up a dedicated plant if farmers paid up unpaid past dues and agreed to metered tariff. However, before the 2004 elections, the chief minister 'waved' past dues of farmers, and the Hoshangabad association disintegrated, its members disillusioned. Orissa organized similar village Vidyut Sanghas in thousands under its reforms; while these lie defunct, Orissa has achieved modest success in improving metered charge collection by using local entrepreneurs as billing and collection agents. It is difficult to foresee if this would work elsewhere because less than 5% of rural load in Orissa is agricultural; it is equally difficult to see what kind of treatment collection agents would receive in Gujarat villages where agricultural load may be 50–80% of total rural load. Although it is early times yet to learn lessons from these, it is all too clear that the old system of metering and billing – in which SEBs employed an army of unionized meter readers – would just not work.[14] That model seems passé; in power as well as surface water, volumetric pricing can work, where needed, only by smartly designed incentive contracts.

From *Degenerate* Flat Tariff to *Rational* Flat Tariff Regime

Flat tariff for farm power is universally written off as inefficient, wasteful, irrational and distortionary, in addition to being inequitable. In the South Asian experience, it has indeed proved to be so. It was the change to flat tariff that encouraged political leaders to indulge in populist whims such as doing away with farm power tariff altogether (as Punjab and Tamil Nadu have) or to peg it at unviably low levels regardless of the true cost of power supply. Such examples have led to the general perception that the flat tariff regime has been responsible for ruining the electricity industry and for causing groundwater depletion in many parts of South Asia.

However, we would like to suggest that flat tariff regime is wrongly maligned; in fact, the flat tariff that South Asia has used in its energy–irrigation nexus so far is a completely *degenerate* version of what might otherwise be a highly rational, sophisticated and scientific pricing regime. Zero tariff, we submit, is certainly not a rational flat tariff; nor is a flat tariff without proactive rationing and supply management. To most people, the worst thing about flat tariff is that it violates the marginal cost principle that advocates parity between the price charged and marginal cost of supply. Yet, businesses commonly price their products or services in ways that violate the marginal cost principle but make overall business sense. Flat rates are often charged to stimulate use to justify the incremental cost of providing a service. In early days of rural electrification, SEBs used to charge a flat-cum-pro-rata tariff to achieve two ends: SEBs wanted each tube well to use at least the amount of power that would justify its investment in laying cable and poles; the flat component of the tariff encouraged users to achieve this level. India's telephone department still provides the first 250 calls for a flat charge even though all calls are metered; the idea here is to encourage the use of telephone service to a level that justifies the incremental cost of providing the service.

In general, however, flat tariff regime is commonly resorted to when saving on the transaction costs of doing business is an important business objective. Organizations hire employees on piece rate when their work is easy to measure; but flat rate compensation is popular worldwide because it is not easy to measure the marginal value product of an employee on a daily basis. Urban public transport systems offer passes to commuters at an attractive flat rate in part because commuters offer a stable business but equally because it reduces queues at ticket windows, the cost of ticketing and collecting fares daily. Cable operators in India still charge a flat tariff for a bunch of television channels rather than charging for each channel separately because the latter would substantially increase their transaction costs. The Indian Income Tax Department a few years ago offered all businesses in the informal sector to pay a flat income tax of Rs 1400 ($30.44)/year instead of launching a nationwide campaign to bring these millions of small businesses within its tax net because the transaction costs of doing so would have been far greater than the revenue realized. A major reason municipal taxes are levied on a flat rate is the transaction costs of charging citizens based on the value they place on the margin of the municipal services.

Are all these businesses that charge for their products or services on a flat rate destined to make losses? No; often they make money because they charge a flat rate. Many private goods share this one feature with public goods like municipal services and defence: the high transaction costs of charging a differential price to different customers based on their use as well as the value they place on the product or service. So they recover their costs through a flat rate and then remain viable through deft supply management. Canal irrigation is a classic example. For ages we have been hearing about the exhortations to charge irrigators on volumetric basis; however, nowhere in South Asia can we find volumetric water pricing practised in canal irrigation. In our view, transaction costs of collecting volumetric charge for canal irrigation become prohibitively high (Perry, 1996, 2001) because: (i) in a typical South Asian system, the number of customers involved per 1000 ha command is quite large; so the cost of monitoring and measuring water use by each user would be high; (ii) once a gravity flow system is commissioned, it becomes extremely difficult in practice for the system managers to exclude defaulting customers from the command area from availing of irrigation when others are; (iii) the customer propensity to frustrate sellers' effort to collect a charge based on use would depend in some ways on the proportion the charge constitutes to the overall scale of his or her income. On all these counts, one can surmise that volumetric pricing of canal irrigation would be far easier in South African irrigation systems serving white commercial farmers; here, a branch canal serving 5000 ha might have 10–50 customers, and charging them based on actual use would be easier than in an Indian system where the same command area would contain 6000–8000 customers (Shah *et al.*, 2002). The only way of making canal irrigation systems viable in the Indian situation is to raise the flat rate per hectare to a level that ensures overall viability.

Supply restriction is inherent to rational flat rate pricing; by the same token, flat rate pricing and on-demand service are incompatible in most situations. In that sense, consumption-linked pricing and flat rate pricing represent two different business philosophies: in the first, the supplier will strive to 'delight the customer' as it were, by providing on-demand service without quantity or quality restrictions of any kind; in the second, the customer has to adapt to the supplier's constraints in terms of the overall quantum available and the manner in which it is supplied. In the case of buffet meals, restaurants give customers a good deal but save on waiting costs, which are a substantial element in the economics of a restaurant. In the Indian *thali* system, where one gets a buffet-type meal served on one's table, the downside is that one cannot have a leisurely meal since the restaurant aims to maximize the number of customers served during a fixed working period and in limited space. Thus, there is always a price for the value businesses offer their customers through products and services offered on flat tariff; but that does not mean that the seller or the buyer is any the worse for flat rate pricing.

The reason why flat rate tariff for power supply to WEMs as currently practised in South Asia is degenerate – and power industry is in the red – is because the power utilities have failed to invest more intelligence in managing rationed power supply. Under flat tariff systems until now, most SEBs have tried to maintain farm power supply at 8–15 h/day right through the year. Raising flat tariff to a level that covers the cost of present levels of supply would be so high that it will send state governments tumbling in the face of farmer wrath.[15] However, we believe that it is possible for the SEBs to satisfy farmer needs while reducing total power supply to farmers during a year by fine-tuning the scheduling of power supply to irrigation needs of farmers. Ideally, the business objective of a power utility charging flat tariff should be to supply the best quality service it can offer its customers consistent with the flat tariff pegged at a given level. The big opportunity for 'value improvement' in the energy–irrigation nexus – and by 'value improvement' we mean 'the ability to meet or exceed customer expectations while removing unnecessary cost' (Berk and Berk, 1995, p. 11) – arises from the fact that the pattern of power demand of the farming sector differs in significant ways from the demand pattern of domestic and industrial customers. The domestic consumers' idea of good-quality service is power of uniform voltage and frequency supplied 24 h/day, 365 days of the year. But the irrigators' idea of good-quality service from power utilities is power of uniform voltage and frequency when their crops face critical moisture stress. With intelligent management of power supply, we argue that it is possible to satisfy irrigation power demand by ensuring a supply of 18–20 h/day for 40–50 key moisture-stress days in *kharif* and *rabi* seasons of the year, with some power available on the rest of the days. Against this, Tamil Nadu supplies power to farmers 14 h/day for 365 days of the year! This is like being in the command area of an irrigation system with all branches and the distribution network operating at full supply level every day of the year.

Groundwater irrigators are always envious of farmers in the command areas of canal irrigation projects. But in some of the best irrigation projects in South Asia, a typical canal irrigator gets surface water for no more than 10–15 times a year. In most irrigation systems, in fact, the irrigator would be happy if he or she got water 6 times a year. In the new Sardar Sarovar project in Gujarat, the policy is to provide farmers a total of 53 cm depth of water in 5–6 installments during a full year. For an irrigation well with a modest output of 25 m^3/h, this would mean the ability to pump for 212 h/ha. In terms of water availability, a WEM owner with 3 ha of irrigable land would be at par with a farmer with 3 ha in the Narmada command if he or she got 636 h of power in a year. The WEM owner would be better off if the 636 h of power came when he or she needed water the most. When the Gujarat government commits to year-round supply of 8 h/day of farm power, it in effect offers tube well owners water entitlements 14 times larger than those that the Sardar Sarovar project offers to farmers in its command area.[16] Under metered tariff, this may not matter all that much since tube well owners would use power and groundwater only when their value exceeded the marginal cost of pumping; but under flat tariff, they would have a strong incentive to use some of these 'excess water entitlements' for low marginal value uses just because it costs them nothing on the margin to pump groundwater. This is why the present flat tariff in South Asia is degenerate.

Rational flat tariff, if well managed, can confer two larger benefits. First, it may curtail wasteful use of groundwater. If farm power supply outside main irrigation seasons is restricted to 2–3 h/day, it will encourage farmers to build small on-farm storage tanks for meeting multiple uses of water. Using progressive flat tariff – by charging higher rates per connected horsepower as the pump size increases – will produce additional incentive for farmers to purchase and use smaller-capacity pumps to irrigate less areas, and thereby reduce overdraft in regions where resource depletion is rampant. Above all, restricted but predictable water supply will encourage water-saving irrigation methods more effectively than raising the marginal cost of irrigation. Second, given the quality of power transmission and distribution infrastructure in rural India, restricting the period of time when the farm power system is 'on' may by itself result in significant reduction in technical and commercial losses of power. The parallel with water supply systems is clear here. In a 1999 paper, for example, Briscoe (1999) wrote that throughout the Indian subcontinent, unaccounted-for water as a proportion of supply is so high 'that losses are "controlled" by having water in the distribution system only a couple of hours a day, and by keeping pressures very low. In Madras, for example, it is estimated that if the supply was to increase from current levels (of about 2 hours supply a day at 2 m of pressure) to a reasonable level (say 12 hours a day at 10 m of pressure) leaks would account for about 900 MLD, which is about three times the current supply in the city.' Much the same logic works in farm power, with the additional caveat that the T&D system for farm connections is far more widespread than the urban water supply system.

Five preconditions for successful rationing

We believe that transforming the present degenerate flat power tariff into rational tariff regime will be easier, and more feasible and beneficial in the short run in many parts of South Asia than trying to overcome farmer resistance to metering. We also believe that doing so can significantly cut the losses of power utilities from their agricultural operation. Five points seem important and feasible.

Separating agricultural and non-agricultural power supply

The first precondition for successful rationing is infrastructural changes needed to separate agricultural power supply from non-agricultural power supply to rural settlements. The most common way this is done now is to keep two-phase power on for 24 h so that domestic and (most) non-agricultural uses are not affected and ration three-phase power necessary to run irrigation pump sets. This is working, but only partially. Farmers' response in states like Gujarat is a rampant use of phase-splitting capacitors with which they can run pumps on even two-phase power. There are technological ways to get around this. It is possible to use gadgets that ensure that the 11 kV line shuts off as soon as the load increases beyond a predetermined level. However, many SEBs have begun separating the feeders supplying farm and non-farm rural consumers. The government of Gujarat has now embarked on an ambitious programme called Jyotirgram Yojana to lay parallel power supply lines for agricultural users in 16,000 villages of the state over the next 3 years at an estimated cost of Rs 9 billion ($195.7 million). In Andhra Pradesh, the process of separation of domestic and agricultural feeders is already 70% complete (Raghu, 2004). This would ensure that industrial users in the rural areas who need uninterrupted three-phase power supply as well as domestic users remain unaffected from rationing of power supplies for agricultural consumers. Another infrastructural change needed would be to install meters to monitor power use so that proper power budgeting can be implemented. For this, meters at transformer level, or even feeder levels, might be appropriate. Many states have already installed meters at the feeder level.

Gradual and regular increase in flat power tariff

Flat tariffs have tended to remain 'sticky'; in most states, they have not been changed for more than 10–15 years while the cost of generating and distributing power has soared. We surmise that raising flat tariff at one go to close this gap between revenue and cost per kWh would be too drastic an increase. However, we believe that farmers would be able to cope with a regular 10–15% annual increase in flat tariff far more easily than a 350% increase at one go as has been proposed by the Electricity Regulatory Commission in Gujarat.

Explicit subsidy

If we are to judge the value of a subsidy to a large mass of people by the scale of popular opposition to curtailing it, there is little doubt that, amongst the plethora of subsidies that governments in India provide, power subsidy is one of the most valued. Indeed, a decision by a ruling party to curtail power subsidy is the biggest weapon that opposition parties use to bring down a government. So it is unlikely that political leaders will want to do away with power subsidies completely, no matter what the power industry and international donors would like. However, the problem with the power subsidy in the current degenerate flat tariff is its indeterminacy. Chief ministers keep issuing diktats to the SEBs about the number of hours of power to be supplied per day to farmers; that done, the actual subsidy availed of by the farmers is in effect left to them to usurp. Instead, the governments should tell the power utility the amount of power subsidy it can make available at the start of each year; the power utility should then decide the amount of farm power the flat tariff and the government subsidy can buy.

Use of off-peak power

In estimating losses from farm power supply, protagonists of power sector reform, including international agencies, systematically overestimate the real opportunity cost of power supplied to the farmers. For instance, the cost of supplying power to the domestic sector – including generation, transmission and distribution – is often taken as the opportunity cost of power to agriculture, which is clearly wrong, since a large part of the high transaction costs of distributing power to the domestic sector is saved in power supply to agriculture under flat tariff. Moreover, a large part of the power supplied to the farm sector is off-peak load power. In fact, but for the agriculture sector, power utilities would be hard-pressed to dispose of this power.[17] More than half of the power supply to farm sector is in the night, and this proportion can increase further. But in computing the amount of power the prevailing flat tariff and prespecified subsidy can buy, the power utilities must use the lower opportunity cost of the off-peak supply.

Intelligent supply management

There is tremendous scope for cutting costs and improving service here. The existing rostering policy in many states of maintaining power supply to the farm sector at a constant rate during prespecified hours is irrational and the prime reason for wasteful use of power and water.[18] Ideally, power supply to the farm sector should be so scheduled as to reflect the pumping behaviour of a modal group of farmers in a given region when they would be subject to metered power tariff at full cost. However, it is difficult to simulate this behaviour because farmers everywhere are subject to flat tariff under which they would have a propensity to use power whenever it is available, regardless of its marginal product. In many states, there is a small number of new tube wells whose owners pay for power on a metered basis; however, they are charged so low a rate that they behave pretty much like flat tariff–paying farmers. Another method is to compare electricity use before and after flat tariff to gauge the extent of overutilization of power and water attributable to flat tariff.[19] However, our surmise is that the pumping behaviour of diesel pump owners, who are subject to full marginal costs of energy, comparable to what electric tube well owners would pay under unsubsidized metered tariff regime, would be a good indicator of the temporal pattern of power use by electric tube wells under metered tariff. Several studies have shown that annual hours of operation of diesel tube wells is often half or less than half compared to flat tariff–paying electric tube wells (Mukherji and Shah, 2002) (Fig. 11.3).[20] Batra and Singh (2003) interviewed approximately 188 farmers in Punjab, Haryana and central Uttar Pradesh to explore if pumping behaviour of diesel and electric WEM owners differed significantly. They did not find significant differences in Punjab and Haryana[21] but their results for central Uttar Pradesh suggested that diesel WEMs are pumped

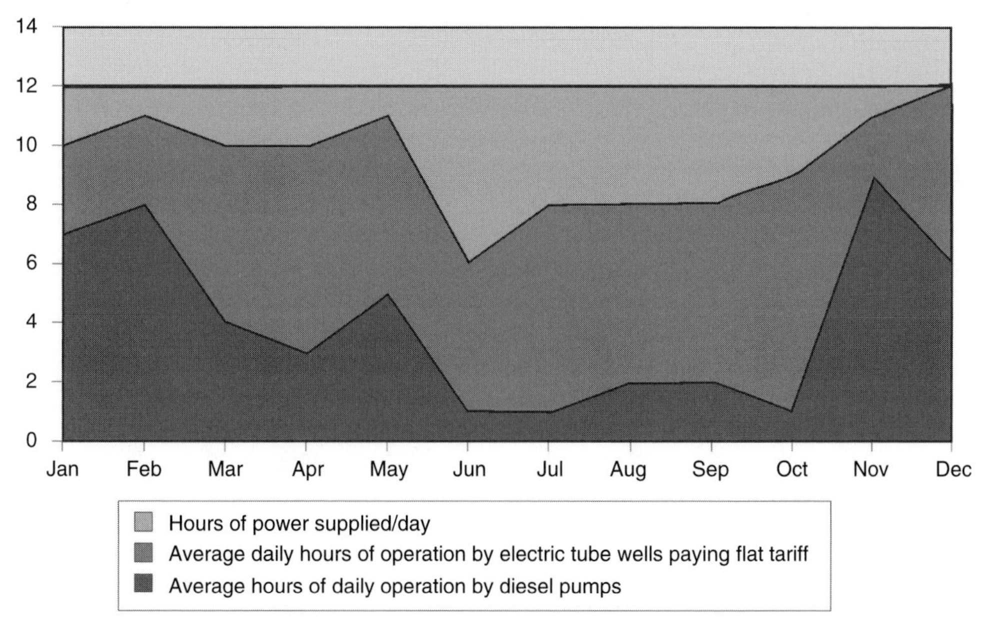

Fig. 11.3. Minimizing waste of power and water through supply management. The numbers shown in this schematic diagram are indicative and not based on actual field data.

when irrigation is needed and electric WEMs are operated whenever electricity is available. Very likely, a good deal of the excess water pumped by farmers owning both electric and diesel pumps is wasted in the sense that its marginal value product falls short of the scarcity value of water and power together.

Figures 11.4 and 11.5 present the central premise of our case: a large part of the excess of pumping by electric tube wells over diesel tube wells is indicative of the waste of water and power that is encouraged by the zero marginal cost of pumping under the present degenerate flat tariff regime. Figure 11.4 presents results of a survey of 2234 tube well irrigators across India and Bangladesh in late 2002, which shows that electric tube well owners subjected to flat tariff everywhere invariably operate their pumps for much longer hours compared to diesel pump owners who face a steep marginal energy cost of pumping (Mukherji and Shah, 2002). It might be argued that diesel pumps on average might be bigger in capacity compared with electric pumps; so we also compared pumping hours weighted by horsepower ratings; and Fig. 11.5 shows that horsepower-hours pumped by flat tariff–paying electric WEMs too are significantly higher than for diesel WEMs everywhere. The survey showed the difference in annual pumpage to be of the order of 40–150%; some of this excess pumping no doubt results in additional output; however, a good deal of it very likely does not, and is a social waste that needs to be eliminated.

Making 'Rational Flat Tariff and Intelligent Power Supply Management' Work

If power utilities undertake a refined analysis of the level and pattern of pumping by diesel pump owners in a region and shave off the potential excess pumping by

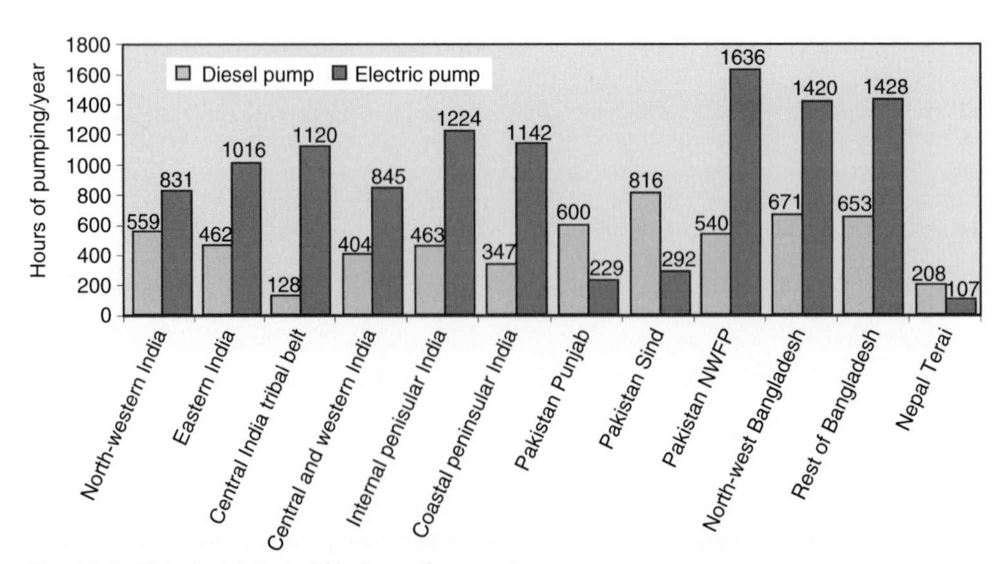

Fig. 11.4. Flat electricity tariff induces farmers to pump more.

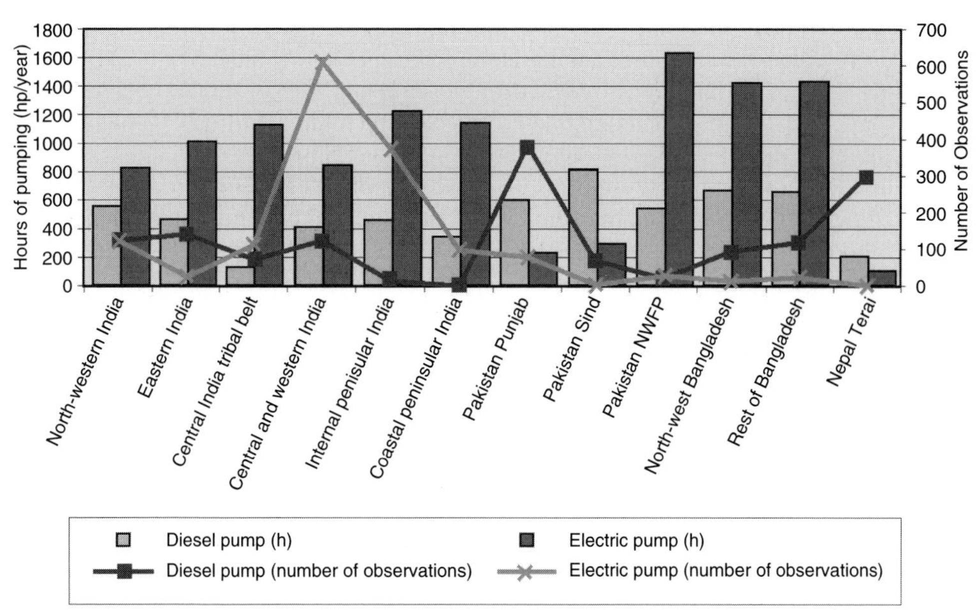

Fig. 11.5. Impact of flat tariff on average annual hours of pumping weighted by pump horsepower.

flat tariff–paying electric tube wells (as shown in Fig. 11.3) by fine-tuning power supply schedule around the year, flat tariff can become not only viable but also socially optimal by eliminating the 'waste'. The average number of hours for which diesel pumps operate is around 500–600/year. At 600 h of annual operation, an electric tube well would use 450 kWh of power per horsepower; if all the power used is off-peak load, commanding, say, 25% discount on a generation cost of Rs 2.5 ($5)/kWh, then farm power supply by the power utility would break even at a flat tariff of Rs 843.75 ($18.34)/hp/year as against Rs 500 ($10.87)/hp/year in force in Gujarat since 1989. The government of Gujarat is committed to raise the flat tariff eventually to around Rs 2100($45.65)/hp/year at the insistance of the Gujarat Electricity Regulatory Commission; however, chances are that if it does so, farmers will unseat the government. A more viable and practical course would be to raise flat tariff to perhaps Rs 900 ($19.57) first and then to Rs 1200 ($26.09), and restrict annual supply of farm power to around 1000–1200 h against the existing regime of supplying farm power for 3000–3500 h/year. A 5 hp pump lifting 25 m^3 of water per hour over a head of 15 m can produce 30,000 m^3 of water per year in 1200 h of tube well operation, sufficient to meet the needs of most small farmers in the region.

Farmers will no doubt resist such rationing of power supply; however, their resistance can be reduced through proactive and intelligent supply management by the following methods:

1. Enhancing the *predictability* and *certainty*: More than the total quantum of power delivered, in our assessment, power suppliers can help the farmers by announcing an annual schedule of power supply fine-tuned to match the

demand pattern of farmers. Once announced, the utility must then stick to the schedule so that farmers can be certain about power availability.
2. Improving the quality: Whenever power is supplied, it should be at full voltage and frequency, minimizing the damage to motors and downtime of transformers due to voltage fluctuations.
3. Better matching of supply with peak periods of moisture stress: Most canal irrigators in South Asia manage with only 3–4 canal water releases in a season; there are probably 2 weeks during *kharif* in a normal year and 5 weeks during *rabi* when the average South Asian irrigation farmer experiences great nervousness about moisture stress to the crops. If the power utility can take care of these periods, 80–90% of farmers' power and water needs would be met. This will, however, not help sugarcane growers of Maharashtra, Gujarat and Tamil Nadu, but then they constitute the big part of the power utility's problems.
4. Better upkeep of farm power supply infrastructure: Intelligent power supply management to agriculture is a tricky business. If rationing of power supply is done by arbitrary increase in power cuts and neglect of rural power infrastructure, it can result in disastrous consequences. Eastern India is a classic example. After the eastern Indian states switched to flat power tariff, they found it difficult to maintain the viability of power utilities in the face of organized opposition to raising flat tariff from militant farmer leaders like Mahendra Singh Tikait. As a result, the power utilities began to neglect the maintenance and repair of power infrastructure; and rural power supply was reduced to a trickle. Unable to irrigate their crops, farmers began en masse to replace their electric pumps by diesel pumps. Over a decade, the groundwater economy got more or less completely dieselized in large regions including Bihar, eastern Uttar Pradesh and North Bengal. Figure 11.6 shows the electrical and diesel halves of India; in the western parts, groundwater irrigation is dominated by electric pumps; as we move east, diesel pumps become more preponderant. The saving grace was that in these groundwater-abundant regions, small diesel pumps, though dirtier and costlier to operate, kept the economy going. But in regions like north Gujarat, where groundwater is lifted from 200 to 300 m, such de-electrification can completely destroy the agricultural economy.

Against this danger, the major advantage the rational flat tariff regime offers is in putting a brake on groundwater depletion in western and peninsular India. Growing evidence suggests that water demand in agriculture is inelastic to pumping costs within a large range. While metered charge without subsidy can make power utilities viable, it may not help much to cut water use and encourage water-saving agriculture. If anything, a growing body of evidence suggests that adoption of water- and power-saving methods respond more strongly to scarcity of these resources than their price. Pockets of India where drip irrigation is spreading rapidly – Aurangabad region in Maharashtra, Maikaal region in Madhya Pradesh, Kolar in Karnataka, Coimbatore in Tamil Nadu – are all regions where water and/or power is scarce rather than costly. Rational flat tariff with intelligent power supply rationing to the farm sector holds out the promise of minimizing wasteful use of both the resources and of encouraging technical change towards water and power saving. Our surmise is that such a strategy can easily reduce annual

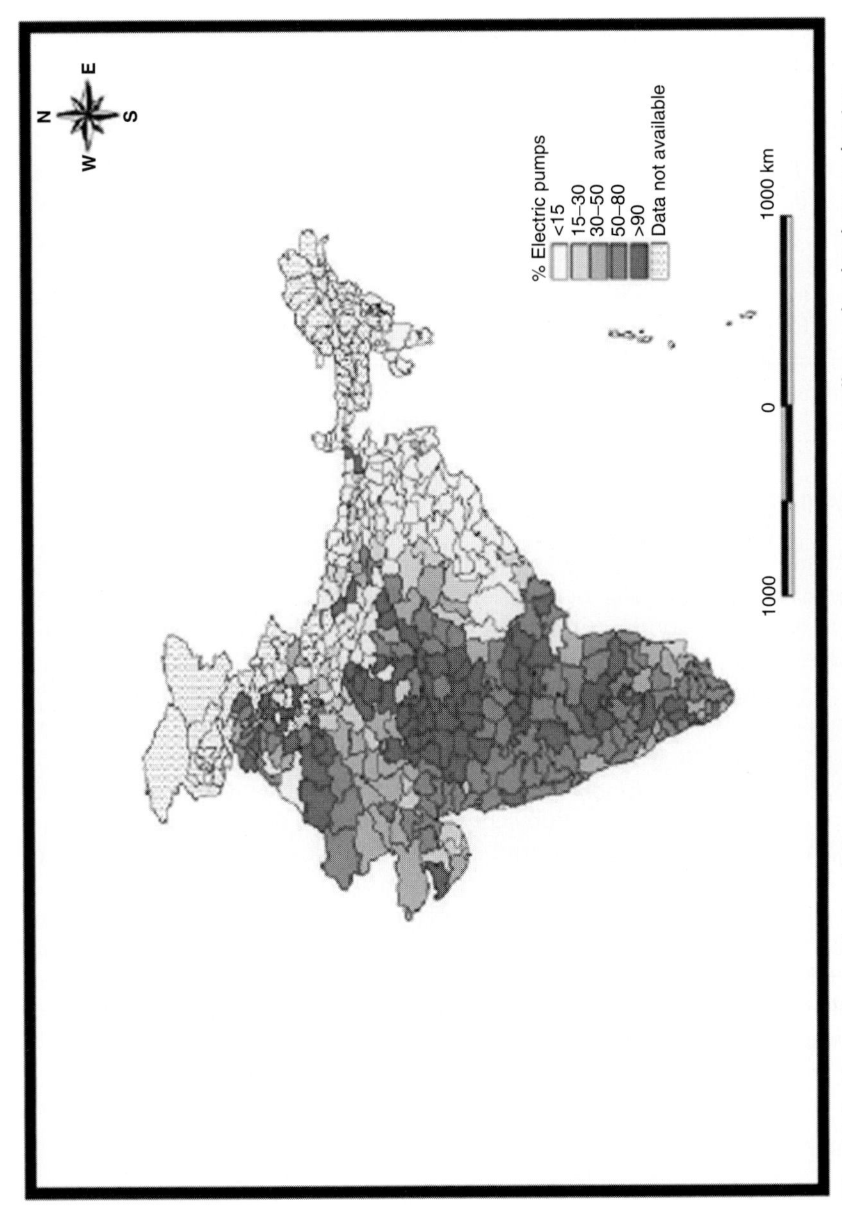

Fig. 11.6. Percentage of electricity-operated groundwater structures to totally mechanized groundwater structures.

groundwater extraction in western and peninsular India by 12–21 km^3/year and reduce power use in groundwater extraction by about 4–6 billion kilowatt-hour of power, valued at Rs 10–15 ($0.22–0.33) billion per year.

Approaches for Rationing

The strongest piece of evidence in support of our argument for intelligent rationing of power supply as the way to go is that intuitively most SEBs in India have already been doing some kind of rationing of farm power supply now for more than a decade. Andhra Pradesh, where the new government announced free power, also announced that farm power supply would henceforth be restricted to 7 h/day. Nobody – farmers included – considers 24 h, uninterrupted power supply to agriculture to be either a feasible proposition or a defensible demand under the flat tariff regime in force. Negotiations between farmer groups and governments almost everywhere in India are carried out in terms of the minimum hours of daily power supply the government can guarantee.

Default system of rationing

Constant hours/day of power supply to farmers, which is the current default, is the least intelligent way of rationing power supply to agriculture because it fails to achieve a good 'fit' between the schedule of power supply and farmers' desired irrigation schedule. It leaves farmers frustrated on days when their crops need to be watered the most; on the other hand, on many other days when the need for irrigation is not high, it leads to wasteful use of power and groundwater. From where the SEBs' present power-rationing practices stand today, they only have to gain by achieving a better fit between power supply schedules and farmers' irrigation schedules. Farmers keep demanding that the 'constant hours/day' must be raised because the default system does not provide enough power when they need it the most. There are a number of ways rationing of power supply in agriculture can be carried out to raise farmer satisfaction and control power subsidies provided (i) it reduces farmers' uncertainty about the timing of power availability; (ii) it achieves a better fit between power supply schedules and irrigation schedules; or (iii) both. We suggest a few illustrative approaches that need to be considered and tried out.

Agronomic scheduling

Ideally, SEBs should aim to achieve the 'best fit' by matching power supply schedules with irrigation needs of farmers. In this approach, the power utility constantly (i) studies irrigation behaviour of farmers in regions and subregions by monitoring cropping patterns, cropping cycles, rainfall events; (ii) matches power supply schedules to meet irrigation needs; and (iii) minimizes supply

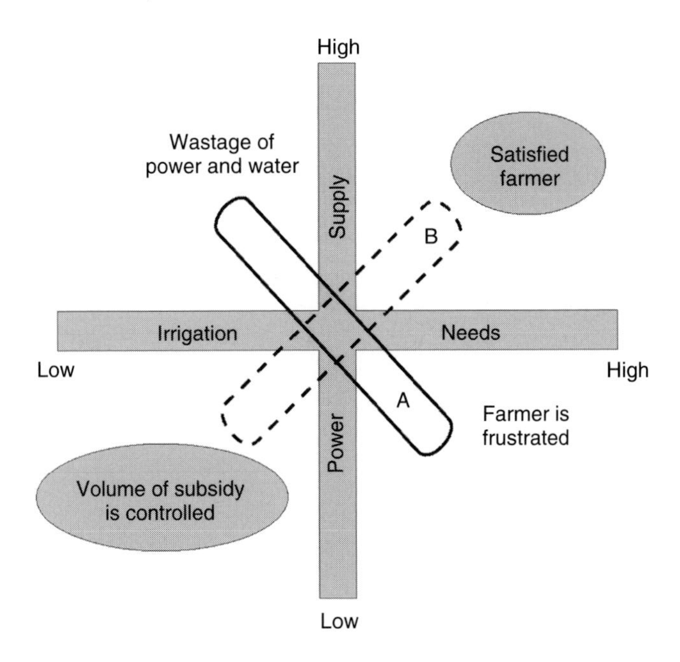

A Mismatch between power supply and irrigation needs; existing system in which the farmer is frustrated.

B A win-win scenario; power supply is good and reliable when the irrigation needs are high (satisfied farmer), and power supply is low when irrigation needs are low (volume of subsidy is controlled).

Fig. 11.7. Improving farmer satisfaction and controlling electricity subsidies through intelligent management of farm power supply.

in off-peak irrigation periods (see Fig. 11.7). The advantages of such a system are that (i) farmers are happier; (ii) total power supply to agriculture can be reduced; (iii) power and water waste is minimized; and (iv) level of subsidy availed is within SEB control. The key disadvantage of this approach is that it is highly management-intensive, and therefore difficult to operationalize.

Demand-based scheduling

In the demand-based approach, feeder-level farmer committees or other representational bodies of farmers assume the responsibility of ascertaining members' requirements of power, and provide a power supply schedule to the utility for a fixed number of allowable hours for each season. This is a modified version of agronomic scheduling in which the power utility's research and monitoring task is assumed by feeder committees. This may make it easier to generate demand schedules but more difficult to serve it. Moreover, the organizational challenge this approach poses is formidable.

Canal-based scheduling

Tube well irrigators outside canal commands justify demands for power subsidies by comparing their lot with canal irrigators who get cheap canal irrigation without any capital investment of their own. However, under the present degenerate flat tariff, tube well irrigators often have the best of both worlds. At 10 h/day of power supply, an Andhra Pradesh tube well irrigator could in theory use 300–500 m^3 of water every day of the year. In contrast, under some of the best canal commands, farmers get irrigation for 10–15 times in an entire year. In this approach, power rationing aims to remove the inequity between tube well and canal irrigators by scheduling power supply to mimic the irrigation schedule of a benchmarked public irrigation system. This can drastically reduce power subsidies from current levels, but for that very reason, will face stiff resistance from tube well–irrigating farmers.

Zonal roster

An approach to rationing that is simpler to administer is to divide the state into, say, seven zones, each assigned a fixed day of the week when it gets 20 h of uninterrupted, quality power throughout the year; on the rest of the days, it gets 2 h. This is somewhat like a weekly turn in the *wara-bandi* system in canal irrigation systems in India Punjab and Pakistan Punjab. The advantages of this approach are: (i) it is easy to administer; (ii) agricultural load for the state as a whole remains constant, so it becomes easy to manage for SEB also; (iii) level of subsidies are controlled; and (iv) power supply to each zone is predictable, so farmers can plan their irrigation easily. Disadvantages are: (i) farmers in deep water table areas or areas with poor aquifers (as Saurashtra in Gujarat) would be unhappy; and (ii) zonal rostering will not mimic seasonal fluctuations in irrigation demand as well as agronomic rationing would do.

Adjusted zonal roaster

The zonal roaster can help farmers plan their cropping pattern and irrigation schedules by reducing uncertainty in power supply but it does not do much to improve the 'fit' between irrigation need and power supply across seasons. In most of India, for instance, following the same zonal roaster for *kharif* and *rabi* seasons makes little sense. Modifying the zonal roaster system so that power supply offered is higher in winter and summer than in monsoon season would improve the seasonal fit as well as reduce uncertainty.

Conclusion

We have argued in this chapter that either a switch to metered tariff regime at this juncture or raising flat tariff fourfold as proposed in Gujarat will very likely backfire in most of the states of India. Metering is highly unlikely to improve the fortunes of the power utilities that have found no smarter ways – than in the 1970s – of dealing with the exceedingly high transaction costs of metered farm power supply, which led them to flat tariff regime in the first place. However, if agriculturally dynamic states like Punjab and Haryana – where non-farm uses of three-phase power supply are extensive and growing in the villages and where productive farmers can afford higher costs of better-quality power supply in their stride – want to experiment with metered power supply, they would be well advised to create micro-entrepreneurs to retail power, meter individual power consumption and collect revenue rather than experiment with wooly ideas of electricity co-operatives that continue to be promoted (Gulati and Narayanan, 2003, p. 129). Despite 50 years of effort to make these work, including with donor support, they have not succeeded in India.[22] The 50-year-old Pravara electricity co-operative in Maharashtra survives but only by owing the SEB several billions of rupees in unpaid past dues (Godbole, 2002). While promoting metering it should also be borne in mind that the component of the transaction costs of metering, which is by far the largest and the most difficult to manage, arises from containing user efforts to frustrate metered tariff regime, by pilfering power, illegal connections, tampering with meters and so on. These costs soar in a 'soft state' in which an average user expects to get away easily even if caught indulging in these.[23] One reason why metering works reasonably well in China is because it is a 'hard state': an average user fears the village electrician whose informal power and authority border on the absolute in his or her domain.[24] The ongoing experiments on privatization of electricity retailing in Orissa will soon produce useful lessons on whether metering-cum-billing agents can drastically and sustainably reduce the cost of metered power supply in a situation in which tube well owners account for a significant proportion of electricity use.

However, with tight and intelligent supply management, in the particular context of South Asia, rational flat tariff (and intelligent power supply management) can achieve all that metered tariff regime can, and more. Flat tariff will have to be raised, but the schema we have set out can cut power utility losses from farm power supply substantially. Total hours of power supplied to farmers during a year will have to be reduced but farmers would get good-quality power aplenty at times of moisture stress when they need irrigation the most. Power supply to agriculture should still be metered at the feeder level so as to be able to measure and monitor the use of power in irrigation in order to manage it well. In this way, the huge transaction costs of metered charge collection would be saved; and if power utilities were to begin viewing farmers as customers, the adversarial relationship between them could even be turned into a benign one. While metered tariff regime will turn groundwater markets into sellers' markets hitting the resource-poor water buyers, rational flat tariff would help keep water markets as buyers' markets, albeit far less so than would be the case under the present degenerate flat tariffs (see Shah, 1993 for a detailed argument). Rational flat tariff – under which power rationing is far more defensible than under metered tariff regime – will make it possible to put an effective check on total use of power and water, and make their use more sustainable than under the present regime or under metered tariff. Moreover, restricting the total hours of operation of farm supply would help greatly curtail technical and commercial losses experienced by SEBs. Above all, rational flat tariff can significantly curtail groundwater depletion by minimizing wasteful resource use. On the basis of an International Water Management Institute (IWMI) survey of 2234 owners of diesel and electric tube wells in India, Pakistan, Nepal Terai and Bangladesh, it was concluded that electric tube well owners subject to flat tariff but unrestricted, poor-quality power supply were worked 40–150% more horsepower-hours compared to diesel tube well owners with greater control over their irrigation schedules. It can easily curtail groundwater draft by 13–14 million electric tube wells at least by 10–14%, i.e. approximately 12–21 km^3/year, assuming electric WEMs pump a total of some 120–150 km^3 of groundwater every year.

Contrary to popular understanding, rational flat tariff is an elegant and sophisticated regime management, which requires a complex set of skills and deep understanding of agriculture and irrigation in different regions. Power utilities in South Asia have never had these skills or the understanding, which is a major reason for the constant hiatus between them and the agriculture sector. One reason is that SEBs employ only engineers (Rao, 2002). In the power sector reforms underway in many Indian states, this important aspect has been overlooked in the institutional architecture of unbundling. Distributing power to agriculture is a different ball game in this region from selling it to townspeople and industry; and private distribution companies will most likely exclude the agricultural market segment in a hurry as being 'too difficult and costly to serve', as Orissa's experience is already showing.[25] Perhaps the most appropriate course would be to promote a separate distribution company for serving the agriculture sector with specialized competence and skill base; and predetermined government subsidies to the farming sector should be directed to the agricultural distribution companies.[26]

Notes

1 According to the Centre for Monitoring Indian Economy, electricity use in Indian agriculture in 2000–2001 was 84.7 billion kWh, much greater than our combined estimate of 69.6 billion kWh equivalent of total energy use for India, Pakistan and Bangladesh where at least one-third of the tube wells are run by diesel pumps. However, we also know that the estimates of agricultural electricity use in India are overestimates (see footnote 5) and include a portion of transmission and distribution losses in non-farm sectors that are passed off as agricultural consumption (CMIE, 2003).

2 Gulati and Narayanan (2003, p. 99) took the difference between the combined cost of supplying power to all sectors and the tariff charged from agriculture sector as a measure of subsidy to agriculture per kWh. Multiplying this with estimated power supply to agriculture, they place power subsidy to agriculture in 2000/01 at Rs 288.14 ($6.26) billion and suggest that these are 78 times more than what they were in 1980/81 but acknowledge that their estimate is likely to be a huge overestimate because of SEB propensity to pass off excessive transmission and distribution (T&D) losses in other sectors as farm consumption.

3 We assumed that an average South Asian tube well uses 4 kWh of electricity equivalent to pump for an hour, which gives us 17.5 billion hours of pumping of groundwater per year. At an average price of Rs 30 ($65)/h, the market value of pump irrigation in the region can be computed at Rs 522($11.34) billion. In many parts of South Asia, water sellers providing pump irrigation service claim one-third crop share; based on this, we computed contribution to farm output as three times the market value of pump irrigation. Alternatively, according to our calculations, a representative South Asian tube well produces around Rs 25,000 ($543.48) worth of irrigation water per year, which helps produce Rs 75,000 ($1,630.44) worth of crops. If we take the World Bank estimate, which places groundwater contribution to India's GDP at 10%, our calculations are severe underestimates of productive contribution of tube well irrigation.

4 Dhawan estimated the net value of marginal product of power in agriculture as Rs 9 ($0.20)/kWh in net terms and Rs 14 ($0.30)/kWh in terms of gross value of output (Dhawan, 1999).

5 Dhawan (cited in Samra, 2002) has asserted that in low rainfall regions of India, 'a wholly [groundwater] irrigated acre of land becomes equivalent to 8 to 10 acres of dry land in terms of production and income'.

6 Shah (2001) analysed this aspect for the Uttar Pradesh SEB and found agricultural power use 35% lower than claimed. Similarly, based on a World Bank study in Haryana, Kishore and Sharma (2002) report that actual agricultural power consumption was 27% less than reported, and the overall T&D losses were 47% while the official claim was 36.8%, making the SEB more efficient than it actually was. Power subsidy ostensibly meant for agricultural sector but actually accruing to other sectors was estimated at Rs 5.50 billion ($0.12 billion) per year for Haryana alone.

7 Farmers are getting away with it in many states. Electricity supply to agriculture became a major issue in India's 2004 parliamentary and state elections. Chief ministers like Chandrababu Naidu of Andhra Pradesh, Narendra Modi of Gujarat and Jayalalitha of Tamil Nadu suffered major electoral reverses arguably on account of farmer opposition to their stand on electricity supply to agriculture. The new chief minister of Andhra Pradesh announced free power to farmers the day after he assumed office; and Jayalalitha, who had abolished free power in Tamil Nadu, restored it soon after the results of election. Gujarat's Narendra Modi softened his hard stand on farm power supply; in Maharashtra, Shiv Sena chief Bal Thakre announced his promise to provide free power to farmers should his party come to power.

8 According to Batra and Singh (2003), farmers resist metering 'because of the prevalence of irregularities in the SEBs'. Complaints of frequent meter burning, which costs the farmer Rs 1000 ($21.74) per meter burnt, false billing, uncertainty in the bill amount, etc., repel farmers from accepting metering. They suggest that farmers also resist metering because of the two-part tariff (energy charge and rental for meter) system offered as an alternative to flat tariff. They are reluctant to pay the minimum bill (rental charge) which they have to pay even if they do not use the pump in a given month.

9 A recent World Bank study for the small state of Haryana estimated that the cost of metering all farm power connections in Haryana would amount to $30 (Rs 1380) million in capital investment and $2.2 (Rs 101.2) million/year in operating them (Kishore and Sharma, 2002). The Maharashtra Electricity Tariff Commission estimated the capital cost of metering the state's farm connections at Rs 11.50 ($0.25) billion (Godbole, 2002).

10 Rao and Govindarajan (2003) lay particular emphasis on geographic dispersion and remoteness of farm consumers in raising transaction costs of metering and billing:

'To illustrate, a rural area of the size of Bhubaneshwar, *the capital of Orissa state*, will have approximately 4000 consumers. Bhubaneshwar has 96000. *The* former will have a collection potential of Rs 0.7 million ($15217) a month; for Bhubaneshwar, it is Rs 22.0 million ($0.48 million) a month.'

11 In states like Gujarat, which had metered tariff until 1987, an important source of opposition to metering is the arbitrariness of meter readers and the power they came to wield over them; in many villages, farmers had organized for the sole purpose of resisting the tyranny of the meter reader. In some areas, this became so serious that meter readers were declared *persona non grata*; even today, electricity board field staff seldom go to villages except in fairly large groups, and often with police escort.

12 The village electrician's reward system encourages him or her to exert pressures to achieve greater efficiency by cutting line losses. In Dong Wang Nu village in Ci county, the village committee's single large transformer, which served both domestic and agricultural connections, caused heavy line losses at 22–25%. Once the Network Reform Program began, the electrician pressurized the village committee to sell the old transformer to the Township Electricity Bureau and raise 10,000 yuan (partly by collecting a levy of 25 yuan per family and partly by a contribution from the Village Development Fund) to get two new transformers, one for domestic connections and the other for pumps. Since then, power losses have fallen to a permissible level of 12% here (Shah *et al*., 2004b).

13 1$ = 8 yuan = Rs 46 (July 2004).

14 A 1997 consumer survey of the power sector revealed that 53% of consumers had to pay bribes to electricity staff for services which were supposed to be free; 68% suggested grievance redressal to be poor or worse than poor; 76% found staff attitudes poor or worse; 53% found repair fault services poor or worse; 42% said they had to make 6–12 calls just to register a complaint; 57% knew of power thefts in their neighbourhoods; 35% complained of excess billing; 76% complained of inconvenience in paying their bills (Rao, 2002).

15 In Madhya Pradesh, the latest state to announce power pricing reforms, the chief minister announced a sixfold hike in flat tariff. No sooner was the announcement made than there was a realignment of forces within the ruling party and seniormost cabinet ministers began clamouring for leadership change. Subhash Yadav, the Deputy Chief Minister, lamented in an interview with India Today: 'A farmer who produces 10 tonNEs of wheat earns Rs 60,000 ($1,304.35) and he is expected to pay Rs 55,000 ($1,195.65) to the electricity board. What will he feed his children with and why should he vote for the Congress?' (India Today, 2002, p. 32). The farmers stopped paying even the revised flat charges in protest; and just before the May 2004 assembly elections, the chief minister announced a waiver of all past electricity dues; yet, he could not save his seat. His Congress government, until now eulogized for a progressive development-oriented stance, was trounced at the polls. Analysts attributed his defeat to the government's failure on three fronts: *bijli, pani, sadak* (electricity, irrigation, roads).

16 At a rate of 25 m^3/h, a tube well can pump 73,000 m^3 of water if it is operated whenever power supply is on. At the water entitlement of 5300 m^3/ha prescribed in the Narmada project, this amount of water can irrigate 13.77 ha of land.

17 The cost of power supply has three components: energy costs, fixed generation costs and T&D costs. The first two account for about 60–80% of the total cost to serve. Energy costs, which is variable, depend on the length of time of power consumption but fixed generation costs depend on how much a consumer uses at peak load. T&D costs depend on where the consumer is connected in the system. Since the

contribution of agricultural power consumption to peak load is often very little, the opportunity cost of power supply to agriculture is lower than the overall average cost of supply. Moreover, agricultural consumption, most of it off-peak, helps smoothen the load curve for the whole system and saves the back-up cost, which are high for coal-based plants and insignificant for hydropower plants.

18 In Tamil Nadu where farm power supply is free, 14h of three-phase power – 6h during the day and 8h during the night – is supplied throughout the year. In Andhra Pradesh, 9h of three-phase power supply is guaranteed, 6h during the day and 3h during the night (Palanisami and Suresh Kumar, 2002); however, it was recently reduced to 7h when the new government announced free power. This implies that in theory, a tube well in Tamil Nadu can run for more than 5000h in a year; and in Andhra Pradesh, it can run for 3200h. If the real cost of power is taken to be Rs 2.5 ($5.4)/kWh, depending upon how conscientious a Tamil Nadu farmer operating a 10hp tube well is, he or she can avail of a power subsidy in the range of Rs 0–93,750 ($0–2038)/year; and an Andhra farmer, Rs 0–60,000 ($0–1304)/year. Moreover, the stories one hears of farmers installing automatic switches that turn on the tube wells whenever power supply starts suggest that a large proportion of farmers are choosing to go overboard in using power and water. Palanisami and Suresh Kumar (2002) mention that many bore well–owning farmers lift water during the night to fill an open well using an automatic switch and then lift water during the day from the open well to irrigate their fields. True, they would not indulge in such waste if they had to pay a metered rate at Rs 2.5 ($5.4)/kWh; but they would not do this either if they got only 3–4h of good-quality power at convenient hours on a pre-announced schedule.

19 An extreme case is Tamil Nadu, where electricity consumption per tube well shot up from 2583kWh/year under metered tariff in the early 1980s to 4546kWh in 1997/98. However, this jump represents three components: (i) increased consumption due to degenerate flat tariff; (ii) increased consumption because of the increased average lift caused by resource depletion; and (iii) T&D losses in other segments that are wrongly assigned to agriculture. Palanisami (2001) estimated that 32% of the increased power use was explained by additional pumping and 68% by increased lift. However, he made no effort to estimate the third point, which we suspect is quite large.

20 We recognize that comparing hours of operation of diesel and electric tube wells is not the same as comparing the quantity of water extracted. However, in understanding the economic behaviour of tube well owners, we believe that comparing hours is more meaningful than comparing water produced. In any case, for the same hours of pumping, an electric pump would produce more water per horsepower compared with a diesel pump *ceteris paribus* due to the former's higher efficiency.

21 Punjab and Haryana have much more productive agriculture compared with other parts of India, with the cost of irrigation being just 8–10% of the gross value of produce. That might explain why the pumping pattern is inelastic to the energy cost. However, this is just a hypothesis and needs to be further confirmed.

22 Thus, Godbole (2002, p. 2197) notes: 'But if co-operatives are to be a serious and viable option (for power distribution), our present thinking on the subject will have to be seriously reassessed. As compared to the success stories of electricity co-operatives (in USA, Thailand and Bangladesh), ours have been dismal failures'.

23 Transaction costs of charge collection will be high even under flat tariff regime if farmers think they can get away. Throughout India and Pakistan, replacing nameplates of electric motors on tube wells has emerged as a growth industry under flat tariff. A World Bank study had recently estimated that in Haryana the actual connected

agricultural load was 74% higher than the official utility records showed (Kishore and Sharma, 2002).

24 Private electricity companies that supply power in cities like Ahmedabad and Surat instill fear of God in their users by regularly meting out exemplary penalties often in an arbitrary manner. The Ahmedabad Electricity Company's inspection squads, for example, are set steep targets for penalty collection for pilferage. To meet these targets, they have to catch real or imagined power thieves; their victims cough up the fine because going to courts would take years to redress their grievances while they stay without power. Although these horror stories paint a sordid picture, the Company would find it difficult to keep its commercial losses to acceptable levels unless its customers were repeatedly reminded about their obligation to pay for the power they use.

25 The Orissa Electricity Regulatory Commission has already opened the gate for the power utility to ask agriculture to fend for itself, when it decided that 'any expansion of the grid which is not commercially viable, would not be taken into account in calculating the capital base of the company. In future unless government gives grants for rural electrification, the projects will not be taken up through tariff route' (Panda, 2002).

26 T.L. Sankar, for instance, has already argued for the need to set up separate supply companies for farmers and rural poor that will access cheap power from hydroelectric and depreciated thermal plants and be subsidized as necessary directly by governments (Rao, 2002, p. 3435).

References

Batra, S. and Singh, A. (2003) Evolving Proactive Power Supply Regime for Agricultrual Power Supply. IWMI-Tata Water Policy Program (Internal report available on request), Anand, Gujarat, India.

Berk, J. and Berk, S. (1995) *Total Quality Management*. Excel Books, New Delhi, India.

Briscoe, J. (1999) The financing of hydropower, irrigation and water supply infrastructure in developing countries. *Water Resources Development* 15(4), 459–491.

CMIE (2003) *Economic Intelligence Service*. Centre for Monitorting Indian Economy, Mumbai, May.

Debroy, A. and Shah, T. (2003) Socio-ecology of groundwater irrigation in India. In: Llamas, R. and Custodio, E. (eds) *Groundwater Intensive Use: Challenges and Opportunities*. Swets & Zetlinger, Lisse, The Netherlands.

Godbole, M. (2002) Electricity regulatory commissions: the Jury is still out. *Economic and Political Weekly* 37(23), 2195–2201.

GOI Ministry of Water Resources (1999) *Integrated Water Resources Development: A Plan for Action*. Report of the National Commission for Integrated Water Resources Development Plan, Vol. I. Ministry of Water Resources, New Delhi, India.

Gulati, A.S. (2002) *Energy Implications of Groundwater Irrigation in Punjab.*, paper for the IWMI-ICAR-Colombo Plan sponsored Policy Dialogue on 'Forward-Thinking Policies for Groundwater Management: Energy, Water Resources, and Economic Approaches' organized at India International Centre, New Delhi, India, September 2–6.

Gulati, A. and Narayanan, S. (2003) *Subsidy Syndrome in Indian Agriculture*. Oxford University Press, Delhi.

Hekmat,0 A. (2002) *Overexploitation of Groundwater in Iran: Need for an Integrated Water Policy*, paper for the IWMI-ICAR-Colombo Plan sponsored Policy Dialogue on 'Forward-Thinking Policies for Groundwater Management: Energy, Water Resources, and Economic Approaches' organized at India International Centre, New Delhi, India, September 2–6.

India Today (2002) Running for cover: demand for a tribal chief minister and a proposed hike in power tariffs pose a serious challenge to Digvijay Singh's leadership. *India Today* 27(47), 32.

Kishore, A. and Sharma, A. (2002) *Use of Electricity in Agriculture: An Overview*, paper for the IWMI-ICAR-Colombo Plan sponsored Policy Dialogue on 'Forward-Thinking Policies for Groundwater Management: Energy, Water Resources, and Economic Approaches' organized at India International Centre, New Delhi, India.

Lim, ER. (2001) presentation made at Conference on Power Distribution Reforms, October 12–13, New Delhi, India.

Mukherji, A. and Shah T. (2002) *Groundwater Socio-ecology of South Asia: An Overview of Issues and Evidence*, paper presented in Conference on Intensive Use of Groundwater: Opportunities and Implications, Valencia, Spain.

North, D. (1997) The Contribution of the New Institutional Economics to an Understanding of the Transition Problem, United Nations University, Wider Annual Lectures 1.

NSSO (1999) *Cultivation Practices in India*, New Delhi, Government of India, National Sample Survey Organisation, 54th round, January–June.

Palanisami, K. (2001) Techno-economic Feasibility of Groundwater Exploitation in Tamilnadu. Presented in ICAR-IWMI Policy Dialogue on Groundwater Management, at CSSRI, Karnal, Haryana, India.

Palanisami, K. and Suresh Kumar, D. (2002) *Power Pricing, Groundwater Extraction, Use and Management: Comparison of Andhra Pradesh and Tamilnadu states*, paper for the IWMI-ICAR-Colombo Plan sponsored Policy Dialogue on Forward-Thinking Policies for Groundwater Management: Energy, Water Resources, and Economic Approaches' organized at India International Centre, New Delhi, India.

Panda, H. (2002) *Power Sector Reform in Orissa and Its Impact on Lift Irrigation; An Assessment and Lessons*, paper for the IWMI-ICAR-Colombo Plan sponsored Policy Dialogue on 'Forward-Thinking Policies for Groundwater Management: Energy, Water Resources, and Economic Approaches' organized at India International Centre, New Delhi, India.

Perry, C.J. (1996) *Alternative Approaches to Cost Sharing for Water Service to Agriculture in Egypt*. Research Report 2. International Irrigation Management Institue (IIMI) Colombo, Sri Lanka.

Perry, C.J. (2001) *Charging for Irrigation Water: The Issues and Options with a Case Study from Iran*. Research Report 52. International Water Management Institute Colombo, Sri Lanka.

Raghu, K. (2004) People's Monitoring Group, Andhra Pradesh, *Power Sector Reforms in Andhra Pradesh*, Presentation made in National Workshop on Electricity Act 2003 for NGOs, June 26–28 July.

Rao, S.L. (2002) The political economy of power. *Economic and Political Weekly*, XXXVII(17), 3433–3445.

Rao, D.N. and Govindarajan, S. (2003) *Community Intervention in Rural Power Distribution*, IWMI-Tata Water Policy Program, Water Policy Highlight No 14, Anand, Gujarat, India.

Repetto, R. (1994) *The 'Second India' Revisited: population, poverty and environmental stress over two decades*, Washington, DC: World Resources Institute.

Samra, J.S. (2002) *Impact of Groundwater Management and Energy Policies on Food Security of India*, paper for the IWMI-ICAR-Colombo Plan sponsored Policy Dialogue on 'Forward-Thinking Policies for Groundwater Management: Energy, Water Resources, and Economic Approaches' organized at India International Centre, New Delhi, India.

Scott, C., Shah, T. and Buechler, S. (2002) *Energy Pricing and Supply for Groundwater Demand Management: Lessons from Mexico*, paper for the IWMI-ICAR-Colombo Plan sponsored Policy Dialogue on 'Forward-Thinking Policies for Groundwater Management: Energy, Water Resources, and Economic Approaches' organized at India International Centre, New Delhi, India.

Shah, T. (1993) *Groundwater Markets and Irrigation Development: Political Economy and Practical Policy*. Oxford University Press, New Delhi, India.

Shah, T. (2001) *Wells and Welfare in the Ganga Basin: Public Policy and Private Initiative in Eastern Uttar Pradesh, India*. Research Report 54. International Water Management Institute (IWMI), Colombo, Sri Lanka.

Shah, T., Koppen, B.V., Merry, D., Lange, M.D., Samad, M. (2002) Institutional *Alternatives in African Smallholder Irrigation*: *Lessons from International Experience with Irrigation Management Transfer*. Research Report 60. International Water Management Institute (IWMI), V. 24p, Colombo, Sri Lanka.

Shah, T., Deb Roy, A., Qureshi, A.S. and Wang, J. (2003) Sustaining Asia's groundwater boom: an overview of issues and evidence. *Natural Resources Forum* 27(2003), 130–141.

Shah, T., Singh, O.P. and Mukherjee, A. (2004a) *Groundwater Irrigation and South Asian Agriculture: Empirical Analyses from a Large-Scale Survey of India, Pakistan, Nepal Terai, and Bangladesh*, paper presented at IWMI Tata Annual Partners' Meeting, Anand, Gujarat, India.

Shah,T., Giordano, M. and Wang, J. (2004b) Irrigation Institutions in a Dyanamic Economy: What is China Doing Differently from India? *Economic and Political Weekly* XXXIX(31), pp. 3452–3461.

Sharma, S.K. and Mehta, M. (2002) *Groundwater Development Scenario: Management Issues and Options in India*, paper for the IWMI-ICAR-Colombo Plan Sponsored Policy Dialogue on 'Forward-Thinking Policies for Groundwater Management: Energy, Water Resources, and Economic Approaches' organized at India International Centre, New Delhi, India.

World Bank and Government of India. (1998) *India-Water Resources Management Sector Review: Groundwater Regulation and Management Report*. The World Bank, Washington DC/ Government of India, New Delhi.

12 To Adapt or Not to Adapt: The Dilemma between Long-term Resource Management and Short-term Livelihood

SRINIVAS MUDRAKARTHA

Vikram Sarabhai Centre for Development Interaction (VIKSAT)
Nehru Foundation for Development, Ahmedabad 380054, India

Introduction

It has been estimated that groundwater contributes 9% to India's gross domestic product (GDP) (Vaidyanathan, 1999). Most of this contribution comes from the use of groundwater in agricultural and livestock production. Put the other way, agriculture and livestock – the two chief sources of livelihood for the masses of India – have come to depend heavily on groundwater use. While this use has brought much benefit to these sectors and the people who depend on them, the historically water-focused, narrow engineering approach of the government, combined with the tendency of people to 'make the most when available, otherwise the neighbour will take it away' has led to secular decline in groundwater levels in many parts of the country (Janakarajan, 1993, 2003). This has resulted in what may be termed the 'tragedy of the open access'. The increasing number of dark and grey zones,[1] and the persisting dependence of millions of farmers on groundwater indicate the chaos that will likely continue in the groundwater sector. The description of groundwater governance in India as a 'colossal anarchy' seems apt (Mukherji and Shah, 2003).

The groundwater problem in India is particularly acute in arid and semi-arid areas. Here, private investment has largely driven the groundwater boom. Farmers now chase the water table by digging and drilling deeper and investing in higher-capacity pump sets. These actions have far-reaching impacts that go beyond the simple economics of groundwater abstraction (Mudrakartha, 2004).

There is already a serious shortage of irrigation water, whether sourced from surface water or groundwater. In many areas, the situation has become so precarious that any shortfall in rainfall even in one season immediately generates a 'drought condition' affecting the lives of people in many ways. The falling groundwater levels also have resulted in drinking water scarcity, in particular where the centralized piped water supply schemes[2] source from groundwater (Mudrakartha and Gupta, 2004). Farmers are compelled to respond and adapt to these changes in a variety of ways to keep the hearth burning, even at the cost of disruption of their social and family life. At the extreme in terms of resource management, farmers sell their topsoil to brick kilns to abet other forms of land degradation (Moench and Dixit, 2004; Mudrakartha, *et al.*, 2004a). This indicates the desperation of some farmers who are not able to adapt.

What we now see is a dilemma between short-term livelihood and long-term resource management, between immediate gains and long-term human welfare as well as resource sustainability. The tendency to obtain short-term benefits even at the cost of resource degradation seems to have set in; the segment of population that depends directly on groundwater for its primary livelihood seems to be facing a constant threat to its conservation and resource management efforts.

This chapter attempts to capture the multifaceted social, physical, cultural, policy and economic dimensions of this dilemma through the study of farmer response to drought, an extreme and compressed example of the general decline in groundwater resources. The study was focused on three arid and semiarid districts in the Indian state of Gujarat, which experienced drought over the period 1999–2002 (Fig. 12.1).

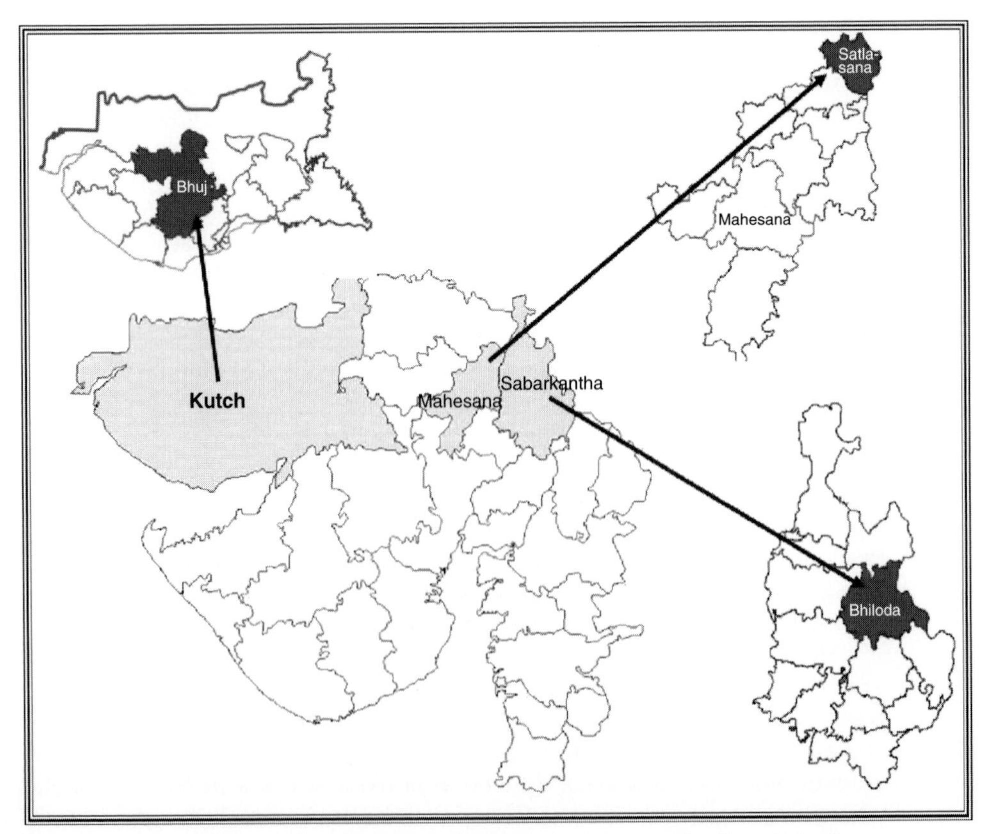

Fig. 12.1. Map of Gujarat showing areas of research study on adaptive strategies.

The chapter first describes the groundwater situation and drought in western India. It then depicts the differential impact of drought on agricultural production and the adaptations farmers have made to respond to new conditions. Finally, it examines how the impact of drought varies across the three study areas, the factors behind this differentiation and what it tells us about policy and practical options for groundwater management.

Drought and Groundwater Hydrologic Response

The definition of drought varies across countries and also within different areas of a country. Half of India, at any point of time, generally suffers from some kind of drought conditions. A meteorological drought[3] is defined as 'a sustained, regionally extensive, deficiency in precipitation condition' (Ramachandran, 2000). The impacts of meteorological drought on water resources, agriculture as well as social and economic activities give rise to what have been called hydrological[4] and, most important for our purposes, agricultural[5] droughts.

Agricultural drought occurs widely in India. About 68% of net sown area in India is highly vulnerable to agricultural drought. Most of this area is located in the 60% of the country that is arid and semiarid (Tenth Five Year Plan, 2002–2007). When drought occurs, there is a loss of biomass along with essential soil-building microorganisms due to the denuded soils being subjected to prolonged periods of dryness. As pressure on resources grows, there is often no time for the land to recover before it is put to use again.

Effects of prolonged agricultural drought, particularly in western India, are manifest in the form of drastic declines in groundwater levels. Out of the 7928 assessment units, 673 units fall under the overexploited category and 425 units under the 'dark' category. Gujarat falls in the highly overexploited category.[6]

As also highlighted by Shah (Chapter 2, this volume), data from the Minor Irrigation Census (Government of India, 1996) have shown that continuous decline of groundwater levels has resulted in a large number of wells and bore wells going dry in many parts of India. In western India, where depletion is the highest, more than 50% of the wells and bore wells are out of commission.

The most recent major drought spell in India was from 1999 to 2002/03, with conditions in 2000 being most severe. In 2000, as can be seen from Table 12.1, almost 55,000 villages or 12% of India's total were affected. The state of our case study region, Gujarat, too suffered from drought during the same period, again with 2000 being the most severe. In fact, the situation was so severe that not only water for agriculture but also drinking water for cattle and human consumption was in extremely short supply.

Description of the Research Study Areas

In the context of the 1999–2002/03 drought, we conducted a study in three areas of Gujarat, to try and understand how people respond differentially to changing resource, in particular groundwater, conditions and what that may suggest

Table 12.1. Losses due to drought 1999–2001. (From Tenth Five-year Plan, 2002–2007.)

Serial number	Year	Districts affected	Villages affected	Population affected (million)	Damage to cropped area (million ha)	Estimated value of damaged crops (million $)	Cattle population affected (million)
1	1999	125	NA	37.00	13.42	1.44	34.56
2	2000	110	54,883	37.81	36.70	79.12	54.17
3	2001	103	22,255	8.82	6.74	NA	3.428

for groundwater management paradigms. The 20 villages studied are located in the Bhuj *taluka* of the arid Kutch district, and in the semiarid Gadhwada[7] region of Satlasana and Bhiloda *talukas* in Mehsana and Sabarkantha districts respectively of the Aravalli Hills region, which forms the uppermost catchment of the Sabarmati river basin. All the study areas are drought-prone where climate is a major factor contributing to regular drought occurrence and desertification processes; in Kutch, there is also a salinization dimension. During the last 50 years, Kutch suffered 30 years of predominantly agricultural drought, while north Gujarat suffers drought 3–5 years in every 10-year period. The key socio-economic and physical aspects of the study sites are given in Table 12.2.

The study was conducted across 400 households spread across 20 villages in the three study sites. The study was conducted over a 2½-year period during 2002–2004. Data from beyond the study duration were also used as these were from regular project villages of Vikram Sarabhai Centre for Development Interaction (VIKSAT).[8] The study was carried out with structured questionnaires and unstructured checklists to capture certain adaptive strategy dimensions through focus groups such as with farmers and women. Since migration formed a key adaptive strategy, there was interaction with families also for understanding stress dimensions, and the extent of their willingness and comfort.

Both north Gujarat and Kutch are drought-prone regions; but the frequency, intensity and type of drought are different and so also is the perception of the people and their adaptive mechanisms. As seen from the following table, the rainfall conditions, social caste composition, natural resource conditions, hydrogeology and livelihood composition are all different.

Groundwater Decline, Drought Conditions and Associated Impacts

The study found that water level declines have been quite drastic in all the three study areas. For example, in Satlasana, the wells were dry with the shallow yielding aquifers totally dewatered. Attempts by farmers to deepen their wells, including drilling of vertical extension bores, met with limited success, as the additional yield did not sustain long. Some farmers took the risk and drilled

Table 12.2. Key characteristics of the study areas.

Serial number	Satlasana	Bhiloda	Bhuj
Literacy (state) 69.97% Male: 80.5% Female: 58.6%			
• *Taluka* (%)	61	70	50
• Male (%)	69	57	56
• Female (%)	31	43	43
Caste composition	Thakore; Chauhan (backward communities) and Patels	Scheduled tribes; Muslims	Rabari; Bharvad; Darbar and others
Livelihood options			
• Primary	Agriculture	Forest products; agriculture; animal husbandry	Animal husbandry; handicrafts
	Animal husbandry	Government service	Agriculture
• Others	Service (mostly private) and small business		
Climatic conditions	Semiarid zone	Semiarid zone	Drought-prone arid zone; disaster-prone (earthquake, cyclone)
Rainfall	650 mm	750 mm	350 mm, erratic
Resource condition (water and soil/land)	Moderate soil fertility; high groundwater depletion and quality deterioration	Moderate soil fertility; groundwater quality medium	Poor to moderate groundwater occurrence; high TDS in ground water; saline soils
Marginal farmers (%)	63	71	(20)
Landless (%)	14	3	57
Women	Practice *purdah* (veil) system	Practice *purdah* (veil) system	Practice *purdah* (veil) system
Others	Improved local breed of livestock; changing agricultural practices	Local breed of livestock; traditional` agricultural practices	Local breed of livestock; traditional agricultural practices

new deep bore wells. Only 5 out of 11 bore wells drilled across four villages yielded a reasonable quantity of water. The rest were dry. By 2001, most of the existing and new wells as well as the bore wells had more or less dried up. Low rainfall did not result in much surface water flows, and hence there was not much recharge to the ground, with the result that the cultivated area and the yields suffered a drastic reduction.

Figure 12.2 for Nana Kothasana is a typical representation of the above scenario for the Gadhwada region, while Table 12.3 presents data on the yield obtained based on focus group discussion in Bhanavas. The comparison was between a normal year (considered here as 1998) and the drought period 1999–2002. It is clearly seen that in Satlasana, the total annual agricultural production was reduced by a drastic 60–70% during *kharif* (monsoon) and 80–95% during *rabi* (winter); for summer crops, the reduction was in the range of 90–95% between 1996/97 and 2002/03. In many cases, the summer crop was almost nil.

Similarly, the impact on livestock was also severe: 10% of the cattle died in Bhiloda, 17% in Bhuj and 16% in Satlasana. The arid Bhuj also witnessed the death of 21% of its camels, in spite of their known resilience and adaptability to water-scarcity conditions (Mudrakartha, 2002 ; Mudrakartha *et al.*, 2004a).

The fall in agricultural output and loss of livestock generally had an adverse impact across all the rural families in the study areas. Rural families, who have a tradition of ensuring their family requirement of food grain through agriculture, were instead forced to purchase their food grain requirements from the market. The much-needed cash flow for this was coming from the animal husbandry, which had assumed greater significance as livelihood realignment took

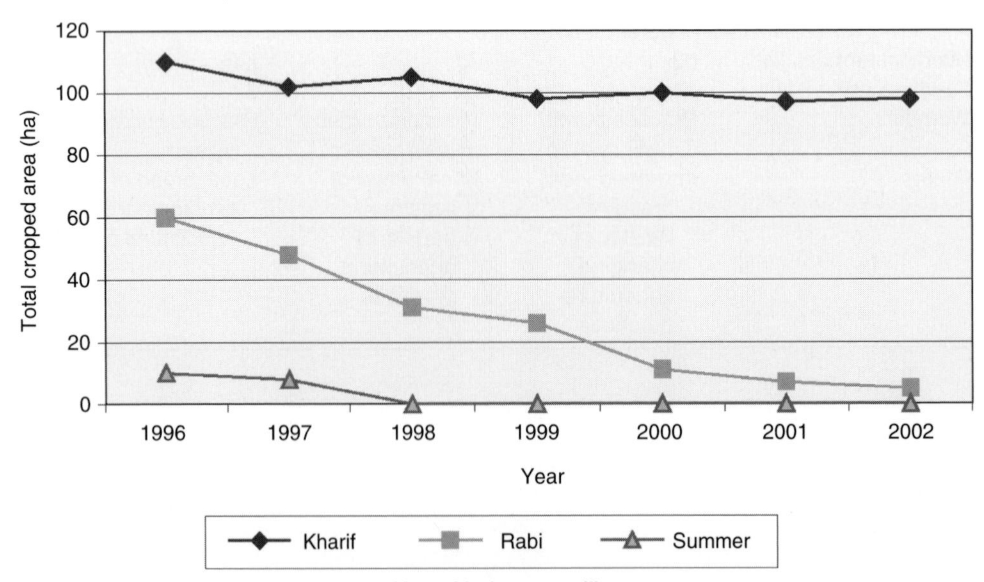

Fig. 12.2. Trends in cropped area in Nana Kothasana village.

Table 12.3. Decline in production of selected crops in Bhanavas village. (From Mudrakartha *et al.*, 2004a.)

Crop year	1998 (%)	1999 (%)	2000 (%)	2001 (%)	2002 (%)
Monsoon crops					
Groundnut	100	50	Did not cultivate	Did not cultivate	Crop failed
Cluster beans	100	30	Did not cultivate	Did not cultivate	Crop failed
Maize	100	50	Did not cultivate	Did not cultivate	Crop failed
Minor millet (*bajra*)	100	70	50	25	Crop failed
Winter crops					
Wheat	100	50	25	10	5
Mustard	100	50	Did not cultivate	Did not cultivate	Did not cultivate
Tobacco	100	10	Did not cultivate	Did not cultivate	Did not cultivate

Note: Year 2000 was the severest of all the 4 years of drought.

place. Figure 12.3 is a typical representation of the livelihood realignment in the study villages. As can be seen from the figure, there was an overall drop in income to 33% of its previous levels by the end of the 4-year drought period. However, this drop and the overall impact of drought were not uniform across study sites. Reduction of income has also led to families spending less on food. While this reduction was 70% in Satlasana and 30% in Bhiloda, Bhuj families ended up spending 9% more than usual. It is interesting to note that the availability of work and cash flow in Bhuj in the years after the earthquake of 2001 helped them to spend money on food.

A few more things were happening on the agriculture front. First, due to the prolonged drought there was total erosion of the well-established agrobiodiversity using local composite seeds and low-chemical fertilizers. Second, since farmers' cash flow was greatly eroded during, or at the end of, the drought period, they bought poor-quality seeds pushed by moneylenders who also sell agricultural inputs. Third, newer seed varieties pushed by the market replaced the conventional, pest-tolerant local varieties.

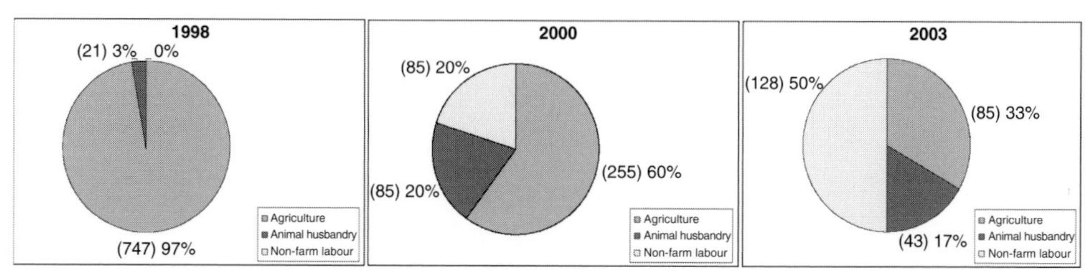

Fig. 12.3. Changes in relative share of livelihood income sources.

Economic Impacts

Figure 12.4 shows the economic impact of drought on the people in the three study sites. As can be seen, there is a movement of families from both higher to lower income levels and below poverty line (BPL)[9] under the influence of prolonged drought conditions. In other words, people have become poorer in

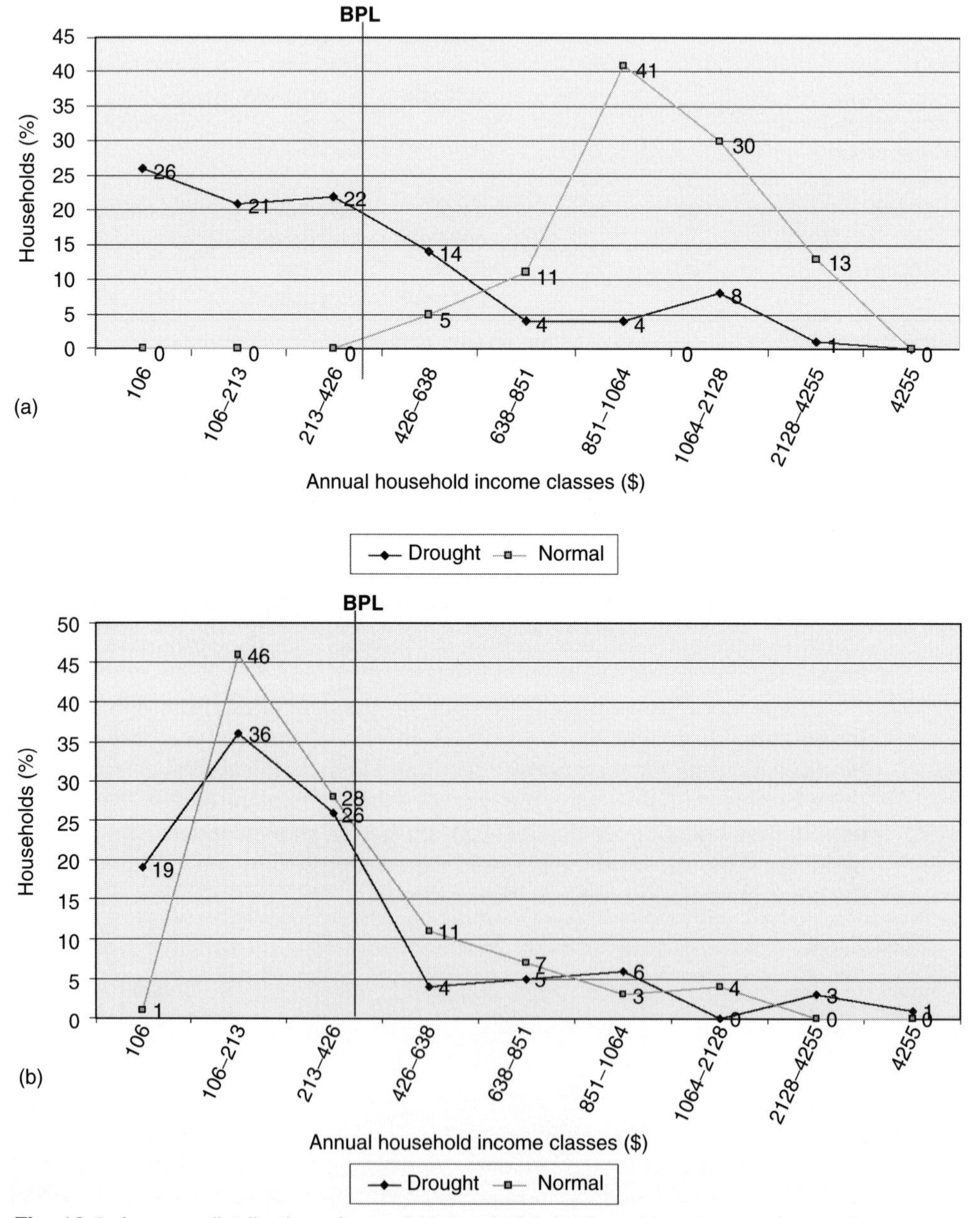

Fig. 12.4. Income distribution of sample households in drought and normal years in (a) Satlasana area, (b) Bhiloda area, and (c) Bhuj area.

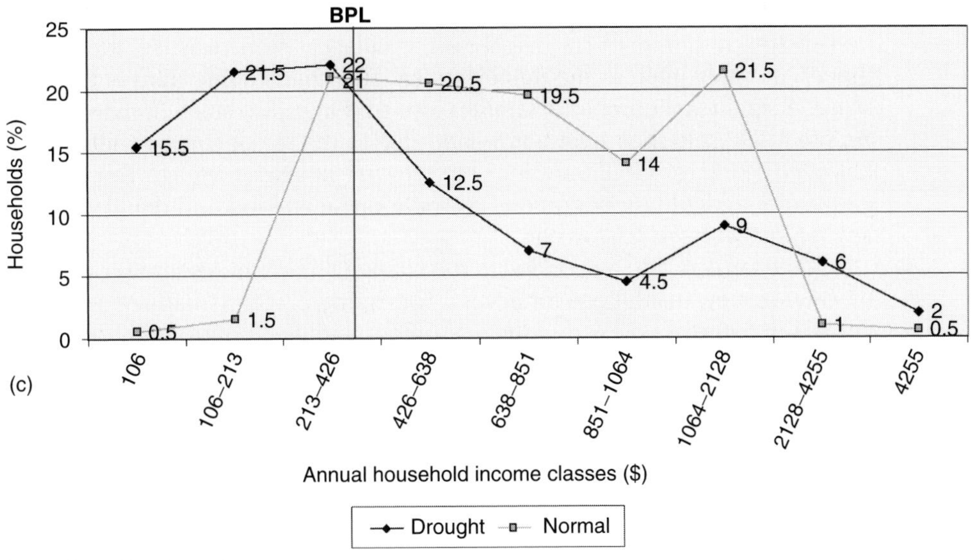

Fig. 12.4. *Continued*

both relative and absolute terms. However, the impacts were not the same in each study area.

Satlasana turned out to be the most vulnerable in spite of people's generally good economic condition. From almost nil during normal years, the number of BPL families during drought swelled to as high as 69%. High drought proneness has made Bhuj families to evolve handicrafts and metalwork, often for the international market, which provide significant income and cash flow throughout the year. Although Bhiloda suffered less relatively, in terms of intensity and magnitude, it was significant. In-depth discussion is available in Moench and Dixit (2004).

Adaptive Responses

How did the people respond to the declining water levels and the drought conditions? In the first instance, those families who had savings tended to use them to cover food expenses and other basic requirements. During the initial years of drought, about 35% of the respondents in Bhiloda and 13.5% in Bhuj, though none in Satlasana, used up their past savings. Many families also resorted to borrowing money: 47% in Bhiloda, 23% in Bhuj and 19% in Satlasana.

In addition to current consumption, farmers also deepened wells, drilled new bore wells and invested on higher horsepower pump sets in an effort to meet with critical irrigation and livestock needs. People sought to raise the required money mostly from traders and moneylenders, often at a very high rate of interest (36–60% per annum depending upon credibility, amount of loan, duration and mortgageability). Interestingly, banks were not willing to finance drilling of bore wells. For example, out of the eight farmers who drilled bore wells in Bhanavas village, five borrowed from moneylenders at a 3%

monthly interest, one sold jewellery to raise money and two others borrowed from better-off farmers in a neighbouring village, Mumanvas. In addition to the 3% monthly interest, the moneylenders also charged one-third of the crop share. A significant number of families also sold trees, livestock or other assets. People rarely sold their agriculture land due to the social status attached to it but often mortgaged it. Interestingly, in the case of land mortgage, the creditor carries out agricultural operations on the mortgaged land and does not share any returns with the landowner.

Some farmers also responded to the challenging water resource conditions by creating new institutions for access. For example, some small and marginal farmers in Satlasana pooled in their resources and went in for joint bore wells in a bid to access groundwater for irrigation. This strategy to share the high cost of investment meant access to water that these farmers could never afford individually (see Box 12.1).

There was a drastic impact on the livelihood occupation scenario between the normal year 1998 and the drought years 1999–2002. As can be seen from

Box 12.1. Groundwater and livelihood change

Chhatrasinh of Bhanavas, high school–educated, married, father of three, has 4 acres of land in two pieces. The 3-acre piece has irrigation facility from a joint well. He followed the general cropping pattern of the area, i.e. groundnut, *gowar* (cluster beans), *bajri* (pearl millet), castor and maize in *kharif*; wheat and castor in *rabi*; and *bajri* in summer. He dug these wells in 1980 and installed electric pump sets of 5 hp each in 1987. Plenty of water was available at a depth of 5–8 m below ground level. Responding to the demand, Chhatrasinh used to sell water to seven farmers to irrigate 10 acres of land. The payment terms varied. Some paid at the rate of one-third of the crop, while others paid $0.25/h. The farmers who used to cultivate castor needed only two irrigations for 4 h in a month, whereas crops such as wheat needed 6–7 irrigations in a month. Hence, the castor cultivators used to pay in cash and the wheat cultivators, in kind. Chhatrasinh used to earn about $425 per year by selling water for winter and summer crops.

Chhatrasinh sold water this way for almost 8 years till the water levels started to dip in 1995. During that year, he deepened both the wells by 8 m each. Within 5 years, i.e. by 2000, both the wells dried up again. Chhatrasinh decided to drill a new bore well. Although drilled to a depth of 80 m the new well struck no water. A few months later, he drilled another bore well of 100 m depth. This well struck water at 65 m, which was enough to irrigate 4 acres of land. For 2 years, the second bore well yielded. By 2003, i.e. within 3 years, the bore well could irrigate just 1 acre of fodder crop (*bajri*). This was a jointly owned well, shared with his cousins. Together they had borrowed $1065 at a monthly interest rate of 3% for drilling this bore well from a private source. So far they could not repay the loan.

After the depletion of groundwater and subsequent collapse of agriculture, for the last 2 years, Chhatrasinh and his son are working as labourers wherever work is available. During 2003, although the monsoon was good after a bad spell of 4 years, he had sown only *bajri* as he did not want to take any major risks with the uncertain monsoon.

Fig. 12.3, in 1998, 97% of the people were engaged in agriculture, which reduced drastically to 33% in 2002. The displaced farmers abandoned agricultural operations temporarily and migrated to urban centres to work as construction labourers, or as agricultural wage labourers in better water-endowed areas (Moench and Dixit, 2004; Mudrakartha *et al.*, 2004a).

Animal husbandry gradually emerged as an important means of livelihood occupation during drought in Mehsana and Sabarkantha districts as it could feed easily into the existing dairy co-operatives (Fig. 12.3). The drought compelled the people to take a re-look at their animal husbandry practices. They abandoned their unproductive cattle, and took better care of the productive ones indicating a significant change in the mindset. This has allowed them to increase their net returns from animal husbandry in spite of animal deaths (Fig. 12.7).

Dairies such as the Mahesana Dairy in Gujarat that have a mandate to take care of the small milk producers take up collection, storage, processing and redistribution of milk to the whole district and beyond. The dairy also manufactures and sells milk products throughout the year. During the drought, the dairy came forward to supply food concentrate for cattle so as to maintain its own production schedules. Since the returns were quick, and the much-needed cash was available in dairy farming, farmers ploughed back some earnings from the milk income for purchasing fodder at higher cost from elsewhere; they also outsourced subsidized fodder supplied by the government as part of the drought relief programme.

Migration (permanent, temporary and commuting to nearby villages and urban areas for work) emerged as another important adaptation strategy. Around 15.5%, 10.8% and 21.4% of the population migrated from Bhiloda, Bhuj and Satlasana, respectively. About 21% of the working population of Satlasana commuted to the nearby town for work on a daily basis. About 2.3% of children below the age of 14 from Satlasana had migrated for work. Child migration also took place either along with parents or individually, which not only affected their education but also exposed them to greater health and security risks (Mudrakartha *et al.*, 2004b; Moench and Dixit, 2004).

The study found that the overwhelming reason for migration was livelihood-related employment. As much as 100% of the migrants in Satlasana, 96% in Bhiloda and 87% in Bhuj migrated in search of employment. In Bhuj, since livestock is a major source of livelihood, 13% of migrants migrated purely for the purpose of grazing cattle.

The caste system and infrastructure development also played an interesting role in facilitating migration. For example, people used their kinship relationship and social networks for obtaining information about the availability of wage labour (civil, construction, semi-skilled and others) and job opportunities through caste members residing in nearby well-endowed villages, cities and towns. The massive expansion of road network, power projects, bridges and communications in recent decades facilitated the movement of information as well as labour force. Although migration was prompted by immediate need, in a number of cases migrants stayed on, leaving agriculture to other family members or leasing away their land.

Further, some farmers have also resorted to the extreme option of selling or leasing away topsoil to manage livelihood stress. This phenomenon is seen in areas with severe water scarcity and dried up aquifers such as in Satlasana.

What role does the forestry management play? The study shows that consistent, longer-term investment on resource regeneration has a positive impact on the environmental flows, and thereby reduces the impact of, and vulnerability to, drought. The following report compares the three study sites from this angle.

Bhiloda villages have invested time and efforts on forest protection and regeneration in thousands of hectares under the inspiration and guidance of VIKSAT and the Bhiloda federation. Regeneration of catchment areas has helped significant surface water conservation, resulting in availability of groundwater throughout the year. A noteworthy difference is that while the impact of rainfall failure is felt immediately in Satlasana, it is felt with a time lag of 1½ years in Bhiloda. In other words, Satlasana was less prepared when a prolonged and intense period of drought occurred recently (1999–2002/03) and therefore had to suffer the most. As negligible forest area exists in Satlasana, people have of late focused more on the non-land-based income-generating activities. One of the most popular alternative options for women is the diamond-polishing industries and private businesses.

It may be mentioned that the forestry programme in Bhiloda has been active for the last two decades supported by an non-governmental organization (NGO)[10] through promotion of effective, robust institutions at village[11] and *taluka*[12] levels and was expanded to the state[13] level. Furthermore, not just a few villages, but most of the Bhiloda *taluka* is engaged in the ongoing successful joint forest management[14] programme, which, in addition to maintaining the environmental flows, also provided them interim forest products. These include non-timber products (*amla*, *timru* leaves, gums and resins, *safed musli* and other herbal products) as well as fuel wood, fodder and grasses; small timber products help them to obtain critical additional cash income. On an average, a family earns $25–110/year from any one product, in addition to fuel wood and fodder collection. Wage labour is also available in forests for plantation and other works regularly provided by the forest department (VIKSAT Annual Reports, 1998–2005).

Further, the tribal job reservation policy has ensured that there is at least one working member from every third family in Bhiloda; the policy of free education for women has encouraged more women to go to schools and colleges in order to improve their chances of obtaining jobs. Finally, prolonged exposure to drought conditions historically has led families in Bhuj to evolve alternative income-generating occupations such as handicrafts and metalwork. They have also developed reasonable links with the international market.

People's Perceptions and Responses to Droughts

People's responses to a particular event have a strong relationship with the social, cultural, climatic, physical and psychological aspects. There has been a perceptible change in the manner in which disasters, in this case droughts, are being viewed and managed. Prolonged and frequent innings with droughts have compelled people to evolve adaptive and coping mechanisms in tune with the changing externalities. Drought is no more considered in its conventional sense, but means different things to different people. For instance, farmers from western Rajasthan and Gujarat feel that drought is when their son loses job in the city (Moench and Dixit, 2004) or when they are forced to employ at least one male member outside of the family avocation. This perception also varies with caste. For Darbars (a forward caste), drought is when the woman is also forced to work as labourer, as happened in the 1999–2002/03 drought spell.

The study found out that about 60% of the population in Satlasana and Bhuj areas believe that the drought is due to insufficient rain while in Bhiloda, only 28% subscribe to this view. On analysis, it is found that the risk-taking ability in Satlasana and Bhuj is low, while it is high in Bhiloda, aided by the confidence derived from healthy management of village forests that yields fuel wood, fodder and non-timber forest products, some of which they sell and obtain reasonable cash flow (VIKSAT Annual Reports, 1998–2005). The tribals of Bhiloda find that investment in forest management would secure their livelihood. Further, resource exploitation and consequent livelihood erosion is not a major issue in Bhiloda due to restrictions by certain tribal-related policies. For example, sale of land beyond tribal families is legally prohibited with a view to protecting their livelihood. However, such policy may also restrict development of tribal areas, although agriculture- and livestock-based livelihood is less threatened. Further, tribal job reservation policy assures government service for at least one member per family; women increasingly participate in small businesses, all of which develops a sense of confidence.

The perception of scarcity of water as a major reason for drought comes out clearly as believed by 80–90% of the respondents from Satlasana and Bhiloda and only 16% from Bhuj.

On the practical front, diversification is emerging as a major strategy to reduce livelihood vulnerability. Diversification is happening on two fronts: one externally, beyond the primary vocations, agriculture and animal husbandry; two internally, within the agriculture sector, for example, by going in for a mix of crops as in the case of Satlasana farmers.

The study also identified some extreme cases of adaptive mechanisms such as sale of assets (land and cattle) and topsoil, which are often difficult to earn back. People are aware that removal of topsoil leads to serious micro-level ecological and soil nutritional imbalance, which has not only immediate effect on yields but also livelihood implications for generations. Although faced with the immediate need of maintaining the families, farmers chose this option because they prefer selling topsoil to land. This option not only jeopardizes the family-level food security system (for both humans and livestock) but also results in loss of contribution to the national food basket.

People have been increasingly adopting external diversification too. Possessing diverse skills is being recognized as a sure way of widening the safety net against drought. For example, in the tribal Bhiloda *taluka*, there has been a perceptible increase in the attendance of school- and college-going children, both boys and girls, primarily motivated by the job opportunities. Every third house has at least one person working in government service, in addition to a significant number working in private establishments and in small-scale industries.

The last 4 years have seen women self-help groups (SHGs) increasingly seeking loans from financial institutions such as the nationalized banks and National Bank for Agriculture and Rural Development (NABARD)[15] for various livelihood purposes, which indicates the increasing role of women in livelihood, and their concern to become 'creditworthy' (Mudrakartha, 2006). Access to funds has helped women members to really consolidate the livelihood options, not only the primary occupations, but also the non-land income-generating activities. This increased basket of options is enhancing people's capabilities to face future drought events with confidence.

Gender Implications of Drought Events

The study shows that female literacy is very low at 31% in Satlasana (a heterogeneous community), compared to the state average 38%, because of the strong perception that girls should take care of household work and siblings. Consequently, 98% of women are engaged in household work (Mudrakartha *et al.*, 2004b; COMMAN, 2005).

Interestingly, the Bhiloda tribal belt and Bhuj have a higher female literacy rate at 43%. The availability of service sector options in Bhiloda has encouraged enrolment of girl children in schools, which is higher compared to other study areas. Women also take up business (22%) in Bhiloda. In contrast, in Satlasana, women do not go in for either service or business. This is because of the sociocultural restrictions on women, especially those of higher castes.

Although Bhiloda and Bhuj show similar female literacy, the business opportunity for females is slightly more in Bhiloda. In contrast to Bhiloda, the service opportunity in Bhuj is found to be nil.

Notably, a disturbing trend is found in the sex composition across the study areas. The overall sex (female/male) ratio (Bhiloda, Bhuj and Satlasana) was 920:1000 as per primary survey (Moench and Dixit, 2004; Mudrakartha *et al.*, 2004b) comparable with official record of 919:1000 (Census, 2001). However, the primary survey threw up the following startling facts:

- Bhiloda: 928:1000; Bhuj: 965:1000; Satlasana: 920:1000.
- Sex ratio of children up to 5 years: Bhiloda: 717:1000 (highly unfavourable to females); Bhuj: 855:1000; Satlasana: 756:1000.
- In Satlasana, the sex ratio in the age group of 6–14 years is 662:1000, which is alarming (due to preference for male children, *inter alia*).

In other words, the overall sex ratio of children up to 5 years of age in the study areas is 789:1000, which is a matter of serious concern. Of much more concern are the Bhiloda and Satlasana areas where the ratio is even more skewed. This scenario projects a great gender and social disparity for the future.

Does drought have an impact on the adverse sex ratio? It was difficult to establish a direct link between the adverse sex ratio and droughts, also because this dimension was beyond the scope of the project. However, indirect evidences include, in addition to the sociocultural beliefs and other practices, the drastic reduction in the expense on food consumption in chronically drought-affected areas, which was 70% in Satlasana and 30% in Bhiloda. In contrast, the expense increased by 9% in Bhuj due to the availability of cash flow because of the large number of post-earthquake relief and rehabilitation programmes. Such an adaptive approach has more inherent sacrifice by womenfolk, who in Indian custom prefer to feed the adult male and the male children first. It was informally gathered that this often led to malnutrition and increased susceptibility to illnesses of mother and child – all of which was beyond the scope of the research study.

Was there an effect on the marriage prospects of girl children in view of the economic and health impacts of drought? The study established a direct link in terms of a rise in the marriage age of girls as prospective families avoided marrying their children into families living in drought-prone areas. Early marriages were also reported as some poor families married off girl children because they were unable to feed them; often, two sisters were married off at a time to save on the marriage costs. The lower dowry demand for younger girls also contributed to early marriage. Ironically, there was also a rise in the marriage age of girls as some parents delayed the marriage of the second daughter because they needed time to gather money. This was particularly observed in families with 2–3 girls or more.

Interestingly, over the last 6–7 years, the Prajapati community of Satlasana *taluka* has evolved a system of 'mass marriages' as a coping mechanism. This is a low-cost marriage arrangement where many girls and boys are married off in a common ceremony, and there is no demand for dowry.

Carrying Forward Adaptive Strategies

In the post-drought spell of 1999–2002/03, specific efforts were made both by the community-based institutions and the local NGO (VIKSAT) to take forward people's adaptive strategies.

- Convinced by the performance of the water-harvesting structures built by VIKSAT in 2001, an increasing number of village institutions are drawing government schemes for construction of check dams and farm ponds under Sardar Jal Sanchay Yojana and Sujalam Sufalam schemes, respectively. They now recognize that in semiarid and arid zones, enhancing water storage is imperative and, if possible, within the subsurface to avoid the high evaporation losses. For example, in Satlasana, more than 100 check dams have been constructed during the last 4 years. The Augmentation of Groundwater Resource through Artificial Recharge (AGRAR)[16] study established that the tanks and check dams, in that order, are the most efficient structures in the given hydrogeological conditions to enhance groundwater recharge and stabilize agricultural yields, and would help reduce vulnerability to rainfall uncertainties (Mudrakartha *et al*., 2005). In Mehsana and Sabarkantha, water-harvesting structures are still less in vogue.
- Initially supported by NABARD, women in Satlasana and Bhiloda have started forming into SHGs 3–4 years ago. Taking bigger strides, they have recently

federated at the *taluka* level to carry forward the process to the large number of villages in the *talukas*. Significantly, in Satlasana, the State Bank of India (SBI) was so impressed with the functioning of the SHGs that it extended loans to these villages previously considered non-creditworthy (Fig. 12.5). Over the last year, these groups have taken loans to the extent of $90,000 with 100% repayment. Interestingly, some groups have also taken loans for the purpose of constructing check dams, which they repaid once the installment from the government scheme was available. This new initiative and noteworthy performance of the Satlasana groups has encouraged the bank to make the check dam construction a bankable scheme, which is a significant policy change.

- Looking at the Satlasana experience, the same bank (SBI) has extended financial support to Bhiloda and Bhuj villages also. Thus, all the three study areas now have access to funds on repayment basis. This also signifies a marked change in the mindset of the people, from expectations of charity or doles to self-reliance with dignity.
- Linkage with banks and access to funds allow people to earn back their lost assets such as livestock and jewellery and strengthen agriculture. More importantly, they only need to pay 8–11% rate of interest per annum (as against 36–60% charged by money lenders). In other words, people are now better equipped to face any future drought thanks to the access to bank loans, which was absent during the 1999–2002 spell.
- Analysis of bank loan utilization indicated that almost 70% of the loan was for agriculture and animal husbandry, while 10% was for releasing land mortgaged during the drought period. This interesting paradigm shift is clear evidence that women's participation in family livelihood has gone up, adding

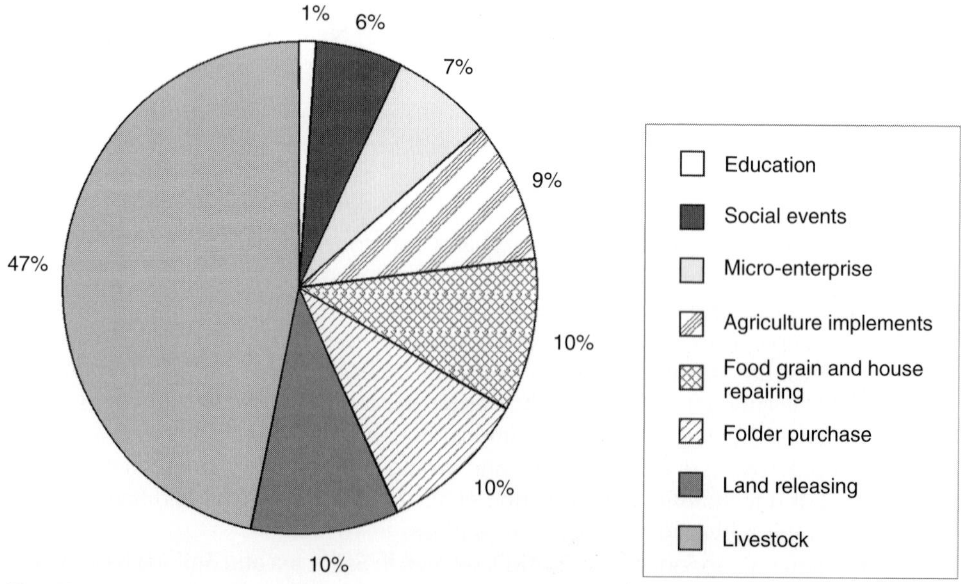

Fig. 12.5. Loan from State Bank of India and purposes.

a new dimension to livelihood management. It has also initiated a direction towards women empowerment evident from their decision-making role at family and village levels. Historically, men used to take 7–12 years to release the mortgaged land or to repay loans taken from moneylenders. Women could hasten livelihood restoration because of their intrinsic risk-taking ability and vision (Mudrakartha and Madhusoodhanan, 2005; Mudrakartha, 2006).

- In addition to banks, many families have taken loans from the revolving fund of the local NGO (VIKSAT) for purposes such as releasing pawned jewellery, setting up small businesses, purchasing food grains and fodder (Fig. 12.6). This fund, operated by a committee comprising representatives of the people's institutions, local leaders and the NGO as per certain norms, catered to those needs that are not covered by the bank, have high interest rate or entail cumbersome procedures. Almost 62% of the fund was used for purchase of seeds and agricultural inputs, while 17% was used for purchase of livestock as it could easily feed into the existing dairy business (Mudrakartha, 2006). Farm-based micro-enterprises such as *amla* products, processed condiments, spices and chilies, as well as non-farm-based enterprise such as handicrafts, bakery, *kirana* shops, flour mill, washing powder preparation and cloth products are slowly picking up. Transactions to the tune of $85,000 are made till date, which indicates vision, commitment and financial management skills of the women's groups and the federations.
- Concepts of seed village and fodder security have been introduced, as a part of which select farmers from within a village were given good-quality improved seeds for multiplication. Beginning with one village in 2003, in 2005/06 almost 200 farmers across eight villages are raising seeds that would meet the requirements of approximately 10–12 villages. Similar procedures are being adopted for fodder, but generally as part of the integrated agriculture approach.

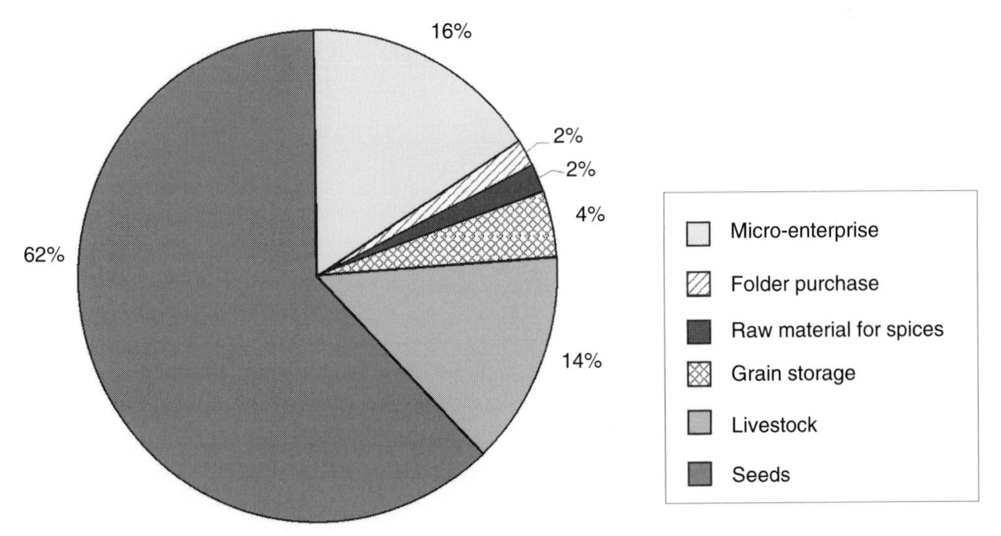

Fig. 12.6. Loan from VIKSAT revolving fund and purposes.

- In the context of institutions, the community-based institutions are more or less in place. Post drought, specific efforts were made to link up with concerned government departments of agriculture, research, extension, seeds, horticulture, livestock and water resources in order to benefit from their technical knowledge as well as draw projects and schemes, including demonstration experiments.
- The federations are taking up a bigger role in terms of sourcing agricultural inputs and fodder in bulk and trading, in the process, providing a decent saving for the farmer as well as improving the federation's financial position. Women's groups and women's federations are making progress in terms of rendering loans accessible to more number of members who are investing in agriculture as well as in livestock purchase. The milk production is linked with the local diary, increasing cash availability (Fig. 12.7).
- Migration resource centres are planned to help migrating families make informed choices so that vulnerability is reduced. In addition to making use of information technology (IT), these centres will be operated by educated members from the villagers themselves. However, the NGOs will help manage and train on the technical part. As part of the activities, information on the menu of adaptive strategies in vogue will be disseminated. This strategy is relevant due to the fact that the understanding of, and response to, drought events is not uniform among villages and within villages.

Implications for Policy

First, there is a need to recognize the fundamental link between groundwater and drought, particularly in the semiarid and arid zones (Mudrakartha

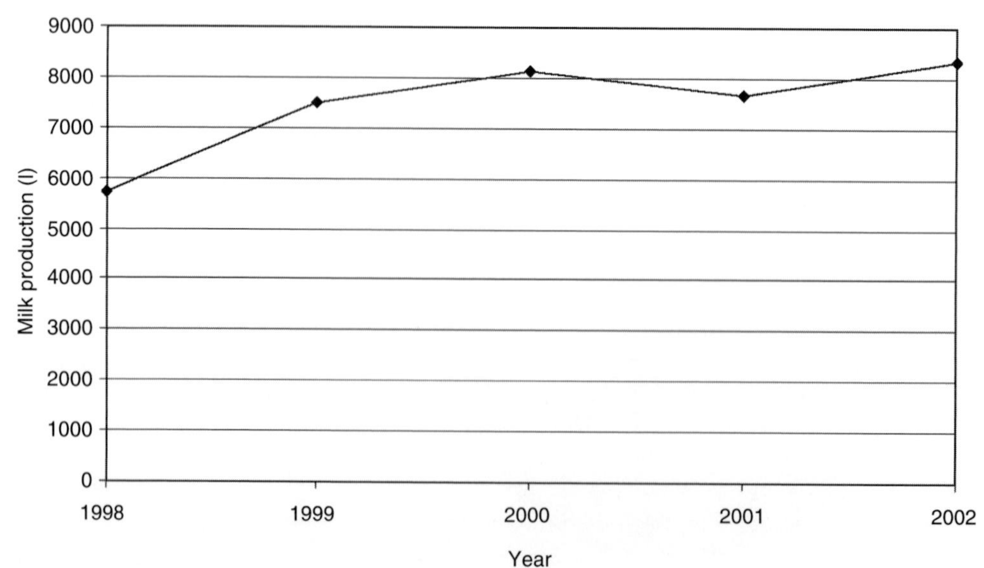

Fig. 12.7. Milk production in Umri village, Satlasana *taluka*.

and Madhusoodhanan, 2005). Two aspects are important: (i) the role of the community in planning, implementation and monitoring, which should include space for community management of groundwater; and (ii) the need for integrating drought interventions into the district perspective plan[17] so that ad hoc spending of large funds during drought events is avoided.

Annual recharge is a crucial element in the context of community management of groundwater and forms the lifeline of the productive systems in drought-prone areas. Therefore, rainwater harvesting for enhancing recharge artificially on a scientific basis should form an integral part of the district perspective plan taking into consideration the site-specific hydrogeology to make the recharge activity effective and meaningful (Mudrakartha and Madhusoodhanan, 2005). Site-specific research is essential to evolve an array of artificial recharge standards to suit different hydrogeological settings (Mudrakartha *et al.*, 2004a).

Second, appropriate policy change to promote community management of ground water is an immediate necessity. Indian experience with forestry management, canal irrigation, watersheds, etc. shows that without people's participation resource protection, development and management is not possible. It is logical that the same principle be extended to the management of groundwater. This is relevant even in the face of the rapidly changing externalities as a majority of the population still depends upon natural resource-based primary productive systems.

Third, there is a need for convergence of institutions. Many institutions are promoted as part of various rural development programmes, often in the same village or in a cluster of villages. While multiplicity of institutions is not an issue, convergence and mutuality constitute the need of the hour. Therefore, an institutional arrangement that coordinates the functions of the various institutions within the perspective plans of an area or an agroclimatic zone needs to be evolved.

Fourth, management of resource through community (e.g. forestry or groundwater) always throws up a variety of management issues. Lack of legal authority, in particular, related with resources, severely hampers their effective functioning. This would also bring about conflicts and litigations leading to an adverse impact on the resource management. Appropriate policy changes are therefore needed to empower the people's institutions. A related concern is the convergence with the *gram sabha* in some way.

Fifth, the changed resource paradigm demands co-management of conventional and people's adaptive strategies (Mudrakartha, 2004). Conventional strategies provide a broader canvas, including linkages with micro-level implementation while adaptive strategies help rooting the conventional strategies in the community domain.

Sixth, livelihood diversification both within and beyond the primary productive sectors needs a stronger push. Although this is happening, the efforts are mostly straitjacketed, i.e. highly sectoral. While the financial institutions push for formation of SHGs and micro-enterprises, the backward and forward linkages are often neglected and, as a result, weak. While a focused, target-oriented approach may help in achieving scale, a broader policy canvas should be spelt out to convey the larger picture.

Seventh, women have proved their skills in strengthening livelihood and also diversification in order to reduce vulnerability. Ensuring backward and forward linkages with women's enterprises for better results will help build stronger and more resilient adaptive mechanisms. Necessary capacity-building strategy should become an integral component of the programme implementation (Mudrakartha, 2006).

Eighth, access to credit is important to help communities come out of the indebtedness trap. A proper combination of community-based institutions, local formal institutions (e.g. dairy) and a committed NGO with active financial agency (e.g. SBI or NABARD) could result in strengthening people's capacities to evolve more effective adaptive strategies.

Communities are increasingly making use of communications technology such as telephones and mobiles through social networks for making informed choices of work during migration. However, since this trend has set in only of late and as it depends upon a variety of complex factors, it may be too premature to expect people to be in complete command and control of their adaptive strategies in the choice of work. The key message is that people are developing confidence both at the family level and the community or village level to face droughts. Their attempt is to develop the ability to maintain the primary livelihood systems, namely agriculture and animal husbandry, mostly in combination. They would like to complement these efforts with non-land-based options to build in the required capacity to adapt to increasing resource challenges. They are also increasingly becoming conscious of the need for an institutional approach to take full advantage of the social capital they have built up over a period of time. This strategy allows them to choose occupations they prefer, and not any occupation under compulsion, as it used to be.

In order that community's efforts are effective, suitable policies to check resource (groundwater, surface water, land, etc.) depletion, degradation and diversion should be formulated carefully. Policies should not be wishful statements but rather those that value processes and are community-centric. They should aim at bridging disconnects – a major lacuna – not only in many of the policy statements but also in implementation. The fact that in India livelihood and natural resources are intricately connected for a vast majority of the population should be borne in mind. Realignment of livelihood and trimming down the huge (and unwieldy) number of agriculture-dependent families up to a certain level may happen naturally because of the availability of an increased range of non-land income-generating options, thanks to IT, infrastructure and communication projects. However, for the adaptive strategies, this is a welcome trend as a wider basket of options reduces vulnerability and increases stability.

Conclusion

Groundwater decline and the associated quality problems in India during the past couple of decades have resulted in severe challenges to the primary productive systems, namely agriculture and dairy. As much as 60% of India's geographic area is under semiarid and arid conditions where the agriculture-based and dairy-based livelihood often gets jeopardized due to long spells of drought conditions.

A study on 400 households in three locations in arid and semiarid regions of Gujarat found that people's response to such drought conditions is not uniform. It varies from a reasonably well-thought-out strategy to ad hoc measures, from household-based to community-based institutions. This variation is found to be dependent upon factors such as social and kinship networks, awareness and education levels, ability to diversify within the primary productive systems and beyond, and non-land-based income options. This is in addition to the economic status of the family and the cohesiveness of the particular caste.

However, it is also found that in spite of the social, economic, caste and gender differences, the presence of a strong and robust institutional mechanism (e.g. dairy, village and *taluka*-level co-operative societies and a committed NGO) goes a long way in providing a complementary, enabling support to families in their adaptive efforts. In particular, during drought conditions, people often made desperate efforts to corner whatever groundwater was available to sustain their *kharif* and *rabi* crops. But long dry spells tended to erode even local seeds and biomass, placing the affected in the hands of the market forces.

Although perceptional differences exist among communities from the three areas as to the causes of drought, a majority believed in the lack of adequate and timely water availability, including from groundwater sources, as a key reason for the livelihood woes. As part of their adaptive strategy, people resorted to borrowing money, selling away jewellery, migration and dairy business. Extreme cases of selling away topsoil for brick making were also identified; farmers who did this were fully aware of the long-term implications on future crop yields. However, they perceived this as a better option than selling away the land, which is linked with the family's social status.

What then is the way out? The study indicates the need for viewing through a livelihood lens, and not through a pure economics lens. The study underlines the dire need for enabling policies and, more importantly, their effective implementation to complement and supplement people's own efforts and adaptive strategies at local level. It also highlights some policies and programmes that have made positive contribution, intended or unintended, to the adaptive strategies of the people. Examples include tribal job reservation policy, free education for girls and the joint forest management programme. Significantly, the regenerated forest cover in Bhiloda has also helped maintain environmental flows that delayed the effects of drought compared to Satlasana and Bhuj areas where such a programme was absent, primarily due to non-availablity of forest land.

Finally, the study emphasizes that adaptive strategies of the people do need to be embedded in the larger conventional resource management systems and welfare measures.

Notes

1 Used in India to denote problem zones in groundwater maps. The zones are categorized based on annual groundwater withdrawals in relation to utilizable recharge: more than 100% of withdrawal to recharge ratio is called 'overexploited'; 85–100% 'dark'; 65–85% 'grey' and less than 65% 'white'.

2 For the past few decades, drinking and domestic water to cities and towns is supplied through pipelines sourcing from either surface water or groundwater.
3 As per the Indian Meteorological Department, a meteorological drought is said to occur when the deficiency of rainfall at a meteorological subdivision level is 25% or more of the long-term average of that subdivision for a given period. The drought is considered 'moderate' if the deficiency is between 26% and 50%, and 'severe' if it is more than 50%.
4 Prolonged meteorological drought causes hydrological drought in the form of scarcity of surface water and declined groundwater levels, resulting in severe shortage of water for both human and animal needs.
5 Agricultural drought is said to occur when soil moisture and rainfall are inadequate during the crop-growing season to support healthy crop growth to maturity. For crops, this causes extreme stress and wilting. Technically, for the purpose of assessment, agricultural drought is defined as a period of four consecutive weeks of severe meteorological drought with a rainfall deficiency of more than 50% of the long-term average or with a weekly rainfall of 5 cm or less during the period from mid-May to mid-October (the *kharif* season) when 80% of the country's total crop is planted, or 6 such consecutive weeks during the rest of the year.
6 According to the Central Groundwater Board, Ministry of Water Resources, Government of India, the dark zones in India are growing at a rate of 5.5% per annum; if corrective measures are not taken, by 2017/18, roughly 36% of the blocks in India will face serious problems of overexploitation of groundwater resources.
7 Gadhwada is a cluster of 27 villages (including the study villages) in Satlasana *taluka* of Mehsana district in north Gujarat with similar sociocultural conditions.
8 VIKSAT Nehru Foundation for Development, Thaltej Tekra, Ahmedabad, is engaged in promoting people's institutions for natural resource management and livelihood enhancement for the last three decades through its five field offices, all located in arid and semiarid regions of Gujarat.
9 The BPL income generally considered as per the World Bank norms is $1/day or Rs 17,000/year. We have used a generous figure of Rs 20,000/year.
10 VIKSAT
11 Tree Growers Co-operative Societies, a common form of village-level institutions in Gujarat registered under the Co-operative Societies Registration Act.
12 Bhiloda *taluka* Lok Van Kalyan Sahkari Sangh Ltd., a *taluka*-level federation of village-level institutions, registered under the Co-operative Societies Registration Act.
13 Sanghathan Kshamata Manch-SAKSHAM; Secretariat at VIKSAT, Ahmedabad. SAKSAM is a state-level federation of federations registered as a Trust and Society.
14 VIKSAT and one of the villages have been awarded the prestigious national award – Indira Vriksh Mitra Award – in 1999 and 2005 respectively.
15 National Bank for Agriculture and Rural Development is established as a development bank, in terms of the preamble of the act: 'for providing and regulating credit and other facilities for the promotion and development of agriculture, small scale industries, cottage and village industries, handicrafts, and other rural crafts and other allied economic activities in rural areas with a view to promoting integrated rural development and securing prosperity of rural areas and for matters connected therewith or incidental thereto'.
16 AGRAR is an international collaborative research project of which VIKSAT is a partner. The project is supported by DFID-UK and coordinated by British Geological Survey, UK. Available at: www.iah.org/recharge/projects/html/.
17 Under the National Food for Work Programme (now renamed National Rural Employment Guarantee Scheme), the Government of India has identified 150 back-

ward districts for preparation of District Perspective Plans, which are underway. VIKSAT has prepared such a perspective plan for Sabarkantha district including artificial recharge activities as one of the major components.

References

COMMAN (2005) Community management of groundwater resources in rural India: background papers on the causes, symptoms and mitigation of groundwater overdraft in India. British Geological Survey Commissioned Report. CR/05/36N. Research Report.

COMMAN (2005) Managing groundwater resources in rural India: the community and beyond. British Geological Survey Commissioned Report. CR/05/36N, Synthesis Document.

Government of India, Minor Irrigation Census, 1996.

Janakarajan, S. (1993) In search of tanks: some hidden facts. *Economic and Political Weekly* 28(26).

Janakarajan, S. (2003) Need to modernize the tradition: changing role of tanks in response to scarcity and variability. Conference on Market Development of Water & Waste Technologies through Environmental Economics. New Delhi, India.

Moench, M. and Dixit, A. (eds) (2004) *Adaptive Capacity and Livelihood Resilience: Adaptive Strategies for Responding to Floods and Droughts in South Asia*. ISET, USA/Nepal, India.

Mudrakartha, S. (2004) Problems, prospects & attitudes in ensuring water security in arid and semi-arid zones. Proceedings of the Regional Workshop on Management of Aquifer Recharge and Water harvesting in Arid and Semi-arid Regions of Asia, Yazd, Iran, 27 November–1 December.

Mudrakartha, S. (2006) Women in livelihoods: SHG as a medium of empowerment. International Conference on Adaptation to Climatic Change and Variability: Emerging Issues in India and the US.

Mudrakartha, S. and Gupta, S.K. (2004) *Ensuring Rural Drinking Water Supply in Gujarat: Resources*. Task force paper on Water Resources for Government of Gujarat.

Mudrakartha, S. and Madhusoodhanan, M.P. (2005) Declining water levels and deteriorating livelihoods. Background Research Papers to Community Management of Groundwater in Rural India: Research Report in British Geological Survey Commissioned Report CR/05/36N. DFID, UK.

Mudrakartha, S., Madhusoodhanan, M.P. and Srinath, J. (2004a) *Coping and Adaptive Response to Drought in Gujarat*. VIKSAT.

Mudrakartha, S., Madhusoodhanan, M.P. and Srinath, J. (2004b) *Community Management of Groundwater Resources in Rural India: A Study of Bhanavas, Samrapur and Nana Kothasana Villages of Satlasana Taluka*, Mahesana District, Gujarat, India.

Mudrakartha, S., Srinath, J. and Pawar, S. (2005) *Augmenting Groundwater Resources through Artificial Recharge: A Case Study of Aravalli Hills*. Satlasana, Gujarat, India.

Mukherji, A. and Shah, T. (2003) *Groundwater Governance in South Asia: Governing a Colossal Anarchy*, No 13. IWMI-Tata Water Policy Programme, Vallabh Vidyanagar, Gujarat, India.

Ramachandran, R. (2000) *Frontline*, 17(12).

Tenth Five-year Plan (2002–2007) *Disaster Management: The Development Perspective*. Government of India. Available at: www.planningcommission.nic.in

Vaidyanathan, A. (1999) Water Resource Management. Oxford University Press, New Delhi, India.

VIKSAT Annual Reports (1998–2005).

13 Lessons from Intensive Groundwater Use In Spain: Economic and Social Benefits and Conflicts

M. Ramón Llamas[1] and Alberto Garrido[2]

[1]*Department of Geodynamics, Complutense University of Madrid, 28040 Madrid, Spain;* [2]*Department of Agricultural Economics, Technical University of Madrid, 28040 Madrid, Spain*

Introduction

Background

Groundwater development has significantly increased during the last 50 years in most semiarid or arid countries of the world (Shah, 2004; Deb Roy and Shah, 2003). This development has been mainly undertaken by a large number of small (private or public) developers, with minor, if any at all, scientific, administrative or technological control. This is why some authors consider this new phenomenon as a *silent revolution* (Llamas and Martínez-Santos, 2005a,b). In contrast, the surface water projects developed during the same period are usually of larger dimension and have been designed, financed and constructed by government agencies, which also take on management and control, whether for irrigation or urban public water supply systems. This historical situation has often produced two effects: (i) most regulators have limited understanding and poor data on the groundwater situation and value; and (ii) in some cases the lack of control on groundwater development has caused serious problems that are later on reviewed in detail.

Spain, as the most arid country in Europe, is no exception to these trends. In Spain, and almost everywhere else, these problems have been frequently magnified or exaggerated by groups with lack of hydrogeological know-how, professional bias or vested interests (Llamas *et al.*, 1992). For instance, the World Water Council (2000, p. 13) states: 'Aquifers are being mined at an unprecedented rate – 10% of world's agricultural production depends on using mineral groundwater'. However, this 10% estimation is not based on any reliable data. In recent decades, the term groundwater overexploitation has become a pervasive and confusing concept, almost a kind of *hydromyth*, that has flooded the water resources literature. A usual axiom derived from this confusing paradigm or *hydromyth* is that groundwater is an unreliable and fragile resource that should only be developed if the conventional large surface water projects are not feasible. This groundwater resource fragility concept has been dominant in Spain during the last 20 years (López-Gunn and Llamas, 2000). In the last decade a good number of authors have also voiced this fragility as a common issue (Seckler *et al.*, 1998; Postel, 1999).

Another usual wrong paradigm or *hydromyth* is the idea that mining non-renewable groundwater is by definition a case of overexploitation, which implies that groundwater mining goes against basic ecological and ethical principles. Some authors (Delli Priscoli and Llamas, 2001; Abderraman, 2003) have shown that in some cases the use of non-renewable groundwater may be a reasonable option. This point of view has been approved by the UNESCO World Commission on the Ethics of Science and Technology (COMEST), as can be read in Selborne (2000).

Purpose

This chapter provides an overview of the positive and negative aspects of the intense groundwater development in Spain during the last 3–4 decades. During this period, Spain has become an industrialized country. The analysis of the changing role of groundwater in Spain's water policy may be useful for other countries that are undergoing or will undergo a similar processes. Llamas and Martínez-Santos (2005a,b) suggest that a worldwide debate on this topic is desirable. One step for this aim has been the organization by the Interacademy Panel (IAP) of an International Symposium on Groundwater Sustainability (Alicante, Spain 24–27 January 2006). The Spanish Water Act of 1985 is one of the few in the world that sets provisions for 'overexploited aquifers'. Relying on the Spanish experience, the main aim of this chapter is to present and discuss: (i) the many meanings of the terms groundwater (or aquifer) overexploitation and sustainability; (ii) the main factors to take into consideration in analysing the pros and cons of intensive groundwater development; and (iii) the strategies to prevent or correct the unwanted effects of intensive groundwater development.

What does intensively used or stressed aquifer mean? During the last decade the expression *water stressed-regions* has become pervasive in the water resources literature. Usually this means that there are regions prone to suffer now or in the near future serious social and economic problems resulting from water scarcity. The usual threshold to consider a region under water stress is 1000 m^3/person/year (United Nations, 1997, pp. 10–13), but some authors increase this figure to 1700 m^3/person/year. If this ratio is only 500 m^3/person/year, the country is considered to be in a situation of absolute water stress or water scarcity (Seckler *et al.*, 1998; Postel, 1999; Cosgrove and Rijsbesman, 2000). This is far too simplistic. Considering only the ratio between water resources and population has meagre practical application. Most water problems are related to quality degradation, and accentuated drought cycles, but not to its relative scarcity. As an example, a good number of Spanish regions with a ratio lower than 500 m^3/person/year enjoy high economic growth and high living standards. Yet, development reinforces itself, and water demand increases, providing rationale for more public investment in water projects. In general, resource scarcity results from economic development, which in turn is endogenous to processes well beyond the boundaries of water policies. Often, the cause–effect direction is mistakenly reversed to conclude that making more water will promote economic development. This causality does not resist close scrutiny.

In its 1997 Assessment of Global Water Resources, the United Nations did a more realistic classification of countries according to their water stress. This assessment considered not only the water/population ratio but also the gross national product per capita (United Nations, 1997, p. 138). Other experts, like Sullivan (2001), have also begun to use other more sophisticated indices or concepts in order to diagnose the current or future regions with water problems. While laudable, even these have yet to prove themselves.

Groundwater development during the past decades has significantly contributed to Spanish agricultural and regional development. These improvements have also taken place in developing countries as particularly highlighted in the cases of South Asia and China (Shah, Chapter 2; and Wang *et al.*, Chapter 3, this volume, respectively). However, there is a pressing need to manage groundwater development and mitigate the externalities of groundwater extraction, accounting for the temporary or intrinsic uncertainties related to water. Sustainable groundwater use requires, sine qua non, the participation of educated and informed groundwater users and other stakeholders. This demands urgently the development of institutional arrangements for groundwater management where users can work jointly with the corresponding water agencies. But close cooperation among individuals does not come naturally, especially when societies face zero-sum gains (Livingston and Garrido, 2004).

What Does Sustainability Really Mean?

Since its early appearance in 1987, the concept of sustainability has been proposed by many as a philosophy to solve most water problems or conflicts. The US Geological Survey (1999) defines groundwater sustainability, though from an exclusively hydrological point of view. The European Union's (EU) Water Framework Directive (WFD), enacted in December 2000, establishes that it is necessary to promote sustainable water use. Probably, most people agree with this general principle, but its practical application in natural resources management is daunting. Shamir (2000) considers that the sustainability concept has up to ten dimensions including hydrology, ecology, economics, policy, intergenerational and intragenerational. It is out of the scope of this chapter to elaborate more on this concept. However, it will be used with specific meaning as much as possible.

In our view, sustainability integrates the concept of future generations. But how many of these should be considered? No scientist is able to predict the situation 1000 years from now, and very few dare to present plausible scenarios for the 22nd century. Most current predictions refer to the needs of humans in one or two generations, i.e. not more than 50 years from now. It is clear that environmental problems have a natural science foundation, but also, and perhaps primarily, a social science foundation. Recently, Arrow *et al.* (2004) have argued that the accumulation of human capital at a faster rate than the consumption of natural stocks could be considered a sustainable growth path. While saving and investment can make growth sustainable, irreversible effects may warrant more precautionary extraction patterns.

The way to solve the existing water problems, mainly the lack of potable water, is not to persist on gloom-and-doom unrealistic campaigns, trying to create *environmental scares* and predicting *water wars in the near future* (see The Economist, 1998; Asmal, 2000; World Humanity Action Trust, 2000) but to improve its management. In other words, the crisis is not of physical water scarcity but of lack of proper water governance, capital and financial resources (Rogers *et al.*, 2006).

The Polysemic and Increasingly Useless Concept of Overexploitation: Overview

This section will consider first the concept of overexploitation from a general perspective, and then the failure of its application in Spain. The term overexploitation has been frequently used during the last three decades. Nevertheless, most authors agree in considering that the concept of aquifer overexploitation is one that resists a useful and practical definition (Llamas, 1992; Collin and Margat, 1993). Custodio (2000, 2002) and Sophocleous (2000, 2003) have most recently dealt with this topic in detail.

A number of conceptual approaches can be found in the water resources literature: safe yield, sustained yield, perennial yield, overdraft, groundwater mining, exploitation of fossil groundwater, optimal yield and others (see glossaries in Fetter, 1994, and Acreman, 1999). In general, these terms have in common the idea of avoiding *undesirable effects* as a result of groundwater development. However, this *undesirability* is not free of value judgements. In addition, its perception is more related to the legal, cultural and economic background than to hydrogeological facts.

For example, in a research study on groundwater-fed catchments, called Groundwater and River Resources Action Programme at the European Scale (GRAPES) (Acreman, 1999), three pilot catchments were analysed: the Pang in the UK, the Upper Guadiana in Spain and the Messara in Greece. The main social value in the Pang has been to preserve the amenity of the river, related to the conservation of its natural low flows. In the Messara, the development of irrigation is the main objective and the disappearance of relevant wetlands has not been a social issue. In the Upper Guadiana the degradation of some important wetlands caused by groundwater abstraction for irrigation has stirred an ongoing conflict between farmers and conservationists (Bromley *et al.*, 2001).

The Spanish Water Act of 1985 does not mention specifically the concept of sustainability in water resources development but indicates that use rates should be in balance with nature. It basically considers an aquifer as overexploited when the pumpage is close to, or larger than, the natural recharge.

The Regulation for the Public Water Domain enacted in pursuant to the 1985 Water Act says that 'an aquifer is overexploited or in risk of being overexploited, when the continuation of existing uses is in immediate threat as a consequence of abstraction being greater or very close to the mean annual volume of renewable resources, or when it may produce a serious water quality deterioration'. According to the law, 14 aquifers have been declared either provisionally or definitively overexploited, for which strict regulatory measures have been designed. However, to a large extent, these measures have not been successfully implemented and a situation of legal chaos still persists in many of these aquifers (MIMAM, 2000).

The misconception of considering that *safe yield* is practically equal to natural recharge, already shown by the late well-known American hydrologist Theiss in 1940, has been voiced by many other hydrogeologists (see Custodio, 2000; Sophocleous, 2000; Hernández-Mora *et al.*, 2001).

Several national and international conferences have been organized by Spanish hydrogeologists over the last two decades to discuss and help dispel the misconceptions related to aquifer overexploitation (see Custodio and Dijon, 1991; Simmers *et al.*, 1992). Nevertheless, the success of these activities was rather limited in Spain and abroad.

It was suggested that a possible definition is to consider an aquifer as overexploited when the economic, social and environmental costs that derive from a certain level of groundwater abstraction are greater than its benefits. Given the multifaceted character of water, this comparative analysis should include hydrologic, ecological, socio-economic and institutional variables. While some of these variables may be difficult to measure and compare, they must be explicitly included in the analysis so that they can inform decision-making processes. Following Hernández-Mora *et al.* (2001), the basic categories of extractive services and *in situ* services are taken into account in the description of costs and benefits of groundwater development. The National Research Council (1997) recognizes that the monetary value of groundwater's *in situ* services (avoiding subsidence, conservation of wetlands or maintaining the base flow of rivers, among others) is a rather complex and difficult task for which there is only limited information. Yet the WFD foresees that Member states must evaluate the environmental and resource costs, providing motivation to environmental economics to build on new applicable methods. Recently Llamas and Custodio (2003) have tried to present 'intensive groundwater use' as a more practical concept. According to the editors of that book 'groundwater use is considered intensive when the natural functioning of the corresponding aquifer is substantially modified by groundwater abstraction'. This concept only describes the physical changes but does not qualify its advantages or disadvantages from the many dimensions of the sustainability concept, including ecological, hydrological, economical, social, intragenerational and intergenerational. On the contrary, other terms such as overexploitation, overdraft and stressed aquifers have a derogatory meaning for most people.

Fortunately, the scientific literature on intensive use of groundwater is increasing rapidly. In the book previously mentioned 20 chapters written by more than 30 well-known authors are included. A second book dealing with intensive use has also been recently published (Sahuquillo *et al.*, 2005).

Water Resources of Spain

Climatic and hydrologic setting

Spain has an area of approximately 505,000 km^2. The average precipitation is 700 mm/year. However, this average has considerable spatial deviation, ranging from 100 mm/year in some islands in the Canary Archipelago to more than 2000 mm/year in the humid north. The average annual temperature is 14°C. The average potential evapotranspiration is about 700 mm/year. In most of Spain potential evapotranspiration is higher than precipitation. The average stream flow is about 110 km^3/year (220 mm/year). From this amount, about 80 km^3/year (160 mm/year) is surface runoff and about 30 km^3/year (60 mm/year) is groundwater recharge. At least one-third of Spain is endowed with good aquifers. These aquifers may be detrital, calcareous or volcanic (Fig. 13.1). The Water Administration has formally identified 411 aquifer systems or hydrogeological units (see Table 4.1 in Llamas *et al.*, 2001), which cover an area of approximately 180,000 km^2. The estimated natural recharge of these aquifers averages 30 km^3/year but varies with weather conditions between 20 and 40 km^3/year (Fig. 13.2).

Water uses

Spain's current total water use is about 36km^3/year or about one-third of the total water resources (110 km^3/year). Use is distributed between irrigation (67%), urban water supply and connected industries (14%) and independent industrial uses and cooling (19%).

Spain has about 43 million inhabitants. This means there is an average usage of almost 3000 m^3/person/year, considering the whole country, but in some areas this indicator is in the range of 200 or 300. Table 13.1 shows the range of groundwater volumes used in Spain in recent years. The higher numbers correspond to dry periods in which groundwater use increases. The dramatic increase in the use of groundwater during the last 40 years is illustrated in Fig. 13.3.

This growth in groundwater use has been the result of groundwater development by individuals, small municipalities and industries. It has not been planned by government agencies. As a matter of fact, Spain is a serious case of *hydroschizophrenia*, i.e. of an almost complete separation of surface and groundwater in the mind of water planners (Llamas, 1985). These water planners have been almost without exception conventional civil engineers working in the Ministry of Public Works (and since 1996, in the Ministry for Environment). The Ministry of Agriculture, independently of the general water resources policy driven by the Ministry of Public Works, promoted the initial use of groundwater for irrigation in Spain in the 1950s. As a result, Spain is among the countries with the highest number of large dams per person: 30 large dams per million inhabitants (Fig. 13.4). The pace of large dam construction in Spain during the last 50 years has been almost 20 large dams per year (Fig. 13.5).

Within the EU, Spain has the lowest percentage (25%) of groundwater use for urban water supply (see Fig. 13.6). The explanation of this anomaly is not the lack of aquifers, but the *hydroschizophrenia* of the government water planners of the water supply systems to large cities and for grand surface water irrigation schemes.

Groundwater ownership and markets

Until the 1985 Water Act came into force, groundwater in Spain was private domain. In contrast, surface water was almost always public domain, ruled by government agencies. Because of the real or imagined problems related to the uncontrolled development of groundwater, the 1985 Water Act declared all

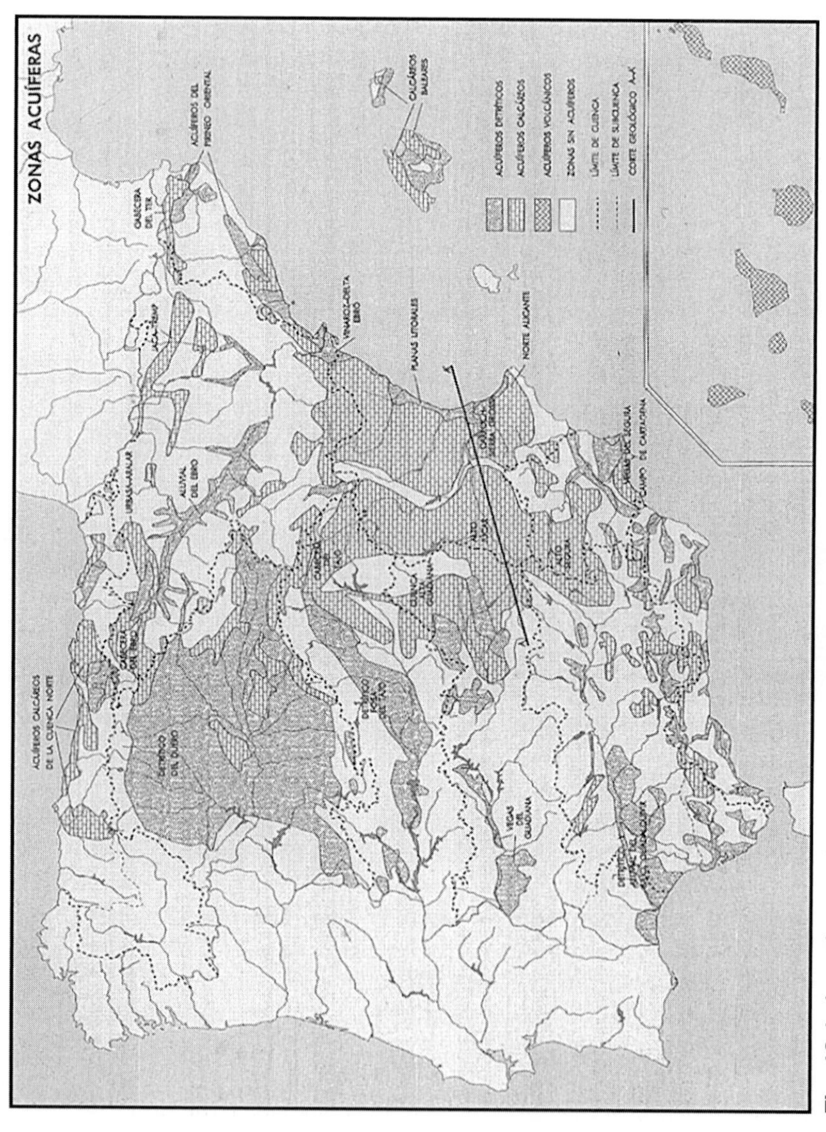

Fig. 13.1. Locations and types of main acquifers in Spain. (From MIMAM, 2000.)

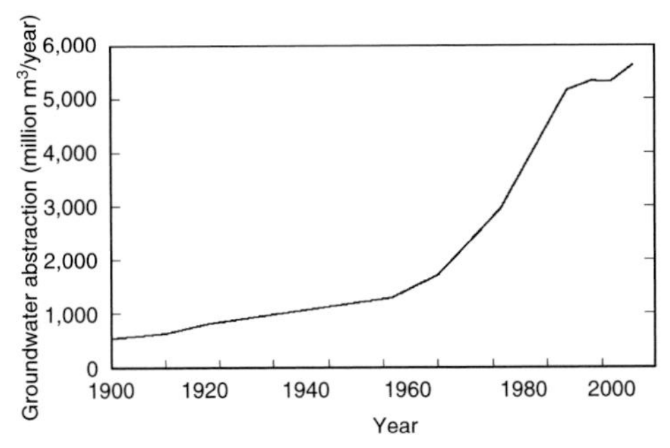

Fig. 13.2. Estimated natural groundwater recharge in Spain, 1940–1995. (From MIMAM, 2000.)

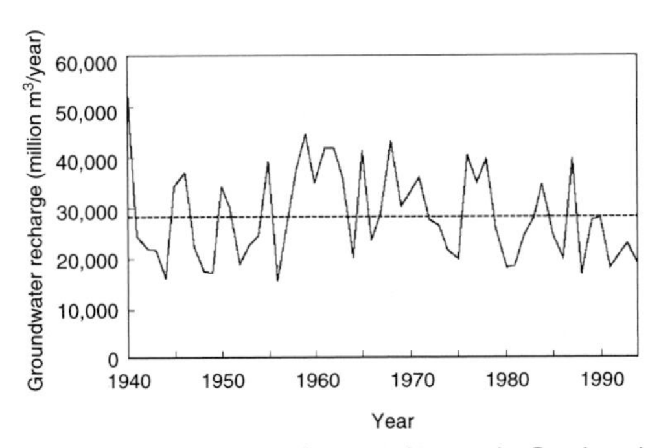

Fig. 13.3. Consumption development of groundwater in Spain, showing a rapid increase from the 1960s to date. (From MIMAM, 2000.)

Table 13.1. Spain's groundwater use summary (estimated from several sources). (From Llamas *et al.*, 2001.)

Activity	Volume applied (million cubic metres per year)	Percentage of total water (surface + groundwater)
Urban	1,000–15,000	~20
Irrigation	4,000–5,000	~25
Industrial and cooling	300–400	~5
Total	5,500–6,500	15–20

groundwater in Spain as public domain. Every new groundwater abstraction requires a permit granted by the corresponding water authority.

The groundwater developments made before 1 January 1986 may continue as private domain, using the same amount of groundwater as before. All these wells, galleries and springs should be inventoried and registered within the basin agencies registries. The main problem is that the legislators and the water authorities underestimated the number of groundwater abstractions and did not provide the economic means to register all the grandfathered groundwater rights. Even

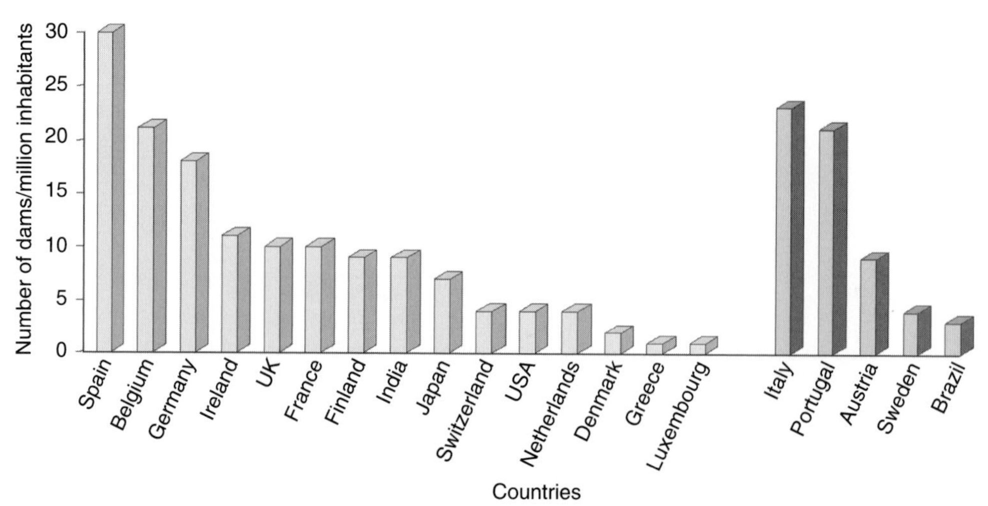

Fig. 13.4. Number of dams per capita in different countries. (From Llamas *et al.*, 2001.)

20 years after the enactment of the 1985 Water Act the number of private groundwater abstraction rights remains uncertain as do, by extension, the pumped volumes. Llamas *et al.* (2001, ch. 8) have estimated that the number of water wells in Spain is between one and two million. This means there are between 2 and 4 wells/km^2; however, this ratio is three times higher if it is applied only to the surface of the 400 aquifer systems. The average groundwater withdrawal from each well is low (between 2500 and 5000 m^3/year), indicating that most are meant for domestic use or small irrigation. The 1985 Water Act states that a permit is not necessary to drill a new well for abstracting less than 7000 m^3/year. Probably 90% of the private groundwater developments have an illegal or alegal status. In order to cope with this complex situation, in 1995 the government began a programme (called ARICA) with a cost of €60 million to have a reliable inventory of the water rights in Spain. The results of the ARICA programme were discouraging and it was practically abandoned. In 2002 the government began another similar programme (this time called ALBERCA) with a budget of €150 million. Detailed information on the progress of the ALBERCA programme is not available yet. However, according to Fornés *et al.* (2005) a larger budget would be necessary to clarify in full the inventory of groundwater rights.

On top of these disappointing results, and prompted to increase the economic efficiency of both surface and groundwaters, the 1985 Water Act was partly amended in 1999, mainly to introduce water markets in some way. This was mainly done to allow greater *flexibility* to sell or buy water rights. In principle, this new flexibility is not relevant to groundwater markets because in Spain most groundwater resources are still in private ownership and they could be sold or bought and leased like any other private asset. The importance of these groundwater markets varies according to the different Spanish regions. In most cases, these are informal (or illegal) markets and the information on them is not reliable (Hernández-Mora and Llamas, 2001). The Canary Islands are an exception to this general situation. This autonomous region of

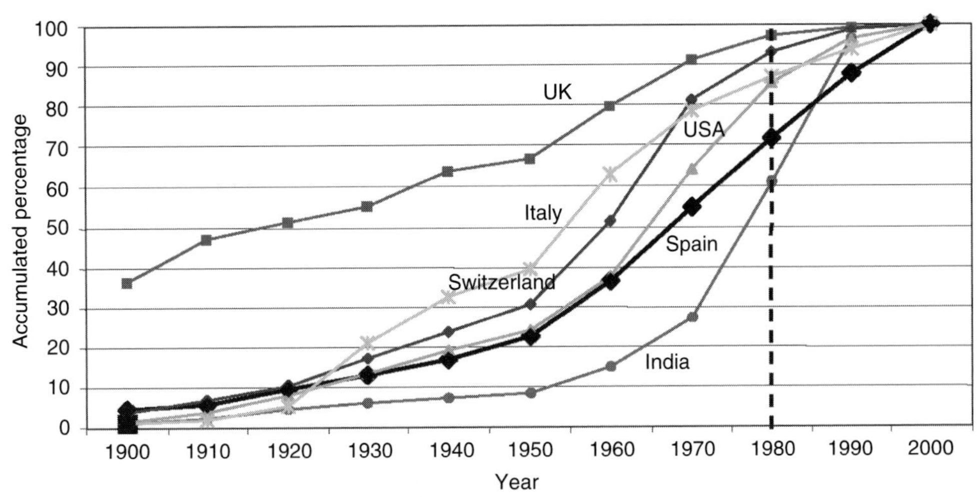

Fig. 13.5. Temporal dam construction rhythm in several representative countries. (From Llamas *et al.*, 2001.)

Spain has a different water code. Almost 90% of the total water uses are supplied by groundwater. Practically all groundwater is in private ownership. Aguilera Klink (2002) has studied the pros and cons of the water markets in this archipelago. Other than this, the 1999 amendment has not produced any substantial water reallocation, even under the 2005 pre-drought conditions prevailing in the country.

Benefits of Groundwater Development

Groundwater sustainability must necessarily take into account the numerous dimensions of this concept, among them the socio-economic and even ecological benefits that result from groundwater use. Socio-economic benefits range from drinking water supply to economic development, as a result of agricultural growth in a region. With respect to the potential ecological benefits, the use of groundwater resources can often eliminate the need for new large and expensive hydraulic infrastructures that might seriously damage the natural regime of a river or stream and/or create serious social problems (World Commission on Dams, 2000).

Drinking water supply

Groundwater is a key source of drinking water, particularly in rural areas and on islands. In Spain, for example, medium and small municipalities (of less than 20,000 inhabitants) obtain 70% of their water supply from groundwater sources (MIMAM, 2000). In some coastal areas and islands the dependence on groundwater as a source of drinking water is even higher. Nevertheless, as it was

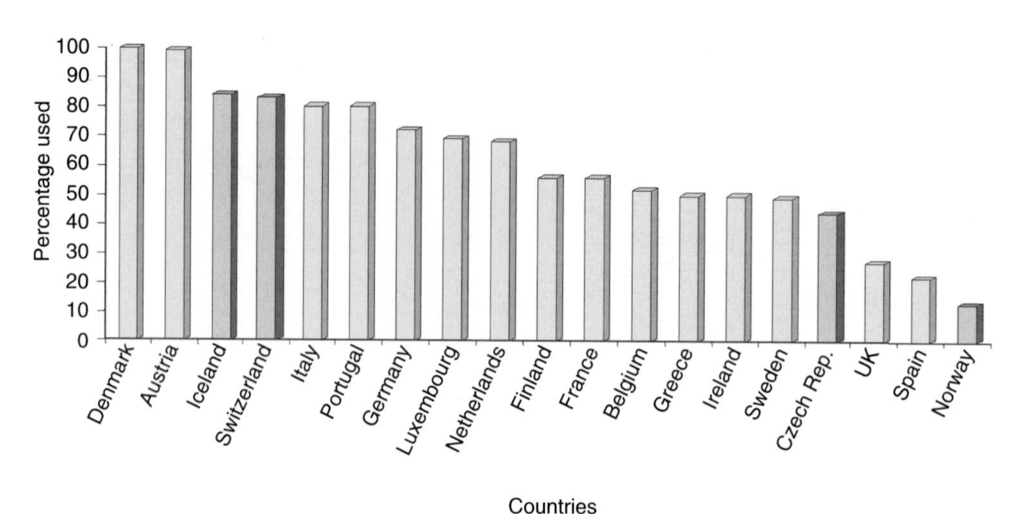

Fig. 13.6. Percentage of groundwater used for urban supply in several European countries. (From Llamas *et al.*, 2001.)

previously mentioned, Spain is one of the European countries with the lowest proportion of groundwater uses for public urban water supply to large cities. Llamas (1985) explains the historical roots of this situation. There were two main causes. The first was that there was a very centralized government system where all the decisions in relation to water policy were taken by a small and selected group of civil engineers working for the Ministry of Public Works. The second was the failure in the 1850s of a proposal of another selected group of mining engineers who also worked for the government. Between the two social groups there existed a certain professional concurrence. Mining engineers supported the use of groundwater to solve Madrid's serious water problems in the latter half of the 19th century. They failed because neither the geology nor the water well technology at that time allowed sufficient understanding about the functioning and potential development of the nearby aquifers.

Irrigation

In Spain, as in many arid and semiarid countries, the main groundwater use is for agriculture. Although few studies have looked at the role that groundwater plays in irrigation, those that do exist point to a higher socio-economic productivity of irrigated agriculture using groundwater than that using surface water. A 1998 study of Andalusia (south Spain) showed that irrigated agriculture using groundwater is significantly more productive than agriculture using surface water, per volume of water used (Hernández-Mora *et al.*, 2001). Table 13.2 shows the main results of the Andalusia study. It is important to note that these results were based on the average water volumes applied in each irrigation unit (or group of fields). The water losses from the source to the fields were not estimated, but are significant in surface water irrigation. Other studies have calculated the volumes used in surface water irrigation as the water actually taken from the reservoirs. For example, the White Paper of Water in Spain (MIMAM, 2000) estimated an average use of 6700 m^3/ha/year and 6500 m^3/ha/year for the two catchments that are the subject of the Andalusia study without differentiating between surface and groundwater irrigation. Using these new figures and the volumes given for irrigation with groundwater in the Andalusia study, a more realistic average volume used for irrigation with surface water of 7400 m^3/ha/year can be estimated. Table 13.2 shows that productivity of groundwater irrigation is five times greater than irrigation using surface water and generates more than three times the employment per cubic metre used. It could be argued that the greater socio-economic productivity of groundwater irrigation in Andalusia can be attributed to the excellent climatic conditions that occur in the coastal areas. While good climatic conditions may influence the results, the situation is similar in other continental regions of Spain (Hernández-Mora and Llamas, 2001). The updated data presented by Vives (2003) about the Andalusian irrigation confirm the previous assessment about the greater social and economic efficiency of groundwater irrigation.

Table 13.3 provides an overview of Spanish irrigation, indicating the water sources and irrigation technologies. In general, drip irrigation and sprinkler systems are more common in the regions where groundwater is used more intensively.

When examining this section it is important to keep in mind the uncertainties of hydrologic data. However, the results are indicative of the greater productivity of irrigation using groundwater. This should not be attributed to any intrinsic quality of groundwater. Rather, causes should be found in the greater control and supply guarantee that groundwater provides mainly during droughts (see Llamas, 2000), and the greater dynamism that has characterized the farmers who have sought their own sources of water and bear the full (direct) costs of drilling, pumping and distribution (Hernández-Mora and Llamas, 2001).

Table 13.2. Comparison of irrigation using surface and groundwaters in Andalusia. (From Hernández-Mora *et al.*, 2001.)

	Origin of irrigation water			
Indicator for irrigation	Groundwater	Surface water	Total	Relation groundwater/ surface water
Irrigated surface (10^3 ha)	210	600	810	0.35
Average use at origin (m^3/ha/year)	4000	7400	6500	0.54
Water productivity (€/m^3)[a]	2.16	0.42	0.72	5.1
Employment generated (EAJ/10^6 m^3)[b]	58	17	25	3.4

[a]€1 ≅ $1.3.
[b]EAJ ~= equivalent annual job, which is the work of one person working full-time for 1 year.

Hydrologic benefits

Another potential benefit of groundwater development is the increase in net recharge in those aquifers that, under natural conditions, have the water level close to the land surface. The drawdown of the water table can result in a decrease in evapotranspiration, an increase in the recharge from precipitation that was rejected under natural conditions and an increase in indirect recharge from surface water bodies. Johnston (1997) analysed 11 American regional aquifers, showing that intensive groundwater development in 9 of these aquifers has resulted in significantly increased recharge.

Shah *et al.* (2003) studied seasonal recovery of groundwater levels for the whole of India after the monsoon rain. They concluded that the depletion of the water table due to groundwater abstraction increases significantly the precipitation recharge. This is quite in agreement with the general hydrogeological principles. There is ample evidence showing that extractions increase the recharge rates augmenting the sustainable use level.

A clear example of this situation is the increase in available resources for consumptive uses that followed intensive groundwater pumping in the Upper Guadiana basin in central Spain (see Bromley *et al.*, 2001). It has been estimated that average renewable resources may have increased between one-third and one-half under disturbed conditions. Figure 13.7 illustrates these results. Prior to the 1970s, groundwater pumping in the Guadiana basin did not have significant impacts on the hydrologic cycle. Intensive pumping for irrigated agriculture started in the early 1970s and reached a peak in the late 1980s. As a result, wetlands that, under semi-natural conditions, had a total extension of about 25,000 ha, cover only 7000 ha today. In addition, some rivers and streams that were naturally fed by the aquifers have now become net losing rivers.

The results of the decline in the water table have been twofold. On one hand, a significant decrease in evapotranspiration from wetlands and the water table, from about 175 million cubic metres per year under quasi-natural conditions to less than 50 million cubic metres per year today. On the other hand, there has been a significant increase in induced recharge to the aquifers from rivers and other surface water bodies. Consequently, more water resources have become available for other uses, mainly irrigation, at the cost of negative impacts on dependent natural wetlands. These impacts are highlighted later.

Disadvantages of Groundwater Use

Groundwater level decline

The observation of a trend of continuous significant decline in groundwater levels is frequently considered an indicator of imbalance between abstraction and recharge. While this may be most frequently the case, the approach may be somewhat simplistic and misguided. Custodio (2000) and Sophocleous (2000) remind us that any groundwater withdrawal causes an increasing piezometric depletion until a new equilibrium is achieved between the pumpage and the

Table 13.3. Descriptive elements of Irrigation in Spain (in hectares). (From MAPYA, 2001.)

Autonomous Community	Surface water	Groundwater	Inter-basin transfers	Water returns	Reuse	Desalinized	Total	Predominant irrigation technique (%)		
								Flood	Sprinkler	Drip irrigation
Andalusia	546,703	224,670	2,783	85	5,639		779,880	42	21	37
Aragón	373,886	20,315		21			394,222	80	18	2
Castilla-León	361,055	113,164		12,428	29		486,676	61	39	–
Castilla-La Mancha	124,262	228,528	1,011				353,801	32	55	13
Cataluña	205,031	53,043		6,377	342		264,793	69	12	19
Extremadura	207,337	3,151					210,488	69	26	5
Galicia	85,061	92					85,153	64	36	–
Murcia	42,553	93,810	51,104	360	1,600	271	189,698	60	3	37
Navarra	79,941	1,682		50			81,673	89	10	1
Rioja	45,771	3,564					49,335	66	29	5
C. Valenciana	146,691	154,821	40,258	4,178	4,534		350,482	80	1	19
Total	2,218,291	896,840	95,156	23,499	12,144	271	3,246,201	59	24	17

new recharge (or capture). This transient situation can be long depending on aquifer characteristics.

With respect to the climatic cycles, in arid and semiarid countries significant recharge can occur only after a certain number of years, which may easily be from 5 to 10 years. Therefore, continuous decline in the water table during a dry spell of a few years, when recharge is low and abstraction is high, may not be representative of long-term trends. Declines in water levels should indicate the need for further analysis. In any case, declines in the water table can result in a decrease in the production of wells and in pumping costs. This economic impact can be more or less significant depending on the value of the crops obtained. For instance, in some zones of Andalusia, the value of crops in greenhouses may reach \$50,000–70,000/ha/year. The water volume used is between 4000 and 6000 m^3/ha. The energy needed to pump 1 m^3 100 m high is 0.3–0.4 kWh. If a \$0.03 kWh energy cost is assumed, this means an increase in energy costs in the order or \$0.01–0.02/m^3.

Analyses on water irrigation costs are rather scarce. The WFD mandates that Member States should collect these and all relevant economic information related to the water services. A preliminary assessment has been done in Spain, which is mentioned by Estrela (2004). Table 13.4 compiles a few studies that have attempted to evaluate the impact of tariff increases as a result of the implementation of Article 9 of WFD.

Table 13.4 provides indication that the application of the full cost recovery (FCR) water rates may have a significant impact in farmers' rents and water demand. Yet, by no means should it be expected that the WFD would entail catastrophic results to the farming sector. This is proven by the evidence supported by the irrigation sector relying on groundwater resources. Generally, water costs are much closer to the FCR prices indicated in Table 13.4 than to the current prices of surface water. Yet, irrigation relying on surface sources will need to adopt more efficient water conveyance and application technologies.

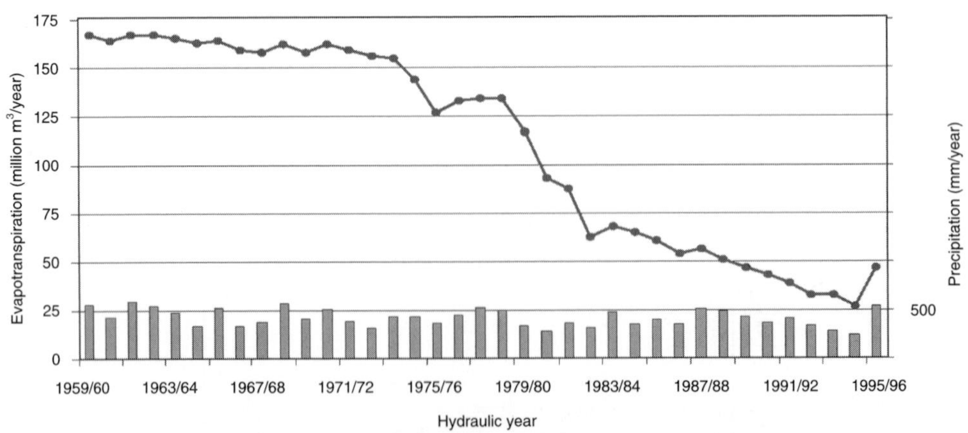

Fig. 13.7. Temporal evolution of evapotranspiration from the water table in the Upper Guadiana basin, caused by water table depletion. (From Martínez-Cortina, 2001, as cited in Llamas *et al.*, 2001.)

Table 13.4 also shows the relevance of water conservation, resulting from FCR prices. Conventional wisdom about water demand for irrigation in Spain should be profoundly revised, in view of the likely reductions that will be achieved by better pricing.

In 2003 the Ministry for the Environment (MIMAM, 2004), in agreement with Article 5 of the WFD, did a preliminary analysis of the cost of groundwater in Spain. This preliminary analysis estimates the cost of groundwater for irrigation and for urban water supply in each of the 400 Spanish hydrogeologic units. The average groundwater irrigation cost for the whole of Spain is about $0.15/m^3, but there exists a great dispersion of values from $0.04/m^3 to $0.40/m^3. This assessment has been done without specific field surveys, and should therefore be considered only as a preliminary approach. The analysis does not include an estimation of the average value of the crops guaranteed with the groundwater abstraction. Experience shows that in Spain the ratio between the value of the crop and the cost of groundwater irrigation is usually very small, usually smaller than 5–10%. In other words the silent revolution of groundwater intensive use is mainly driven by output markets (Llamas and Martínez-Santos, 2005a,b), and will not be deterred by the increasing pumping costs of lower water tables.

A significant fact is that a large sea water desalination plant (40 million cubic metres per year) has been completed in Almería (south Spain) in 2004. The main use of this treated water was supposed to be greenhouse irrigation. The price of this desalinated sea water offered by the government to the farmers is in the order of $0.40/m^3. This is a political price subsidized by the EU and by the Spanish Government. The real cost might be about double. The farmers are reluctant now to accept the price of $0.40/m^3, although in that area the value of the greenhouse crops obtained is in the order of $60,000/ha/year and the cost of the necessary 4000–6000 m^3/ha/year would be smaller than $2000–3000/ha/year or less than 5% of the crop value. Probably the main reason for their reluctance is that in some cases they can buy or obtain groundwater at an even lower price. They do not care about the right to abstract such groundwater because, as it has been previously stated, the administrative and legal situation of groundwater rights is usually chaotic; in other words, most of the water wells in operation are illegal and the government is unable to control them. The preliminary economic analyses in the Jucar basin (Estrela, 2004) seem to confirm that the market drives the silent revolution of groundwater intensive use. For instance, in the small aquifer Crevillente in that basin, farmers are pumping their groundwater from a depth of almost 500 m at a cost of $0.40/m^3. They grow special grapes for export with a value of about $20,000–30,000/ha. They use groundwater at about 3500 m^3/ha/year. Therefore the ratio of groundwater irrigation cost to the crop is less than 5%. The sustainability of this groundwater abstraction is not threatened by the groundwater level depletion but because of groundwater quality degradation.

Degradation of groundwater quality

Groundwater quality is perhaps the most significant but not the most urgent challenge to the long-term sustainability of groundwater resources. Yet, accord-

Table 13.4. The effects of the WFD on the irrigation sector. (From Garrido and Calatrava, 2006.)

	Present rate		Tariff increase		Results		
RBA	Type	Levels[a](€/cm)	Medium	FCR[b]	Farm income	Water demand	Other results
Duero	Per hectare	0.01	0.04	0.06	–40% to –50%	–27% to –52%	Great influence of agricultural policies
Guadalquivir	Per hectare and volume	0.01–0.05	0.05	0.1	–10% to –19%	0% to –10%	Same
Duero	Per hectare and volume	0.01	0.04	0.1	–10% to –49%	–5% to –50%	Technical response
Guadalquivir	Per hectare and volume	0.01–0.05	0.06	0.12	–10% to –40%	–1% to –35%	Technical and crop response
Guadalquivir	Per hectare and volume	0.01–0.05	0.03	0.09	–16% to –35%	–26% to –32%	Technical and crop response
Guadiana	Per hectare	0.005	0.03	0.06	–15% to –20%	–30% to –50%	Technical and crop response
Júcar	Per hectare, volume and hourly rates	0.03–0.15	0.06	0.15	–10% to –40%	0% to –40%	Technical response
Segura	Per hectare, volume and hourly rates	0.05–0.30	0.10	0.25	–10% to –30%	0% to –10%	Very inelastic demand

[a]Equivalent measure.
[b]FCR = Full cost recovery rates.

ing to the Kuznet's curve, countries implicitly accept a degradation of their environmental quality in return of higher living standards, up to a point where the preferences are reversed. This point has been empirically estimated to be in the range of $6000–10,000 of per capita income.

Restoration of contaminated aquifers can be a very costly and difficult task. Most often, degradation of groundwater quality is primarily related to as point or non-point source pollution from various sources such as return flows from irrigation, leakage from septic tanks and landfills or industrial liquid wastes. These problems are not exclusive to industrialized countries but also may be serious in developing countries. The WFD emphasizes the recovery of groundwater quality in the EU but pays little attention to groundwater quantity problems. This situation may be caused by the insufficient participation of European Mediterranean experts in the preparation of the WFD. Therefore, other arid and semiarid countries should be very prudent in taking the WFD as a good paradigm for their water policy. Groundwater abstraction can also cause changes in groundwater quality. Some indicators of the susceptibility of an aquifer to water quality degradation are given in Custodio (2000). Although groundwater pollution is possibly the most serious problem from a long-term perspective, the quantitative issues may be the more urgent and politically pressing ones. In these cases, the problem is often related to inadequate well field location and not necessarily to the total volumes abstracted. Technical solutions to deal with problems of saline or lower-quality water intrusion have been developed and applied successfully in some places such as California and Israel. Unfortunately, the public awareness in Spain about groundwater pollution problems is still weak, mainly due to the scarcity of government reports and action to assess and to abate groundwater pollution.

Susceptibility to subsidence and/or collapse of the land surface

Aquifers in young sedimentary formations are prone to compaction as a result of water abstraction, and therefore the decrease in intergranular pore pressure. For example, this has been the case in the aquifers underlying Venice or Mexico City. More dramatic collapses occur commonly in karstic landscapes, where oscillations in water tables as a result of groundwater abstractions can precipitate the occurrence of karstic collapses. In both cases, the amount of subsidence or the probability of collapses is related to the decrease in water pressure. Fortunately, these types of geotechnical problems are not relevant in Spain.

Interference with surface water bodies and streams

Decline in the water table as a result of groundwater withdrawals can affect the hydrologic regime of connected wetlands and streams. Loss of base flow to streams, desiccation of wetlands and transformation of previously gaining rivers to losing rivers may all be potentially undesirable results of groundwater abstraction and serve as indicators of possible excessive abstraction. The already mentioned Upper Guadiana catchment in Spain is a typical example of this type of situation, which will be dealt with in more detail later. According to the WFD most groundwater abstraction in the Upper Guadiana basin in Spain must be cancelled because of its evident interference with the surface waters and its ecological impacts. However, the WFD states that when this solution implies serious social problems, the corresponding Member State may ask for derogations based on the hydrological, economic and social consequences. Most likely Spain will request derogations to the EU not only in the Upper Guadiana basin but also in many other aquifers where an intensive use of groundwater exists.

Ecological impacts on groundwater-dependent ecosystems

The ecological impacts of drawdown of the water table on surface water bodies and streams are increasingly constraining new groundwater developments (Llamas, 1992). Drying up of wetlands, disappearance of riparian vegetation because of decreased soil moisture and alteration of natural hydraulic river regimes can all be used as indicators of overexploitation. Reliable data on the ecological consequences of these changes are not always available, and the social perception of such impacts varies in response to the cultural and economic situation of each region. The lack of adequate scientific data to evaluate the impacts of groundwater abstraction on the hydrologic regime of surface water bodies makes the design of adequate restoration plans difficult. For instance, wetland restoration programmes often ignore the need to simulate the natural hydrologic regime of the wetlands, i.e. not only restore its form but also its hydrological function. Similar problems result in trying to restore minimum low flows to rivers and streams. Oftentimes minimum stream flows are determined as a percentage of average flows, without emulating natural seasonal and year-to-year fluctuations to which native organisms are adapted.

The social perception of the ecological impacts of groundwater abstraction may differ from region to region and result in very different management responses. GRAPES, an EU-funded project previously mentioned, looked at the effects of intensive groundwater pumping in three different areas: Greece, Great Britain and Spain (Acreman, 1999). In the Pang River in Britain, conservation groups and neighbourhood associations with an interest in conserving the environmental and amenity values of the river that had been affected by groundwater abstraction mainly drove management decisions. In the Upper Guadiana basin, dramatic drawdown in the water table (30–40 m) caused jointly by groundwater abstraction and drought (see Fig. 13.8) resulted in intense conflicts between nature conservation officials and environmental non-governmental organizations (NGOs), irrigation farmers and water authority officials. The conflicts have been ongoing for the last 20 years and have not yet been resolved. Management attempts to mitigate the impact of water level drops on the area's wetlands have so far had mixed results (Fornés and Llamas, 1999; Bromley *et al.*, 2001). On the other hand, in the Messara Valley in Greece, the wetland degradation caused by decline in the water table has not generated any social conflict. This situation seems to confirm that ecological awareness is deeply related to economic value of water and to the cultural background of each region.

Stakeholders' Participation in Groundwater Management

Spain has a long tradition of collective management of common pool resources. Probably the *Tribunal de las Aguas de Valencia* (Water Court of Valencia) is the most famous example. This Court has been meeting at noon every Thursday for many centuries at the entrance of Cathedral of Valencia to solve all the claims among the water users of a surface irrigation system located close to Valencia. All the members of the Court are also farmers. The decisions or settlements are oral and cannot be appealed to a higher court. The system has worked and it is a clear proof that 'the tragedy of commons' is not always true. Further evidence of social cooperation in Spain is the nearly 6000 *Comunidades de Regantes* (Irrigation Communities of Surface Water Users Associations). Some of them have been in operation for several centuries. Currently these communities are legally considered entities of public right. They are dependent on the Ministry for the Environment and are traditionally subsidized with public funds, mainly for the maintenance of the irrigation infrastructures.

The 1985 Spanish Water Act preserved the traditional *Comunidades de Regantes* that existed before its enactment and recommended these institutions for surface water management. It also extended this type of collective institution to groundwater management, and required the compulsory formation of *Comunidades de Usuarios de Aguas Subterráneas* (Groundwater Users Communities) when an aquifer system was legally declared overexploited. A short description of these institutions is contained in Hernández-Mora and Llamas (2000).

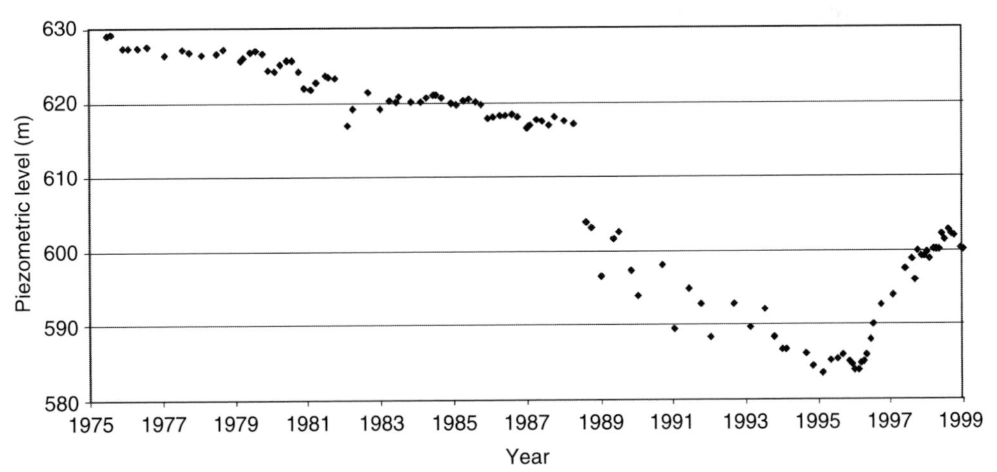

Fig. 13.8. Water table evolution in Manzanares (Upper Guadiana catchment, Spain). (From Martínez-Cortina, 2001, as cited in Hemández-Mora *et al.*, 2001.)

A more detailed description of the nature and evolution of some of these new groundwater user associations and communities can be found in López-Gunn (2003). The current situation can be summarized as follows:

1. It seems clear that the key issue for the acceptable functioning of these institutions is a bottom-up approach from the outset on the part of the governmental water authorities. This explains the almost perfect functioning of the Llobregat delta groundwater user association, which has been in operation since the 1970s under the previous 1879 Water Act. In that Water Act groundwater was legally privately owned but the corresponding Water Authority officers and the groundwater users (mainly water supply companies and industries) were able to work jointly. Something similar has occurred in the implementation of the Groundwater User Community for the eastern La Mancha aquifer located in the continental plateau. In this case, the groundwater users are mainly farmers and the irrigated surface covers about 900 km^2. This aquifer has never been declared legally 'overexploited' by the corresponding Water Authority.
2. In two important aquifers the situation has been the opposite. The western La Mancha and the Campo de Montiel aquifers are also located in the continental plateau in Spain. Their total area is about 7500 km^2 and their irrigated area is about 2000 km^2. The Guadiana River Water Authority legally declared both aquifers overexploited in 1987, in a typical top-down and control-and-command approach. Only in 1994 the corresponding groundwater user communities were implemented. And this was only possible thanks to a generous economic subsidies plan (paid mainly by the EU) to compensate the decrease in the groundwater abstraction. Nevertheless, a good number of farmers have continued to drill illegal water wells and they are not decreasing their pumpage. On the other hand, after 10 years the economic incentives from the EU have been discontinued. The Spanish Parliament asked the Government in July 2001 to present a plan for the sustainable use of water in this area by July 2002. This requirement has not been accomplished yet by Sophocleous (2000).

As the chief engineer for groundwater resources in the Ministry for Environment stated, serious difficulties have been faced in enforcing the setting up of the groundwater users associations in the aquifers legally declared 'overexploited' (Llamas, 2003). Only 2 out of 17 groundwater user communities that have to be implemented in the corresponding legally declared 'overexploited aquifers' are operative (Hernández-Mora and Llamas, 2001; Llamas *et al.*, 2001, ch. 9). As recognized in the White Paper on Water in Spain (MIMAM, 2000), the main cause is that these new groundwater user communities were established top-down, i.e. the water authorities imposed their implementation without the agreement of the farmers who are the main stakeholders. The 1999 amendments to the 1985 Water Act and the 2001 Law of the National Water Plan have provisions to overcome these difficulties and to foster the implementation of institutions for collective management of aquifers with ample participation of the stakeholders under a certain control of water authorities. It is too early to assess the results of these provisions.

In Spain, in addition to the communities born under the auspices of the 1985 Water Act, there are a large number of private collective institutions or associations to manage groundwater. Only a few years ago a group of them set up the Spanish Association of Groundwater Users. This is a civil (private) association that is legally independent of the Ministry for the Environment. Despite the wide recognition of the benefits of this type of associations, it is early to ascertain whether the needed economic, tax, and operational incentives are in place.

Hydrosolidarity and Groundwater Management in Spain

Overview

In Spain, like everywhere else, ethical factors play a crucial role in water uses and water management. Several recent publications address this topic (see Delli Priscoli and Llamas, 2001; Selborne, 2000; Llamas, 2003b). Human solidarity is one of the ethical principles that underlay most water policy agreements or treaties. One of the meanings of the concept of solidarity, as it applies to the use of natural resources, is that a person's right to use those resources should be constrained or limited by the rights or needs of other human beings now or in the future, including protection of the natural environment.

Nowadays, few people would dare to speak openly against hydrosolidarity (the need to share water resources). In practice, however, it might be difficult to find constructive ways to facilitate an equitable and fair sharing of water resources among concerned stakeholders, particularly in densely populated arid and semiarid regions. Lack of knowledge, arrogance, vested interests, neglect, institutional inertia and corruption are some of the obstacles frequently encountered to achieve hydrosolidarity (Llamas and Martínez-Santos, 2005b). The noble and beautiful concept of hydrosolidarity may also be used in a corrupt or unethical way by some lobbies in order to pocket *perverse subsidies*, which are bad for the economy and the environment (Delli Priscoli and Llamas, 2001). An example of the improper use of hydrosolidarity is that of the Segura catchment area. It has influenced the approval of the large aqueduct for the Ebro River water diversion included in the first National Water Plan in Spain, which was approved as a Law in July 2001 by the Spanish Parliament, and rebuffed after the general election of March 2004.

The Segura catchment

This section is mainly taken from an invited paper presented by Llamas and Pérez Picazo in the 2001 Stockholm World Water Week. The term 'hydrosolidarity' has been coined mainly by professors Falkenmark and Lundquist who were the organizers of this water week.

Hydrology

The Segura catchment is located in south-eastern Spain. Its main features are: (i) surface area 19,000 km^2; (ii) average annual precipitation: 400 mm, ranging from 800 mm in the headwater to 200 mm in the coastal plain; (iii) annual potential evapotranspiration: 800–900 mm; (iv) average streamflow: 1000 million cubic metres. The relief is abrupt with mountains that reach an altitude of 2000 m. The geology is complex with numerous faults and thrusts. Calcareous aquifers cover about 40% of the catchment's surface. Natural groundwater recharge is estimated at about 600 million cubic metres per year (about 60% of the total stream flow). The climate is typical of Mediterranean regions: hot summers, frequent flash floods and long droughts.

Water development until the 1960s

As much as 60% of the Segura River basin is within the Murcia Autonomous Region and the remaining 40% divided between the autonomous regions of Valencia and Castilla-La Mancha. The mild climate and the important base flow (typical of a karstic catchment) of the Segura River encouraged the development of an important agricultural economy in the region. It was based on an irrigation network on the flood plains of the middle and lower part of the catchment area, which dates as far back as the Muslim occupation 1200 years ago. Vegetables, citrus and other fruits have been cultivated in the region for many centuries. Agro-industry (food processing) has also been significant at least since the beginning of the 20th century. Collective systems to manage surface irrigation were implemented several centuries ago.

Until recently, agriculture was the main revenue-generating activity in the Segura catchment area. Murcia was considered the orchard of Spain. Since the integration of Spain in the EU (1986), the demand for its agricultural products increased significantly. The scarcity and/or variability in the availability of surface water resources have motivated the construction of 24 reservoirs that provide total storage of about 1000 million cubic metres. Although good at preventing floods, they have not satisfied the farmers' water demands for irrigation at a nominal price. Politicians and engineers who have advocated for the transfer of water resources from 'humid' Spain to 'dry' Spain have backed the old paradigm, with intense reliance on subsidies. In 1933, the first formal proposal to transfer water from the Tagus River headwaters (in central Spain) to the Segura River was formally made, but became operative only in 1979.

Groundwater abstraction boom

In the 1950s and 1960s, the Spanish Ministry of Agriculture launched a significant effort to promote groundwater irrigation in Spain. This promotion can be said to be totally independent of the National Water Policy that, as mentioned earlier, was driven by the corps of civil engineers of the Ministry of Public Works. This initial activity, heavily subsidized with public funds, soon became a catalyst that promoted intensive water well drilling by many private farmers in many regions of Spain. The most active region in this respect was the Segura catchment area. There were several reasons for the special development of groundwater abstraction in this region: (i) the area had a long tradition of irrigation with surface water and a traditional capacity to market high-value crops in Spain and abroad; (ii) many farmers had the expectation that these groundwater-irrigated areas would have some kind of preference in the allocation of surface water coming from the Segura reservoirs, from the Tagus River water transfer or from the future Ebro River water transfer project. In 1976, several years before the arrival of the first Tagus water, the new areas irrigated with groundwater required more water than the total theoretical volume to be transferred to the Segura catchment in the 1980s.

In Spain, according to the Water Law of 1879, groundwater was private-owned. The landowner could drill a water well in his or her land and pump as much groundwater as he or she wished, unless a third person was affected. Nevertheless, in the 1950s special regulations were enacted by the government that theoretically made groundwater a part of the public domain in the *Vegas del Segura* (Segura flood plains). The lack of experts in hydrogeology in the Segura Water Authority made this regulation difficult to enforce.

Even after the enactment of the 1985 Water Act the control of the old and new water wells in the Segura catchment area is rather scarce. The situation can accurately be described as one of administrative and legal 'chaos' (see Llamas and Pérez Picazo, 2001). For example, the official White Paper on Spain's Water (MIMAM, 2000, p. 343) admits that in this region only about 2500 water wells out of more than 20,000 drilled are legally inventoried by the Segura Water Authority.

The Tagus River water transfer and the future Ebro River water transfer

In 1979, almost 50 years after the first formal proposal, water from the Tagus River was transferred to the Segura catchment through a 300 km long aqueduct. The capacity of this aqueduct is about 33 m^3/s or 1000 million cubic metres per year, but the maximum volume approved for transfer during the first phase was only 600 million cubic metres per year. The reality is that the average volume transferred during the first two decades of operation of the aqueduct has been about 300 million cubic metres per year. The theoretical 600 million cubic metres to be transferred was distributed thus: 110 million cubic metres for urban water supply, 400 million cubic metres for irrigation and 90 million cubic metres as estimated losses during transfer. It was also stipulated that when the volume of water transferred is smaller that this theoretical amount, urban water supply had a clear priority. One interesting aspect of this project is that the beneficiaries of the transferred Tagus water pay a tariff for the water that is significantly higher than that usually paid by surface water farmers in Spain (approximately €0.005/m^3). In this case, they pay an average of about €0.1/m^3, although water for urban supply has a higher tariff than water for irrigation. The Law of the National Water Plan enacted in 2001 approved a new water transfer of 1050 million cubic metres per year from the Ebro River in northern Spain to several regions along the Mediterranean coast. Almost 50% of this volume was for delivery to the Segura catchment area. The planned aqueduct was almost 900 km long. Out of the total volume transferred, about 50% is for urban water supply and the rest for supplying water to areas in which groundwater abstraction has been excessive and has impacted the storage and groundwater quality of the aquifers. The Ebro water transfer met strong opposition among many different groups, parties and area-of-origin regional governments. Demonstrations summoned hundreds of thousands of people Valencia (for) and Zaragoza (against) the transfer. According to the government, the real cost of the Ebro water transfer would be about €0.30/m^3, but analysts argued it would be much higher.

The conflict about the Ebro water transfer: lack of hydrosolidarity or false paradigms?

In 2001 a poll was held to assess the social perception of the Ebro water transfer. Of those interviewed, 50% were in favour of the transfer, 30% were against it and the remainder had no opinion on the issue. One could conclude that those who were against the Ebro transfer lacked solidarity with the Mediterranean regions because they denied water to *thirsty areas*, while the Ebro River has a surplus of water, which is 'wasted uselessly' into the Mediterranean Sea. Most people, in every culture or religion think that it is a good action to give fresh water to the thirsty. In Western civilization this is a biblical tenet. But are the people in the Segura catchment region really thirsty? Certainly not. Almost 90% of the water used in this area is for irrigation of high-value crops and not for urban water supply. The irrigation economy in Segura is flourishing and very efficient. Table 13.5 shows the evolution of irrigated lands in the Segura catchment region, which has almost tripled since 1933, when the use of surface water reservoirs and groundwater was minimum.

The second old and current false paradigm is that farmers cannot (and should not) pay the *full cost* of the infrastructures to bring them water from the Ebro River. Most authors consider that if the full cost of the transfer were passed on to the farmers and urban users through water use fees, they would not support the Ebro water transfer or be willing to pay for it, since there are cheaper and faster solutions to meet their water needs. As discussed earlier, detailed studies undertaken in Andalusia, Spain, have shown clearly that groundwater irrigation is much more efficient than surface water irrigation: it produces about five times more cash per cubic metre used, and three times more jobs per cubic metre. The analysis done for Andalusia (a sample of almost one million hec tares), and the conclusions drawn from it, can be applied to most irrigated areas of Spain (3.5 million hectares). Other studies shown in Table 13.4 support this conclusion.

Llamas and Pérez-Picazo (2001) considered that both paradigms are now obsolete. However, some time will be necessary to change the mentality of the general public. These false paradigms are also frequent in other countries, as it is mentioned in Llamas and Martínez-Santos (2005b). It seems probable that the conflicts between the farmers and the conservation lobbies will increase in the near future. To avoid or mitigate such conflicts a stronger policy of transparency, accountability and general education (without obsolete paradigms) seems important.

Table 13.5. Evolution of irrigated area in the Segura catchment area. (From Llamas and Pérez Picazo, 2001.)

Year	Area (ha)
1933	90,000
1956	104,000
1963	115,000
1983	197,000
1993	235,000
2000	252,000

Conclusions

In Spain, like everywhere else, complexity and uncertainty characterize water management problems in general, and more so in the case of groundwater. Uncertainty is an integral part of water management. This uncertainty relates to scarcity of data, strong non-linearities in groundwater recharge values, scientific knowledge and changing social preferences. Honesty and prudence in recognizing current uncertainties is necessary. At the same time, there needs to be a concerted effort to obtain more and better hydrological data on which to base management decisions.

Intensive groundwater development is a new situation in most arid and semiarid countries. Usually, it is less than 30–40 years old. Four technological advances have facilitated this: (i) turbine pumps, (ii) cheap and efficient drilling methods, (iii) scientific hydrogeology advance, and (iv) cheap and accessible energy. Full cost (financial, operation and maintenance) of groundwater abstraction is usually low in comparison to the direct benefits obtained.

Mainly individual farmers, industries or small municipalities have carried out groundwater development. Financial and technical assistance by conventional Water Authorities has been scarce. This is why the new situation can be properly described as a *silent revolution* by a great number of modest farmers at their own expense.

The lack of planning and control of groundwater development has resulted in ecological or socio-economic impacts in a few regions. Property rights and institutional uncertainty is now worrying the beneficiaries; despite this none seems to be withdrawing and many others risk becoming users beyond the law and the public control.

Aquifer overexploitation is a complex concept that needs to be understood in terms of a comparison of the social, economic and environmental benefits and costs that derive from a certain level of water abstraction. It is meaningless and misleading to define overexploitation in purely hydrogeological terms given the uncertainties in recharge and abstraction values and the fact that the amount of available resources in a catchment area is variable and can be influenced by human actions and management decisions. The assumption that a long trend (e.g. 10 years) of decline in groundwater levels implies real overexploitation or overdraft may be too simplistic and misleading. This concept has been used in Spain to provide grounds for public action, igniting a top-down sort of policy that has failed to deliver significant benefits.

Increasing emphasis on cost-effective and environmentally sensitive management practices places a new thrust on broad public involvement in any water management decision-making process. But guaranteeing effective public participation in management processes requires informing and educating the public on increasingly complex scientific and technical issues. Effective information and education campaigns are therefore essential. The conflicts that are often a part of water management processes require the use of innovative conflict resolution mechanisms, which will allow for the discovery of feasible solutions that are accepted by all and can be successfully implemented. Up to now very little has been done in this direction in Spain.

Because of the persistence of obsolete paradigms, the wonderful concept of hydrosolidarity was recently improperly used in Spain to promote perverse subsidies mainly through the Ebro River water transfer to the Mediterranean regions. In the opinion of these authors, fortunately, the construction of the Ebro River diversion has been cancelled because it would be a wasteful use of public money. However, the initial solution proposed by the new government is equally prone to 'perverse subsidies'. The difference is that the public funds will be employed in the construction of more than a dozen large desalinating plants. The probability that farmers accept this solution is small. The main reason for this rejection is that abstracting or buying groundwater is significantly cheaper than paying for desalinized water. Probably, in most cases this abstraction of groundwater may not be sustainable and it is against the spirit and the provisions of the WFD. However, logically under the current administrative and legal chaos in groundwater development farmers are not very concerned about the need for achieving an environmentally sustainable groundwater development. They are much more concerned with the economic and social sustainability of groundwater development. Yet the amended law of the National Water Plan includes a certain number of articles, which, if actively enforced, would contribute efficiently to introduce a new water culture in Spain.

References

Abderraman, W.A. (2003) Should intensive use of non-renewable resources always be rejected? In: Llamas, M.R. and Custodio, E. (eds) *Intensive Use of Groundwater*. A.A. Balkema, Lisse, The Netherlands, pp. 191–206.

Acreman, M.C. (compiler) (1999) Guidelines for the sustainable management of groundwater-fed catchments in Europe. Report of the Groundwater and River Resources Action Programme on a European Scale (GRAPES), Institute of Hydrology, Wallingford, UK.

Aguilera K.F. (2002) *Los Mercados de Agua en Tenerife* (*Water Markets in Tenerife*). Bakeaz, Bilbao, Spain.

Arrow, K., Dasgupta, P., Goulder, L., Daily, G., Ehrlich, P., Heal, G., Levin, S., Mäler, K.-G., Schneirder, S., Starrett, D. and Walker, B. (2004) Are we consuming too much? *Journal of Economic Perspectives* 18(3), 147–172.

Asmal, K. (2000) Water: from casus belli to catalyst for Peace. Address in the opening session in the Stockholm Water Symposium. 14 August 2000.

Bromley, J., Cruces, J., Acreman, M., Martínez, L. and Llamas, M.R. (2001) Problems of sustainable management in an area of overexploitation: the Upper Guadiana catchment

Central Spain. *Water Resources Development* 17, 379–396.

Collin, J.J. and Margat, J. (1993) Overexploitation of water resources: overreaction or an economic reality? *Hydroplus*, 36, 26–37.

Cosgrove, W.J. and Rijsberman, F.R. (2000) *World Water Vision*. Earthscan, London.

Custodio, E. (2000) The complex concept of groundwater overexploitation. Papeles del Proyecto Aguas Subterráneas de la Fundación M. Botín, A 1. Santander, Spain.

Custodio, E. (2002) Overexploitation: what does it mean? *Hydrogeology Journal* 10, 254–277.

Custodio, E. and Dijon, R. (1991) Groundwater overexploitation in developing countries. Report of a UN Interregional Workshop, UN.INT/90/R43.

Deb Roy, A. and Shah, T. (2003) Socio-ecology of groundwater irrigation in India. In: Llamas, M.R. and Custodio, E. (eds) *Intensive Use of Groundwater*. A.A. Balkema, Lisse, The Netherlands, pp. 307–336.

Delli Priscoli, J. and Llamas, M.R. (2001) International perspective in ethical dilemmas in the water industry. In: Davis, C.K. and McGinn, R.E. (eds) *Navigating Rough Waters*. American Water Works Association, Denver, Colorado, pp. 41–64.

Estrela, T. (coordinator) (2004) The Jucar Pilot Basin. Provisional Article 5 Report. Pursuant to the Water Framework Directive. Confederación Hidrográfica del Jucar.

Fetter, C.W. (1994) *Applied Hydrogeology*, 3rd edn. Prentice-Hall, New York.

Fornés, J. and Llamas, M.R. (1999) Conflicts between groundwater abstraction for irrigation and wetlands conservation: achieving sustainable development in the La Mancha Húmeda Biosphere Reserve (Spain). In: Griebler *et al.* (eds) *Groundwater Ecology: A Tool for Management of Water Resources*. European Commission, Environmental and Climate Programme, pp. 227–236.

Fornés, J.M., de la Hera, A. and Llamas, M.R. (2005) The silent revolution in groundwater intensive use and its influence in Spain. *Water Policy*, 7(3), 253–268.

Garrido, A. and Calatrava, J. (2006) Recent and future trends in water charging and water markets. In: Garrido, A. and Llamas, M.R. (eds) *Water Policy in Spain: Resources for the Future*. Washington DC (in press).

Hernández-Mora, N. and Llamas, M.R. (2000) The role of user groups in Spain: participation and conflict in groundwater management. CD-ROM of the 10th World Water Congress, Melbourne, 12–16 March 2000, International Association of Water Resources.

Hernández-Mora, N. and Llamas, M.R. (eds) (2001) *La economía del agua subterránea y su gestión colectiva* (*Groundwater Economics and its Collective Management*). Fundación Marcelino Botín y Mundi Prensa, Madrid, Spain.

Hernández-Mora, N., Llamas, M.R. and Martínez, L. (2001) Misconceptions in aquifer overexploitation: implications for water policy in southern Europe. In: Dosi, C. (ed.) *Agricultural Use of Groundwater: Towards Integration between Agricultural Policy and Water Resources Management*. Kluwer Academic Publishers, Dordrecht, The Netherlands, pp. 107–125.

Johnston, R.H. (1997) Sources of water supplying pumpage from regional aquifer system of the United States. *Hydrogeology Journal* 5(2), 54–63.

Livingston, M.L. and Garrido, A. (2004) Entering the policy debate: an economic evaluation of groundwater policy in flux. *Water Resources Research*, 40(12).

Llamas, M.R. (1985) Spanish water resources policy: the illogical influence of certain physical and administrative factors. *Member of the 18th International Congress of the International Association of Hydrologists* 18(2), 160–168.

Llamas, M.R. (1992) Wetlands: an important issue in hydrogeology. In: Simmers *et al.* (eds) *Selected Papers on Aquifer Overexploitation*, Vol. 3. Heise, Hannover, pp. 69–86.

Llamas, M.R. (2000) Some lessons learnt during the drought of 1991–1995 in Spain. In: Vogt and Somma (eds) *Drought and Drought Mitigation in Europe*. Kluwer Academic Publishers, Dordrecht, The Netherlands, pp. 253–264.

Llamas, M.R. (2003a) Lessons learnt from the impact of the neglected role of groundwater in Spain's water policy. In: Al Sharhan and

Wood (eds) *Water Resources Perspectives: Evaluation, Management and Policy*. Elsevier Science, Amsterdam, pp. 63–81.

Llamas, M.R. (2003b) Ethical considerations in water management systems. *Water Nepal* 9/10(1/2), 13–27.

Llamas, M.R. and Custodio, E. (eds) (2003) *Intensive Use of Groundwater: Challenges and Opportunities*. A.A. Balkema, Lisse, The Netherlands.

Llamas, M.R. and Pérez-Picazo, M.T. (2001) *The Segura Catchment Management and the Debate on Hydrosolidarity in Spain*. 2001 Seminar of the Stockholm International Water Institute, Stockholm.18 August 2001.

Llamas, M.R. and Martínez-Santos, P.M. (2005a) The silent revolution of intensive ground water use: Pros and Lons. Guest Editorial. *Ground Water* 43(2), 161.

Llamas, M.R. and Martínez-Santos, P. (2005b) Intensive groundwater use: Silent revolution and potential source of conflicts, invited editorial for *Journal of Water Resources Planning and Management* (American society of Civil Engineers) 131, 337–341.

Llamas, M.R., Back, W. and Margat, J. (1992) Groundwater use: equilibrium between social benefits and potential environmental costs. *Applied Hydrogeology* (Heise–Verlag) 1(2), 3–14.

Llamas, M.R., Fornes, J., Hernández-Mora, N. and Martínez-Cortina, L. (2001) *Aguas subterráneas: retos y oportunidades* (*Groundwater: Challenges and Opportunities*). Fundación Marcelino Botín y Mundi-Prensa, Madrid, Spain.

López-Gunn, E. (2003) The role of collective actions in water governance: a comparative study of groundwater user associations in La Mancha aquifer in Spain. *Water International* 28(3), 367–378.

López-Gunn, E. and Llamas, M.R. (2000) New and old paradigms in Spain's water policy. In *Water Security in the Third Millenium: Mediterranean Countries Towards a Regional Vision*. UNESCO Science For Peace Services 9, 271–293.

Ministerio de Agricultura, Pesca y Alimentación (MAPYA) (2001) Plan Nacional de Regadíos (National Irrigation Plan), Madrid, Spain.

Ministerio de Medio Ambiente (MIMAM) (2000) Libro Blanco del Agua en España (The White Paper on Water in Spain), Secretaría de Estado para Aguas y Costas, Madrid.

Ministerio de Medio Ambiente (MIMAM). (2004) *Valoracion del costo del Uso del Agua subterránea en España* (*Valuing the Cost of Groundwater Use in Spain*) Technical Assistance, electronic versión, April 2004.

National Research Council (1997) *Valuing Ground Water*. National Academy Press, Washington, DC.

Postel, S. (1999) *The Pillar of Sand*. W.W. Norton, New York.

Rogers, P., Llamas, M.R. and Martínez-Cortina, L. (eds) (2006) Foreword. In: Water Crisis: Myth or Reality? A.A. Balkema, Lisse, The Netherlands.

Sahuquillo, A., Sánchez-Vila, X. and Martínez-Cortina, L. (eds) (2005) *Groundwater Intensive Use*. International Association of Hydrogeologists, Selected Papers, No. 7, A.A. Balkema, Leiden, The Netherlands.

Seckler, D., Amarashinge, U., Molden, D., de Silva, R. and Barker, R. (1998) World water demand and supply, 1990 to 2025. *Scenarios and Issues*, Research Report 19, International Water Management Institute, Colombo, Sri Lanka.

Selborne, L. (2000) *The ethics of Freshwater Use: A Survey*. COMEST, UNESCO, Paris.

Shah, T. (2004) *Groundwater and Human Development: Challenges and Opportunities in Livelihoods and Environment*, Paper presented at the 14th Stockholm Water Symposium, World Water Week, Stockholm, 16–20 August.

Shah, T., Debroy, A., Qureshi, A.S. and Wang, J. (2003) Sustaining Asia's groundwater boom: an overview of issues and evidence. *Natural Resources Forum* 27, 130–140.

Shamir, U. (2000) Sustainable management of water resources, in transition towards sustainability. Interacademy Panel Tokyo Conference, pp. 62–66.

Simmers, I., Villarroya, F. and Rebollo, L.F. (eds) (1992) Selected papers on overexploitation. *Hydrogeology*, Selected Papers, Vol. 3, Heise, Hannover.

Sophocleous, M. (2000) From safe yield to sustainable development of water resources – the Kansas experience. *Journal of Hydrology* 235, 27–43.

Sophocleous, M. (2003) Environmental implications of intensive groundwater use with special regard to streams and wetlands. In: Llamas and Custodio (eds) *Intensive Use of Groundwater*. A.A. Balkema, Lisse, The Netherlands, pp. 93–112.
Sullivan, C.A. (2001) The potential for calculating a meaningful water poverty index. *Water International* 26, 471–480.
The Economist (1998) *Environmental Scares*. 20 December 1998, 21–23.
United Nations (1997) Assessment of Global Water Resources, pp 10–13.
US Geological Survey (1999) *Sustainability of Ground-Water Resources*. Circular 1186.
Vives, R. (2003) Economic and social profitability of water use for irrigation in Andalusia. *Water International* 28(3), 326–333.
World Commissions on Dams (2000) *Dams and Developments: A New Frame for Decision-Making*. Earthscan, London.
World Humanity Action Trust (WHAT) (2000) Governance for a sustainable future, Part IV. Working with Water, *Reports of WHAT*, London, 145–197.
World Water Council (2000) A Water Secure World, The Commission Report.

14 Groundwater Management in the High Plains Aquifer in the USA: Legal Problems and Innovations

John C. Peck

Professor of Law, University of Kansas School of Law Lawrence, KS 66045; Special Counsel, Foulston Siefkin LLP, Overland Park, KS 66210, USA

Introduction

Description of the aquifer and the region

The High Plains area of central USA stretches from the Rio Grand River on the south to the Canadian border on the north, and from the 'humid prairie plains' (Kromm and White, 1992, p. 1) on the east to the Rocky mountains on the west. The imprecise eastern boundary runs generally along the eastern portions of the tier of states extending from North Dakota to Texas on the south. It is generally a level, treeless, grassland surface, except along watercourses, with a windy and subhumid climate (Kromm and White, 1992, p. 1). The High Plains aquifer system, including the Ogallala and Equus Beds, is the 'largest underground reservoir in the country' (Kromm and White, 1992, p. 3; Sophocleous, 2005), contains approximately 4000 km^3 of water and underlies parts of eight states. As shown in Fig. 14.1, the 570,000 km^2 aquifer mostly underlies parts of three states: Nebraska has 65% of the aquifer's volume, Texas 12% and Kansas 10% (Kromm and White, 1992, p. 15). Yet, due to the varying thickness in the aquifer, only 37% of Nebraska overlies the aquifer. The saturated thickness can range from 365 m (Kromm and White, 1992, p. 16) to less than 1 m (Groundwater Atlas, 1995, p. 4). Recharge of the aquifer through precipitation is slight, ranging from a high of 15.25 cm annually to a low of 0.06 cm (Kromm and White, 1992, p. 16).

The total population in the High Plains aquifer region has hovered around 2 million since 1960, with small growth areas in some states, but with overall declines showing up in the past few years. With approximately 3.5 people/km^2, this region is very sparsely populated. The High Plains aquifer provides 30% of the groundwater used for irrigation in the USA, and approximately

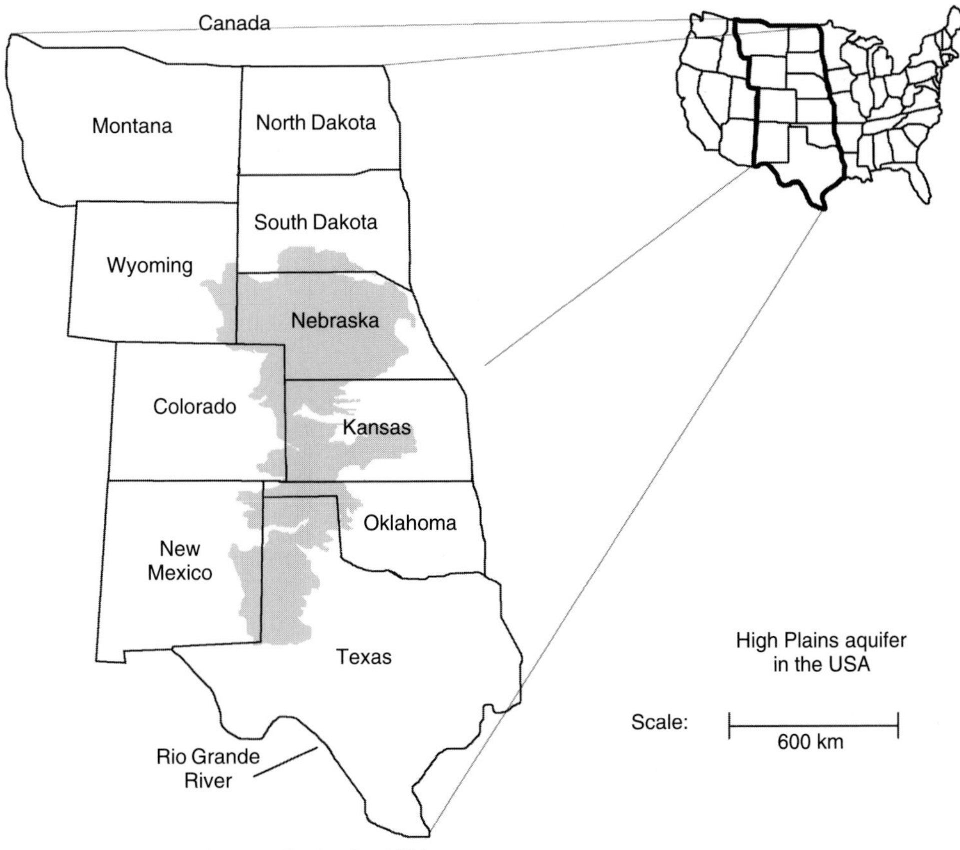

Fig. 14.1. High Plains aquifer in the USA.

20% of the irrigated land in the USA is located in this region (Sophocleous, 2005, pp. 352–353). Agriculture dominates the economy, with virtually all land in some form of agricultural use and with the related agribusiness devoted to seed, fertilizer, pesticide, herbicide, machinery and credit (Kromm and White, 1992, pp. 17, 20). The main irrigated crops are: maize, wheat, sorghum and cotton; others include lucerne, potatoes, vegetables, soybeans and pinto beans. Centre pivot irrigation systems, typically covering circles of 52.6 ha on a square field of 64.8 ha, are common. Farm sizes typically range from 1300 to 2300 ha.

Chapter overview

This chapter concentrates on Kansas, Nebraska and Texas – the three predominantly agricultural states overlying most of the groundwater in the High Plains aquifer. American states are autonomous and able to devise their own water allocation laws, except as constrained by the US Constitutional provisions that protect property from being taken by the government without compensation and that delegate specific powers to the federal government, such as under the 'interstate commerce clause'. Each of the three states has a unique groundwater law, recognizes water rights as property rights and faces unique issues and problems. They have also developed innovations in groundwater management. Because relevant state boundaries are not based on the boundaries of aquifers or river basins, some interstate tensions have arisen. Moving groundwater from its source to other locations within a state or beyond a state's borders often creates disputes, leading to concern by the public, protective legislation and litigation. However, although Kansas and Nebraska are contiguous states, and Kansas, Nebraska and Texas share the High Plains aquifer, the focus of this chapter is not primarily interstate sharing or disputes over the High Plains aquifer. That issue is broached only briefly in the section on interstate conflict with the state of Kansas, which treats the Republican River Compact conflict.[1]

This chapter describes a range of state-level issues and responses to groundwater law. It begins with a description of groundwater law, both for the USA in general and for the three states in particular. It then follows with an account of groundwater allocation law problems faced by Kansas due to groundwater mining. The next section addresses recent groundwater issues and innovations in each state: (i) in Kansas, a water reuse project and an aquifer storage and recovery(ASR) project; (ii) in Nebraska, two types of interstate conflicts, one dealing with antiexportation statutes and the other dealing with allocation of an interstate river and the surface water–groundwater interaction with that river; and (iii) in Texas, questions about the continued efficacy of its Rule of Capture groundwater doctrine and about the advisability of moving groundwater long distances within the state. The chapter finally summarizes and draws some conclusions.

Groundwater Law

American water allocation law, in general

Water vs. water rights

In American law, water is deemed 'personal property'. Personal property includes 'goods', i.e. things that are movable. The Uniform Commercial Code covers the law of contracts for the sale of goods and would therefore cover water purchase contracts. In contrast, water rights are deemed 'real property'. A water right is a right to use a certain annual quantity of water at a certain place, diverted from a specific point of diversion at a certain rate, and in perpetuity – as long as the water right holder follows the law and the prescribed conditions of the water right. Typical real property concepts and documents apply to sales of water rights just as they apply to sales of land: the deed is the conveying instrument, the mortgage is the security document and the statute of frauds requires that contracts of sale be in writing. But water rights are not exactly like land rights: a water right is a right to use the water, not ownership of the water; some types of water rights may be lost by non-use; and state constitutions or statutes may declare that the state's water resource is owned by the public or dedicated to the use of the public.

The law of groundwater allocation: various methods[2]

Several doctrines have developed in the USA for groundwater allocation. Some states use the Rule of Capture (also known as the English Rule or Absolute Ownership Doctrine), which holds that the owner of a tract of land owns all water underneath that land and can pump water without limit, except for prohibitions on malicious or wasteful use. Similarly, the Reasonable Use Doctrine (also known as the American Rule) permits unrestricted pumping, except that the use of the water must be for a reasonable purpose and be used on the landowner's land. The Correlative Rights Doctrine holds that landowners overlying an aquifer must share the aquifer. The Prior Appropriation Doctrine applies the principle of 'first in time, first in right' to groundwater; the earlier 'senior user' may enjoin a later 'junior' right holder who impairs the 'senior' right holder's use. The Restatement of Torts §858 rule combines elements of the doctrines of Reasonable Use and Correlative Rights.

Groundwater law in the states of Kansas, Nebraska and Texas

Kansas

The Absolute Ownership Doctrine prevailed in Kansas until 1945 when the state enacted the Kansas Water Appropriation Act (Kansas Statutes Annotated, 2005, §§82a–701, et seq.), adopting the Prior Appropriation Doctrine for groundwater. Persons wanting to divert water since 1945 have had to obtain a permit from the chief engineer of the Division of Water Resources (DWR) before diverting water. The Act allowed people who were using the water on the date the Act became effective to claim 'vested rights'. People who owned water rights by virtue of landownership alone but who were not diverting water lost their rights. From 1945 on, they had to apply for appropriation rights.

Groundwater pumping from the numerous permits granted from 1945 through the 1970s resulted in serious groundwater mining. In response the legislature enacted the Groundwater Management District (GMD) Act. Five GMDs have been established, and they have the power to enact management programmes and recommend regulations to the DWR. These regulations cover matters such as well spacing and overall aquifer withdrawal policy. For example, Southwest Kansas GMD No. 3 has a 'depletion' formula allowing a regulated lowering of the water table. The Equus Beds GMD in central Kansas has adopted 'safe yield'[3] regulations. The GMD Act provides that in cases of serious groundwater mining, the chief engineer of the DWR may establish intensive groundwater use control areas (IGUCAs) following a public hearing. It also provides that if the chief engineer establishes an IGUCA, he has extraordinary powers of regulation, including the power to reduce the annual quantity of water rights within the IGUCA.

Some quantity of groundwater in Kansas is connected hydrologically with neighbouring streams. While several states like Wyoming have water right dispute resolution statutes that expressly recognize this interconnection (Wyoming Statutes, 2005, §41-3-916), and some states have defined this interconnection,[4] Kansas law is less clear (Peck and Nagel, 1989, pp. 199, 281–300). Yet, the chief engineer has recognized the interconnection in some situations such as in establishing IGUCAs.

Nebraska[5]

Nebraska uses a hybrid of the Reasonable Use Doctrine and the Correlative Rights Doctrine for groundwater rights. The right to use groundwater in Nebraska comes from ownership of the overlying land. No permit is required to drill wells except in groundwater management areas (GMAs), but owners must register them. A preference statute favours domestic use over all other uses and agricultural use over industrial or manufacturing uses. Statutes also regulate the location of wells with respect to nearby streams and other wells.[6] The owner may not use more than a reasonable quantity and may have to share it with others if the groundwater supply is insufficient for all owners. While the Reasonable Use Doctrine generally prohibits the user from using water off the overlying land, Nebraska permits public water suppliers to do so, with compensation to injured overlying landowners, and also permits water use offsite for agricultural uses if it does not adversely affect other users and is deemed in public interest.

Unlike Kansas, which has five special districts devoted exclusively to groundwater management, Nebraska is divided into 23 natural resource districts (NRDs) based on river basin boundaries covering the entire state.[7] Each NRD has its own priorities and programmes, covering matters such as erosion prevention, flood prevention and control, water supply, conservation of surface and groundwaters, drainage, recreation and forest management. Under the Nebraska Ground Water Management and Protection Act (Nebraska Revised Statutes, 2005, §§46–701, et seq.), groundwater management is a local rather than a state responsibility. NRDs develop management plans, which must be approved by the state director of the Department of Natural Resources. To protect the quality and quantity of water and to prevent conflicts between users of groundwater and appropriators of surface water, NRDs may establish GMAs (Nebraska Revised Statutes, 2005, §46–712) inside of which they may implement controls (Nebraska Revised Statutes, 2005, §46–739). The Act permits NRDs to regulate and control groundwater in the GMA with well spacing, pumping restrictions, rotation requirements, metering and reduction of irrigated areas.

Legislative amendments to the Act in 2004 have drawn attention to the issues of hydrologically connected surface and groundwaters (Nebraska Revised Statutes, 2005, §§46–703(2), 46–713, 46–715 through 46–718). They require evaluation of 'the expected long-term availability of hydrologically connected water supplies' (Nebraska Revised Statutes, 2005, §46–713) and create the possibility of different types of management, through the development of 'integrated management plans' (Nebraska Revised Statutes, 2005, §46–715), when the groundwater is not connected with surface water. A Nebraska Supreme Court case in 2005 recognized the right of surface water users to sue alluvial groundwater pumpers for damages, if the groundwater pumping causes unreasonable harm (*Spear T Ranch v. Knaub*, 2005).

Texas

For groundwater allocation, Texas employs the Rule of Capture, a common-law rule (judge-made, not legislated). The Rule of Capture provides that the landowners may 'take all the water they can capture under their land and do with it what they please, and they will not be liable to neighboring landowners even if in so doing they deprive their neighbors of the water's use' (Potter, 2004, p. 1). Indeed, Texas landowners own the underlying groundwater (Texas Water Code, 2005, §36–002). Texas is the only western state that follows the Rule of Capture (Potter, 2004, p. 1).

The Texas Supreme Court adopted the Rule of Capture in a 1904 case (*Houston and Texas Central Railroad Company v. East*, 1904), choosing to apply that rule instead of the Reasonable Use Doctrine, and in doing so cited two public policy considerations – the unknown and uncertain character of groundwater, and the fact that choosing another doctrine would generally interfere with agriculture, industry and hence the development of the state. Thus, the Texas Rule of Capture exists as a common-law rule, and court decisions have modified the Rule of Capture to prevent '(1) willful waste, (2) malicious harm to a neighbor, and (3) subsidence' (Potter, 2004, p. 9). In 1917, Texas amended its constitution to add the following Conservation Amendment:

> The conservation and development of all of the natural resources of this state ...including...the conservation and development of its...water...and the preservation and conservation of all such natural resources...are each and all declared public rights and duties; and the Legislature shall pass any such laws as may be appropriate thereto.
>
> (Texas Constitution, 2005, Art. 16, §59)

By way of this constitutional section, the legislature has the power and duty to change the Rule of Capture if necessary.

The legislature has provided for the creation of groundwater conservation districts (GCDs) to conserve, preserve and protect groundwater (Texas Water Code, 2005, §36–011)[8]; 87 GCDs have been confirmed or are pending in the state, and 11 GCDs overlie portions of the Ogallala aquifer within the High Plains aquifer system. The GCD legislation, however, expressly recognizes that landowners own the groundwater (Texas Water Code, 2005, §36–002). GCDs are required to adopt management plans to address goals[9] and regulate well drilling (Texas Water Code, 2005, §36–113), and are also empowered to enact and enforce rules that regulate well spacing, limit groundwater production and conserve groundwater.

Texas classifies its waters as surface water, diffused surface water and groundwater (*Waters and Water Rights*, 1991 and 2004 Cumulative Supplement, v. 6, p. 774). Statutes do not expressly cover the interrelationship of surface and groundwaters or provide for the conjunctive use between the two classes. Texas court decisions seem to maintain these two as distinct and separate classes by creating a presumption that groundwater pumped near streams causing an effect on the streams is nonetheless deemed groundwater and thus governed by the Rule of Capture (*Waters and Water Rights,* 1991 and 2004 Cumulative Supplement, v. 6, p. 774).

Groundwater Allocation Law in Kansas: Property Rights and the Problem of Claims of 'Takings'

The legal problem: Is compensation required when the state restricts groundwater pumping?

Like most states in western USA, Kansas follows the Prior Appropriation Doctrine for both surface and groundwaters (*Waters and Water Rights,* 1991 and 2004 Cumulative Supplement, v. 6; Kansas Statutes Annotated, 2005, §§82a–701, et seq.). When Kansas adopted that doctrine in 1945, replacing the Absolute Ownership Doctrine, it continued to recognize the rights then used, as 'vested rights', but eliminated unused rights without compensating the holders of those rights. Landowners not using their underlying groundwater challenged the constitutionality of the Act on the basis of an 'unconstitutional taking' for which compensation should be due from the state. The basis of such a constitutional challenge was that the US Constitution's Fifth Amendment requires the government to compensate people when it takes their property. They claimed that by eliminating their unused water rights, the state had 'taken' their water rights from them. They argued that even an unused water right was a property right. The courts, however, have upheld the Act against such challenges (*Williams v. City of Wichita,* 1962).

Another potential challenge arises when the state does not eliminate water rights entirely, but merely restricts groundwater pumping by water right holders to levels below their permitted annual quantities. Extensive regulatory reduction of pumping is arguably tantamount to a 'taking' of a property right even though the government is not technically acquiring title to the water right. While American water rights have generally been viewed as property rights, the original version of the Kansas Water Appropriation Act did not expressly define a water right as a property right. The legislature amended the Act in 1957 to define a water right to be a property right:

> [A] water right is a 'real property right appurtenant to and severable from the land ... [and it] ... passes ... with a conveyance of the land by deed, lease, mortgage, will, or other voluntary disposal, or by inheritance'.
>
> (Kansas Statutes Annotated, 2005, §§82a–701 (g))

The proliferation of irrigation water rights in Kansas from the 1950s through the 1970s led to a serious groundwater mining problem. To slow pumping in the Walnut Creek Basin in west-central Kansas and thereby to protect the Cheyenne Bottoms Wildlife Preserve (an important migratory bird stopover point) from the pumping by the basin's more than 800 irrigation water users, the chief engineer held hearings in 1990 to establish an IGUCA. Following several weeks of hearings and testimony, the chief engineer issued an IGUCA order that established an IGUCA and recognized the interconnection of the groundwater and the Arkansas River and its tributary, the Walnut Creek. After finding that the annual basinwide 'safe yield' (sustainability[10]) was 27,753,792 m^3 and that irrigators and others were pumping almost twice that quantity, the chief engineer instituted 'safe yield' in the river basin and along with it substantial reductions in irrigation pumping. The order divided the water rights into two large groups – 'senior rights' and 'junior rights' – and cut back annual quantities for both groups, but much more significantly for junior rights. The irrigators appealed the order, claiming an unconstitutional taking of property, but eventually dropped their appeal. Thus, Kansas courts have still not decided the 'takings' issue.

Whether such governmentally imposed curtailments are constitutional is an open question in Kansas. Generally in the USA, the western states by court decision are moving away from the view that a water right is an immutable property right to be treated just like a land property right.[11] California, for example, has upheld the 'public trust doctrine' for water rights (*National Audubon Society v. Superior Court of Alpine County*, 1983), meaning that the state is viewed as holding the water resource as trustee for the people and as not having the power to grant unrestricted, unchangeable rights to its water users. The state has not only the power but also the duty to periodically review water rights in light of current conditions, not conditions existing at the time of permit issuance. States have also recognized that water use quantities may be curtailed when waste is occurring.[12] Of course, the nature of the Prior Appropriation Doctrine itself requires the recognition that junior rights must be curtailed when senior rights are impaired (i.e. injured or damaged).[13] But in Kansas, impairment is usually claimed in cases of alleged direct impairment (lowering of the water table or reduction of the pump rate) by the pumping of one well that adversely affects another, not by a general lowering of the water table caused by general aquifer pumping throughout the area. This view, then, might prohibit a senior well owner, whose water table is dropping, from enjoining other junior irrigators in the region where pumping generally causes areawide water table declines but does not directly impair the senior well owner's water right.

The public policy issue: Should the present generation preserve groundwater for future generations?

Aligned with the legal question of whether states like Kansas *can* legally restrict groundwater pumping without having to compensate the affected water right holders is the ethical and public policy question of whether Kansas *should* restrict pumping to preserve groundwater resources for future generations.[14] Most prudent policymakers and socially conscious citizens would say that a society should not waste water, and Kansas has implemented measures to encourage conservation, such as requiring conservation plans for various water users (Kansas Statutes Annotated, 2005, §82a–733). It is one thing to require conservation measures, especially as express conditions on water rights permits for prospective water users; it is quite another to require reductions by current water users for the purpose of 'saving' water for future generations. Even if it were constitutional to do it without compensation, a proposition that is debatable in Kansas, the answer to the ethical question is not obvious:

> The ethical question of imposing safe yield [sustainability[15]] is intriguing no matter which way one resolves the legal question – if no compensation is required, the water user suffers the immediate economic loss; if compensation is required, the taxpayer loses; in either case, forced curtailments will cause someone to suffer and sacrifice for the future.
>
> (Peck, 2004, pp. 349, 351)

This generation's policymakers deciding the issue could consider statements of preserving water for future generations found in statutes, political platforms, the media and literature – popular, environmental and philosophic.

From the ethical arena, several rules come into play:

- The Golden Rule – 'Do unto others as you would have them do unto you.'
- Frankl's rule of logotherapy – 'Live as if you were living already for the second time and as if you had acted the first time as wrongly as you are about to act now!'
- Kant's categorical imperative – 'Act only on that maxim whereby thou canst at the same time will that it should become a universal law.'
- Rawls' principle – '[T]he correct principle is that which the members of any generation (and so all generations) would adopt as the one their generation is to follow and as the principle they would want preceding generations to have followed (and later generations to follow), no matter how far back (or forward) in time'.
- The simple solution to the problem of dividing a piece of pie[16] (Peck, 2004, pp. 352–253).

Deciding whether to adopt strict controls on aquifer pumping to conserve water for the future is a very difficult issue. Current irrigation water users are making a 'beneficial use' of the aquifer, as defined in the current Kansas administrative regulations. Opponents to that view deem it wasteful to pump large quantities of groundwater for irrigated crops not normally grown in the otherwise dry-land wheat-farming area of western Kansas, with the resulting crops used for feeding cattle to satiate the nation and the world's hunger for beef. If Kansas were to restrict current agricultural groundwater users from pumping for the benefit of future generations, a serious disruption of the present economy of western Kansas would result.[17] Moreover, it is likely that the groundwater saved and conserved for the future would eventually be pumped for municipal use, not irrigation.

Recent Issues and Innovations in Groundwater Use and Management

Kansas

Dodge City's water reuse project

The beef cattle industry is very important to the state of Kansas as a whole[18] and to Dodge City, located in south-western Kansas, in particular. Confined feed yards near Dodge City fatten cattle for slaughtering, rendering, packing and shipping, both for domestic and international purposes. The city relies on groundwater for its municipal water supply, which covers household, commercial and industrial (beef plant) uses. Faced with the cost of constructing an expensive wastewater treatment plant in the 1980s, Dodge City opted instead to pipe its municipal wastewater 17 km south of the city to large ponds, where the wastewater undergoes aerobic and anaerobic treatment. A farming operation then applies this wastewater for irrigating maize, milo and lucerne, thus saving the farmers the cost of buying nitrogen and other plant nutrients.

Before this water reuse project was constructed, the participating farmers had drawn freshwater from the aquifer for irrigation purposes. The city constructed the wastewater ponds near the farmers' wells and irrigated fields. The 1987 agreement between the farmers and the city provided that in exchange for the use of wastewater for a 40-year term, the farmers would lease their groundwater rights to the city, except for small amounts of water needed to dilute the treated wastewater. The city, however, did not use the groundwater under its lease from 1987 to 2004.

Recent growth in population and industry has caused the city to increase the wastewater treatment capacity at the ponds and commence use of the groundwater rights under lease for municipal purposes. For this expansion, the project participants face several legal and scientific issues and challenges to insure continued success. The change in groundwater rights is one such problem. Permission is required from the DWR to change the type of use (irrigation to municipal), place of use (farms to city) and points of diversion (old irrigation wells to new municipal wells) (Kansas Statutes Annotated, 2005, §82a–708b). A water quality challenge is to avoid applying low-quality water that would harm the plants and pollute the groundwater. Optimizing the quality and quantity of the irrigation water insures high crop yields and high-quality groundwater over time. Measures are used prior to pond treatment to remove some pollutants. Cropping choices and wastewater application schedules by the farmers are important. They must observe a fine balance: on the one hand they must maintain profitability in their farm operations, but on the other hand they must minimize pollutant migration to the groundwater by optimizing nutrient uptake and usage by the plants. As the city and its industries grow, so does the contractual obligation of the farmers to accommodate more wastewater for irrigation, requiring the farmers to acquire additional crop land.

Wichita's Equus Beds aquifer storage and recovery project

Wichita is the largest city in Kansas and is located in the south-central part of the state, just south of the Equus Beds aquifer, the 'eastern most extension of the High Plains aquifer system' (Equus Beds Information Resource, 2005). In the 1930s, Wichita established wells in the Equus Beds and began pumping groundwater for municipal use. Running through the Equus Beds area is the Little Arkansas River, which joins the Arkansas River at Wichita.

Until the early 1990s, Wichita drew heavily from the Equus Beds aquifer. Extensive groundwater use by Wichita and irrigating farmers drew down the aquifer approximately 13 m in some locations, with a total loss of approximately 24.6 million cubic metres of water from aquifer storage from the time heavy pumping began in the 1940s. Irrigators with water rights junior to Wichita's water rights may have to shut down their wells if the water table keeps dropping. In addition to the lowering of the water table, the other problem in the region is a large, underground saltwater plume located north-west of the Wichita wells, migrating towards the city's well field.

Wichita is working on an ASR project to replenish the Equus Beds for the benefit of both Wichita and irrigators as well as to provide a hydraulic barrier to impede the migration of the saltwater plume moving towards the Wichita well field. The basis of the ASR project is that flood and other higher-than-normal-flow water seeps down into the banks of the Little Arkansas River, is held there and can be withdrawn for recharge into the deeper Equus Beds aquifer. The plan is to refill the depleting aquifer with flood water.

A demonstration project from 1995 to 2004 showed that engineering aspects of the ASR project were feasible, but there were legal problems because of inadequate statutes and regulations. In response, the DWR worked with Wichita and promulgated a new set of regulations designed explicitly for 'aquifer storage and recovery permitting' (Kansas Administrative Regulations, 2005, §§5-12-1, et seq.). Each applicant for an ASR project must file applications for two types of appropriation permits: (i) to divert water either directly from the river or from bank storage; and (ii) to divert water from the Equus Beds aquifer for its ultimate use. The applicant must also comply with relevant regulations of the Kansas Department of Health and Environment regarding the quality of the injected or artificially stored water.

In applying for the first permit to divert water from the river or from bank storage, the applicant must describe the volumetric area in which the water will be stored. The bottom of the basin storage area is the lowest level that has occurred within 10 years of the application; the top is the elevation representing the maximum storage potential, i.e. the pre-development water table elevation. The applicant must also include a methodology of accounting for the water stored on an annual basis to enable 'recharge credits' to be calculated. The regulation seeks an accounting system that sets up a 'water balance' for the water entering and leaving the storage area, considering recharge, groundwater inflow and outflow, evapotranspiration, groundwater pumpage of recharge credits, and all non-domestic wells in the basin storage area.

Wichita's ASR project covers four phases to be completed in 2015, with a goal of 378.5 million litres per day capacity. Phase I, scheduled for completion in 2007, will have a capacity of 37.85 million litres per day. At a public hearing held by the DWR in December 2004 to consider Wichita's Phase I permit applications, the public expressed concerns about the unknowns – the effect of the ASR project on groundwater quality, downstream water right holders and downstream riparian owners. The interested participants in the project (Wichita, the GMD, DWR and the public) must continually review the data and analyse the goals, objectives and performance of the project, and modify it when necessary if its twin goals of recharging the Equus Beds aquifer and halting movement of the saltwater plume for the benefit of Wichita and area irrigators are to be met.

Nebraska

Protecting the state's groundwater from interstate exportation: antiexportation statutes and the Sporhase case

American state legislatures have occasionally sought to protect natural resources from export to other states, and these restrictions have been fought in court.[19] In 1967, Nebraska enacted a statute regulating the movement of groundwater out of state (Nebraska Revised Statutes, 1978, §46-613.01). It provided that groundwater could be withdrawn for use in another state if the Director of Water Resources granted a permit after finding that the withdrawal was 'reasonable,... not contrary to the conservation and use of ground water, and... not otherwise detrimental to the public welfare' (Nebraska Revised Statutes, 1978, §46-613.01). Such withdrawals were prohibited outright, however, unless the destination state granted a 'reciprocal right to withdraw and transport ground water from that state' to Nebraska (Nebraska Revised Statutes, 1978, §46-613.01).

The case of *Sporhase v. Nebraska* involved a farmer who owned contiguous tracts of land in Nebraska and Colorado. He irrigated both tracts from his Nebraska well without obtaining the required permit. Nebraska sought an injunction on the basis that Colorado totally banned groundwater exports and thus could not reciprocate as required by the Nebraska statute. The state was successful in the Nebraska courts. The US Supreme Court reversed the Nebraska court's decision. The Court based the decision on the Commerce Clause of the US Constitution, which gives Congress the power to regulate interstate commerce.

The case involves what is commonly called the 'negative Commerce Clause' because, while Congress has the power to enact relevant legislation dealing with the interstate movement of groundwater, it had not done so. The Court first held that water is an 'article of commerce', thus implicating the Commerce Clause. The Court then noted that 'the exercise of unexercised federal regulatory power does not foreclose state regulation of its water resources' (*Sporhase v. Nebraska*, 1982, p. 954), as long as the statute 'regulates evenhandedly to effectuate a legitimate local public interest, and its effects on interstate commerce are only incidental' (*Sporhase v. Nebraska*, 1982, p. 954).[20] The Court also found that the first three aspects of the statute (it must be reasonable, not contrary to conservation and not detrimental to the public welfare) were permissible, but that the reciprocity clause was unconstitutional as being too broad a restriction:

> Even though the supply of water in a particular well may be abundant, or perhaps even excessive, and even though the most beneficial use of that water might be in another State, such water may not be shipped into a neighboring State that does not permit its water to be used in Nebraska.
>
> (*Sporhase v. Nebraska*, 1982, p. 958)

A reciprocity clause might be permissible only if

> it could be shown that the State as a whole suffers a water shortage, that the intrastate transportation of water from areas of abundance to areas of shortage is feasible regardless of distance, and that the importation of water from adjoining States would roughly compensate for any exportation to those States.
>
> (*Sporhase v. Nebraska*, 1982, p. 958)

The Court further stated that an arid state might justify a complete ban on exports by demonstrating a close relationship between the ban and conservation. The arid state of New Mexico, for example, might justify a ban if it could show that the very water it was prohibiting from export could be used to alleviate water shortages in New Mexico by piping water to those areas.

In 1984, following the Sporhase case, Nebraska amended the statute to remove the reciprocity language (Nebraska Revised Statutes, 2005, §46-613.01), but the amended statute retained the protection of the health, safety and welfare of its citizens.[21]

Interstate conflict with the state of Kansas: Nebraska groundwater pumping affects the Republican River

Disputes among American states over interstate rivers have been common. Three methods of dispute resolution have evolved: (i) a state may sue another state in the original jurisdiction of the US Supreme Court, which will apply the doctrine of Equitable Apportionment (*Kansas v. Colorado*, 1907); (ii) Congress may allocate the water[22]; and (iii) the states may settle their differences with interstate compacts, as is illustrated by the Republican River Compact entered into by Kansas, Colorado and Nebraska in 1942 (Kansas Statutes Annotated, 2005, §§82a-518).

Draining a 64,491 km^2 watershed, the Republican River begins in Colorado, runs eastward into Kansas, turns northward into Nebraska and then south-east running back into Kansas. Because of the interstate nature of the river and the potential for conflict, Kansas, Colorado and Nebraska signed the Compact in 1942 with a view of equitably dividing the waters of the river and its tributaries and of avoiding future conflict. The Compact provided the name and location of each basin and subbasin, defined the 'virgin annual water supply' as 'the water supply within the Basin undepleted by the activities of man' (Kansas Statutes Annotated, 2005, Art. II) and allocated to each state a portion of the virgin annual water supply. The Compact runs in perpetuity.

In the 1990s, Kansas claimed that Nebraska was using more of its share of water by allowing unregulated pumping of alluvial groundwater. After unsuccessful facilitation talks, Kansas sued Nebraska and Colorado in the US Supreme Court in 1999. A threshold issue involved alluvial groundwater. Nebraska denied that the Compact covered groundwater pumping, in that the language of the Compact did not expressly address groundwater in its allocation scheme. The Supreme Court ruled against Nebraska on that issue, holding that '[t]he... [c]ompact restricts a compacting State's consumption of groundwater to the extent the consumption depletes stream flow in the Republican River Basin' (*State of Kansas v. State of Nebraska and State of Colorado*, 2002, Special Master's First Report and Case Management Order). In 2003, the states settled the other issues in the case. Some of the settlement topics included treatment of groundwater pumping (including the use of computer modelling of the groundwater system as a means of accounting for the consumption of groundwater), dispute resolution, a moratorium on the construction of new groundwater wells, formulas for determining future compact compliance, use of 5-year running averages for accounting and compliance and a framework for working together 'to improve operational efficiencies and the usable water supply in the lower Republican River basin' (Testimony of David L. Pope, 2003).

Texas

Rethinking the Rule of Capture

The century-old Texas Rule of Capture is undergoing evaluation and debate (100 Years of Rule of Capture, 2004).[23] Professor Corwin Johnson said:

> All that can be said in favor of the rule of capture is that it leaves the market free to allocate water to uses regarded by the market as most valuable... [but that]... eventually its lack of restraint leads to diminishing, and eventual depletion, of the available supply of aquifers... [and that] it not only threatens the supply of water in Texas, but also deprives Texas landowners of rights they might otherwise have [because] [t]hey have no legal remedy for dewatering of their wells by others.
> (Johnson, 2004, p. 11)[24]

In a paper prepared for the Texas Public Policy Foundation, water resources economic consultant Clay Landry stated:

> [T]he rule of capture makes it extremely difficult for landowners to conserve and manage their groundwater assets... [because]... the only way they can protect their claim is by pumping the water... [resulting in]... a race to the pumphouse.
> (Landry, 2000, p. 1)

Support exists, however, for retaining the Rule of Capture in Texas. Those supporting the rule argue:

> [T]he rule of capture in combination with regulation by local option groundwater conservation districts [GCDs] has proven to be an effective means of developing

> and managing Texas' groundwater resources... [and that]... [a]s a practical matter, the days of operating under an unrestricted rule of capture in Texas are past... [because]... [t]he vast majority of production occurs from resources that are included within GCDs where the rule of capture is significantly limited by district rules and permitting requirements.
>
> (Caroom and Maxwell, 2004, pp. 41, 55)

Professor Johnson recommended that the courts adopt the Restatement of Torts §858 to replace the Rule of Capture (Johnson, 2004, p. 15). Alternatively, the Texas Legislature could adopt one of the various groundwater allocation doctrines used by other states,[25] or 'ignore the rule of capture, and continue on its present course of addressing directly groundwater problems' (Johnson, 2004, p. 16). While adopting the Prior Appropriation Doctrine 'would be helpful' because of the quantification of the rights, integration with surface water and preservation of historic use, that doctrine too would have disadvantages, as noted elsewhere in this chapter (Johnson, 2004, pp. 16–17). Water resources economist Landry concluded in his paper that because '[s]trong markets make for good markets' (Landry, 2000, p. 3), '[p]roperty rights and water markets offer the best hope among all other options for efficiently and equitably allocating this precious resource to its most highly valued uses' (Landry, 2000, p. 8). Supporters of the Rule of Capture argue that refinement, not replacement, would be preferable.

Moving Ogallala aquifer groundwater to other uses and places in Texas

Nebraska's attempt to prevent the interstate movement of water has been discussed earlier. However, some states place limits on the intrastate movement of water. The Kansas Water Transfer Act, for example, regulates water diversions exceeding 2.36 million cubic metres per year transported 56 km or more, with special permitting requirements (Kansas Statutes Annotated, 2005, §§82a-1501, et seq.).

In contrast, the Rule of Capture in Texas permits landowners to pump water and use it on or off the land overlying the aquifer. Diversions of the Ogallala aquifer groundwater already exist in Texas, and more are planned. For example, the Canadian River Municipal Water Authority (CRMWA), which supplies water to almost 500,000 people in 11 cities, draws water from Lake Meridith and Ogallala wells in the Texas Panhandle. The CRMWA has obtained permits for 49.32 million cubic metres of water per year from the Panhandle Groundwater Conservation District #3. The City of Amarillo has also purchased water rights for 177,840 ha in Roberts County (Water Ranching, 2002).

The Mesa Water Project (MWP), a large project proposed in 1999, would pump and move 246.6 million cubic metres of Ogallala aquifer water per year to municipalities in the state. The MWP involves 200 landowners in the Texas Panhandle and initially includes approximately 988,000 ha in Roberts County, one of four counties involved in the project (Mesa Water, 2005). These four counties cover 6,125,000 ha, with 247,000 ha now in irrigation. The project sponsors hope to help meet Texas' water needs over the next 125–200 years by constructing an extensive pipeline from the source wells to various reservoirs associated with the Brazos River and using the river itself as a conduit, thus making conjunctive use of surface and groundwaters. Ultimate water purchasers include the Dallas-Fort Worth metropolitan area, San Antonio and other cities.

Such large diversions of groundwater over long distances in Texas are not without controversy. The concerns involve matters such as the privatization of water supplies (Knickerbocker, 2002); claims that the withdrawals may greatly exceed recharge (Eaton and Caplan, 2003), leaving no water for the children and grandchildren of the local people (McKenzie, 2004) and the otherwise adverse effects on rural communities (Water and the Future of Rural Texas, 2001); the failure of these water marketing projects to take third party effects into account (Water and the Future of Rural Texas, 2001); the lack of a state groundwater policy (Water and the Future of Rural Texas, 2001) and water quality, wildlife and environmental issues when fresh groundwater is mixed with salty river water (Ostdick, 2004).

Conclusion

Introduction

The High Plains aquifer region is an agricultural area of modest precipitation, sparse population, relatively large farms and abundant but declining groundwater resources. Ironically, while the landforms, land use, demography and water resources are fairly uniform across the entire region, what varies among the states are the laws and the legal institutions regulating the water resources. The question is whether the ideas presented in this chapter involving American water law have relevance and applicability in other regions of the world.

Water rights law and water rights doctrines

Each legal doctrine involving groundwater allocation and use discussed in this chapter has merits and demerits. The Rule of Capture applied in Texas, giving landowners ownership of underlying groundwater, provides great freedom of use by the landowner, but gives little protection against impairment by neighbours and little control by the state over the declining water table. The same holds true with the doctrines of Reasonable Use and Correlative Rights employed by Nebraska. With its requirements of permits prior to use, Kansas' Prior Appropriation Doctrine applied to groundwater provides a greater level of state control and protection of the water rights from other users. The doctrine's disadvantages are the lack of freedom of groundwater use by landowners and a heavy requirement of state resources (money and personnel) necessary to administer the complex system of water rights. However, once water rights are obtained under the various doctrines, all three states recognize them as property rights protected against government takings without compensation by the US Constitution's Fifth Amendment.

While the three states apply different allocation doctrines to its groundwater resources, a common and important element is that in each state the legal doctrine was applied early on, and it developed along with the growth of the state's population and water use. Even in Kansas, which changed from the Rule of Capture to the Prior Appropriation Doctrine in 1945, the predominant period of groundwater development occurred after 1945. Thus, there has been no need to superimpose a doctrine on a state having no prior existing water allocation law.

If a country has a problem of extensive exploitation of groundwater, it may benefit from having laws in place to administer, control and limit groundwater pumpage (Singh, 2002).[26] Water allocation law could help if a country has the power to gain control over the groundwater resource and to provide some system of controlling further water use. However, having groundwater management laws in place does not necessarily insure groundwater conservation or prevent groundwater mining, as is shown by the declining groundwater problems in the High Plains aquifer states (Peck, 2003).[27] Enacting groundwater management laws prior to the onset of intensive groundwater exploitation is preferable to waiting until exploitation occurs, but many countries in the world already find themselves dealing with aquifers that have declining yields, water quality or both.

Choosing a water allocation method is difficult, and the methods used in America are, of course, not the only choices. Some of the selection factors to be considered by a country include the type of legal doctrine already in place, if any, including constitutional protection of property against government takings without compensation; the extent to which groundwater resources are already being overused and the current rate of growth of groundwater use; the density of population and water wells; the strength and viability of the judiciary, administrative agency system and legal system in general to resolve water disputes expeditiously; and the availability of public funds and hydrologic and other scientific and legal expertise and data available to administer the system.[28] A country having areas with large numbers of groundwater irrigation users per unit area might find the costs of administration of the Prior Appropriation Doctrine prohibitive. Moreover, superimposition of strict regulations might result in serious unrest or even revolt among water users. It might be preferable to have a system in which water rights are clearly defined, but little or no continuous administration is involved. For example, an alternative to the US doctrines might involve a hybrid system that would establish new rights and recognize existing ones, but having them last for a term of approximately 20 years as opposed to having them last in perpetuity.[29] Water could be reallocated at the end of the term. The rights could be freely transferrable, and disputes could be resolved by arbitration or other alternative forms of resolution.

Water conservation, water reuse, water recharge and recovery

This chapter has described various private projects and government actions that attempt to conserve groundwater in central USA. While technological advances continue to improve efficiencies in irrigation and other water uses, legal and economic problems can arise when the government imposes conservation on existing water users. In the USA, change has been slow but sure: as technology improves, the law has slowly increased its involvement in overall management of groundwater. Claims of compensation for 'taking' of property when the state reduces water right quantities are weaker when the state gradually puts conservation measures in place. But the threat or perceived inevitability of such regulation combined with the need for additional water can produce innovation in conserving existing supplies and in acquiring new ones, as demonstrated by Dodge City's reuse project and Wichita's recharge and recovery project.

Laws limiting the movement of water across political boundaries

A country with its internal state boundaries overlying groundwater aquifers may face situations similar to that of Nebraska in the Sporhase case. The Supreme Court's decision resulted in a limitation on state power to prevent the movement of groundwater to points outside its state boundaries. From a policy standpoint, this decision seems to strike a suitable balance between the needs of the state in protecting the health and welfare of its citizens in times of crises and the needs of a free flow of commerce. Other countries may have to make this policy decision based on other principles and considerations, but it would seem that the Sporhase balance might be universally relevant, at least in considering these two factors.

The proposal of the MWP to make large intrastate diversions of Texas groundwater presents different legal and policy issues than those of interstate diversions. Even intrastate water diversions cause public concern about the disruption of the economies of the places of origin as well as environmental, human displacement and other costs. If a state government has power over its water resources, it can make necessary policy judgements about the costs and benefits to the exporting and importing areas in the state, and can take into account relevant externalities. If a government employs the Rule of Capture, however, it has tacitly left such decisions in the hands of private enterprise.

Conflict resolution between political entities

It may be preferable for countries to have conflict resolution procedures in place before disputes arise, whether the conflicts are among individual water users, states within countries or neighbouring countries. The water law allocation doctrines applied by the three American states are mainly applicable to individual water users. Of the three methods of interstate water conflict resolution in the USA mentioned earlier, the interstate compact is theoretically preferable, as the states have agreed to the allocation in advance. Compacts sometimes come about only after one state has had to resort to a lawsuit seeking equitable apportionment in the Supreme Court, and recent litigation indicates that having a compact does not insure against further disputes. Having two states to recognize, discuss, negotiate and resolve their disputes by compact, however, seems preferable to the uncertainty of equitable apportionment decided by the Supreme Court. Interstate water dispute resolution involves a complexity and level of detail that makes it generally unsuitable for Congressional allocation.

Notes

1 To date, interstate water compacts have involved primarily rivers and only tangentially groundwater. Litigation on one such compact is discussed in the section on interstate conflict with the state of Kansas. However, a model interstate compact specifically involving groundwater is being discussed (The Utton Center, 2004), and Oklahoma legislation in 2001 proposed a multistate groundwater compact (The Bimonthly Newsletter, 2001).

2 For surface water allocation, the USA is divided into two regions and two doctrines. In the eastern states, those states lying east of the High Plains region, precipitation is abundant. These states use the 'Riparian Doctrine'. By virtue of owning land adjacent to a river, riparian landowners have the right to use a reasonable amount of water on their riparian tracts, but their rights are shared with other owners. They neither gain their rights by using the water nor lose them by ceasing to use the water. Courts settle disputes.

In the west, the 'Prior Appropriation Doctrine' holds that 'first in time is first in right'. The first person to use water along a stream gains the 'senior right' to a reasonable quantity for that type of use. Each right that follows is 'junior' to the senior right, but senior to those that follow still later. In times of water shortage, junior rights may be shut down in favour of more senior rights. Most western states now have elaborate administrative systems requiring permits prior to diversion. Water use is not restricted to riparian land. Rights not used are lost by abandonment. Either courts or administrative agencies settle disputes.

3 Apparently, terms such as 'safe yield' and 'aquifer overdraft' have fallen out of favour with groundwater hydrologists. 'Sustainable use' has replaced 'safe yield', and 'intensive groundwater exploitation' has replaced 'aquifer overdraft'. Nevertheless, this chapter uses the term 'safe yield' throughout, because Kansas regulations continue to use the term. Kansas Administrative Regulations §5-1-1 (mmm) defines 'safe yield' as 'the long-term sustainable yield of the source of supply, including hydraulically connected surface or groundwaters.'

4 For example, the final report of the special master in *Nebraska v. Wyoming* (No. 108, Original, US Supreme Court) contains this statement: 'The settlement negotiations, therefore, specifically addressed that groundwater pumping concern, and the parties agreed on a definition of a 'hydrologically connected groundwater well' as a well 'so located and constructed that if water were intentionally withdrawn continuously for 40 years, the cumulative stream depletion would be greater than or equal to 28% of the total groundwater withdrawn by that well.' NPDC Charter, Ex. 4, para. III.D.2.b' (*Nebraska v. Wyoming*, 2001, p. 31).

5 Professor Norm Thorson and others have provided summaries of Nebraska water law (Thorson, 1991, pp. 494–496; Nebraska Water Policy Task Force, 2004).

6 The Nebraska legislature has expressly found that pumping water for irrigation from wells located within 50ft of the bank of a stream may have a direct effect on the stream (Nebraska Revised Statutes, 2005, §46–636), requiring a permit in such cases (with some exceptions) (Nebraska Revised Statutes, 2005, §46–637). Another legislative section prohibits the drilling of irrigation wells within 600 ft of a registered irrigation well (Nebraska Revised Statutes, 2005, §46–609).

7 The Nebraska Association of Resource Districts is the trade association for the NRDs (Nebraska Association of Resource Districts, 2005).
8 The Edwards aquifer supplying San Antonio with municipal water supply is not part of the High Plains aquifer. The legislature has treated the Edwards aquifer differently by empowering the Edwards Aquifer Authority to regulate and restrict Edwards aquifer use (*Waters and Water Rights*, 1991 and 2004 Cumulative Supplement, v. 6, pp. 787–792).
9 These goals include the following: (i) providing the most efficient use of groundwater; (ii) controlling and preventing waste of groundwater; (iii) controlling and preventing subsidence; (iv) addressing conjunctive surface water management issues; (v) addressing natural resource issues; (vi) addressing drought conditions and (vii) addressing conservation (Texas Water Code, 2005, §36–1071).
10 See footnote 3.
11 The classic statement of this trend appears in a California case: 'All things must end, even in the field of water law. It is time to recognize that this law is in flux and that its evolution has passed beyond traditional concepts of vested and immutable rights' (*Imperial Irrigation District v. Water Resources Board* (1990), pp. 250, 267). To some extent, Texas is an exception. In 1999, the Texas Supreme Court refused to abandon the Rule of Capture in favour of the Reasonable Use Doctrine (*Sipriano v. Great Springs Waters of America, Incorporated*. 1999) discussed earlier.
12 In the *Imperial Irrigation District* case, the California court did not define waste, but concluded that 'wasteful practises' included 'canal spills, excess tailwater (the water running off the "tail" of a farm as the result of excess water being introduced at the "head" of the system), and…canal seepage' (*Imperial Irrigation District V. Water Resources Board*, 1990, p. 258). In Kansas, DWR regulations define 'waste of water' as 'any act or omission that causes any of the following: (i) The diversion or withdrawal of water from a source of supply that is not used or reapplied to a beneficial use on or in connection with the place of use authorized…(ii) the unreasonable deterioration of the quality of water…thereby causing impairment…(iii) the escaping and draining of water intended for irrigation use…or (iv) the application of water…in excess of the needs for this use' (Kansas Administrative Regulations, 2005, §5-1-1 (cccc)). GMD regulations prohibit waste of water (Kansas Administrative Regulations, 2005, §§5-21-2 and 5-22-3).
13 Kansas statutes do not expressly define the term 'impair', but K.S.A. §82a-711 states that 'impairment shall include the unreasonable…lowering of the static water level… beyond a reasonable economic limit.' A 1973 Kansas district court case held that impairment had occurred when 'plaintiff's authorized diversion rate is decreased by at least 20% in addition to the rate reduction caused by the pumping of plaintiff's irrigation well' (*File v. Solomon Valley Feedlot, Incorporated*, 1973, para. 5).
14 Other authors raise the same issue (Llamas, 2004, p. 9).
15 For the use of the term 'safe yield', see footnote 3.
16 The example of dividing a piece of pie requires one child to cut the larger piece into two parts and then permits the other child to pick which piece he or she wants.
17 Llamas presents a similar view: 'Fossil groundwater has no intrinsic value if left in the ground except as a potential resource for future generations, but are such future generations going to need it more than present ones?' (Llamas, 2004, p. 9)
18 'Kansas ranked second nationally with 6.65 million cattle on ranches and feed yards as of January 1, 2004. * * * Cattle represented 61% of the 2002 agricultural cash receipts. * * * Kansas ranks second in commercial cattle processed with 8.9 million head in 2003…second in value of live animals and meat exported to other countries at $822.2 million in 2001…second in fed cattle marketed with 5.5 million in 2003 …[which] represents 23.2% of all cattle fed in the USA' (Economic Impact of the Kansas Livestock Industry, 2005).

19 Oklahoma, for example, once sought to prohibit interstate transfer of minnows seined from waters of the state (*Hughes v. Oklahoma*, 1979). Other examples include natural gas (*West v. Kansas Natural Gas Company*, 1911; *Pennsylvania v. West Virginia*, 1923); game birds (*Geer v. Connecticut*, 1896); river water (*Hudson County Water Company v. McCarter*, 1908) and groundwater (*City of Altus v. Carr*, 1966).
20 Citing *Pike v. Bruce Church, Incorporated*, 1970.
21 The Nebraska Supreme Court upheld the constitutionality of the new section (*Ponderosa Ridge LLC v. Banner County*, 1996). Similarly, Kansas amended Kan. Stat. Ann. §82a-726 in 1984 to remove a comparable reciprocity provision and to protect the 'public health and safety' of its citizens.
22 In *Arizona v. California*, 1963, the US Supreme Court held that the Boulder Canyon Project Act 1928, enacted by the US Congress, represented a comprehensive scheme of apportioning waters of the Colorado River.
23 The Texas Water Development Board held a symposium on the subject in June 2004.
24 The late Professor Corwin Johnson taught at the University of Texas School of Law and was a leading authority on Texas water law.
25 See page 99, above.
26 The following is a contrasting position: 'Many policy researchers, including the IWMI-Tata researchers...believe the case for direct regulation hopeless in south Asian settings, not because it is unnecessary or undesirable but on the grounds of administrative feasibility and costs.' The authors, however, point out that 'China's experience with direct management (including well and withdrawal permits) ... has at least shown some positive signs' (Shah *et al.*, 2004, p. 3456).
27 That article came out of a paper delivered at the World Water Council 3rd World Water Forum in Kyoto, Japan, in March 2002.
28 In my talk on groundwater doctrines at the 3rd World Water Forum, I described the Cheyenne Bottoms dispute, which involved 800 irrigators in a dispute with 15 lawyers taking over 2 years to resolve. See pages 302–303 on the necessity of compensation. Dr Singh responded to that in contrast, the comparable situation in India would involve 8000 irrigators in an even smaller geographical area, with perhaps only one or two lawyers and requiring more than 20 years for the judicial system to resolve the issue. In 'irrigation institutions', the authors state that India has '20 odd million pump owners, a number that is growing at the rate of 0.8–1 million per year' (Shah *et al.*, 2002, p. 3456).
29 South Dakota, for example, provides that permits for works to withdraw water from the Madison formation in certain counties are limited to 20 years, unless the Water Management Board determines that there would be no adverse effects on other Madison formation users (South Dakota Codified Laws, 2005, §46-2A-20).

References

Arizona v. California, 373 U.S. 546, 83 S.Ct. 1468, 10 L.Ed.2d 542 (1963).

Boulder Canyon Project Act of December 21, 1928, 45 Stat. 1057.

Caroom, D.G. and Maxwell, S.M. (2004) The Rule of Capture – If It Ain't Broke...., *100 Years of the Rule of Capture: From East to Groundwater Management*. Available at: http://www.twdb.state.tx.us/publications/reports/GroundWaterReports/GWReports/Report%20361/4%20CH%20Caroom.pdf.

City of Altus v. Carr, 255 F. Supp. 828 (W.D. Texas, summarily aff'd, 385 U.S. 35 (1966)).

Eaton, J.M. and Caplan, R. (2003) Water for People and Nature: The Story of Corporate Water Privatization, Available at: http://www.theallianceforddemocracy.org/pdf/wpn01.pdf.

Economic Impact of the Kansas Livestock Industry (2005), Kansas Livestock Association. Available at: http://www.kla.org/economics.htm

Equus Beds Information Resource, A partnership of the Kansas Department of Health and Environment, GIS Policy Board, Data Access and Support Center and Information Network of Kansas, Inc. (2005) Available at: http://www.equusinfo.org/tech_narrative.shtml.

File v. Solomon Valley Feedlot, Incorporated, No. 8831, para. 5 (District Court, Mitchell County, Kansas, 1973).

Geer v. Connecticut, 161 U.S. 519, 16 S.Ct. 600, 40 L.Ed. 793 (1896).

Groundwater Atlas of the United States. (1995) *High Plains Aquifer*, U.S. Geological Survey. Available at: http://capp.water.usgs.gov/gwa/ch_c/C-text5.html.

Houston and Texas Central Railroad Company v. East, 98 Tex. 146, 81 S.W. 279 (1904).

Hudson County Water Company v. McCarter, 209 U.S. 349, 28 S.Ct. 529, 52 L.Ed. 828 (1908).

Hughes v. Oklahoma, 441 U.S. 322, 99 S.Ct. 1727, 60 L.Ed.2d 250 (1979).

Imperial Irrigation District v. Water Resources Board, 225 Cal.App.3d 548, 275 Cal. Rptr. 250 (Cal. App. 4 Dist.) (1990).

Johnson, C.W. (2004) What Should Texas Do About the Rule of Capture? *100 Years of Rule of Capture: From East to Groundwater Management*, Texas Water Development Board, Report 361. Available at: http://www.twdb.state.tx.us/publications/reports/GroundWaterReports/GWReports/Report%20361/2%20CH%20Johnson.pdf.

Kansas Administrative Regulations (2005) Office of Revisor of Statutes of Kansas. Topeka, Kansas.

Kansas Statutes Annotated (2005) Office of Revisor of Statutes of Kansas, Topeka, Kansas. Kansas Legislature 2005–2006. Available at: http://www.kslegislature.org/legsrv-statutes/index.do.

Kansas v. Colorado, 206 U.S. 46, 27 S.Ct. 655, 51 L.Ed. 956 (1907).

Knickerbocker, B. (2002) Privatizing water: a glass half empty? *Christian Science Monitor* 24 October. p. 1. Available at: http://www.csmonitor.com/2002/1024/p01s02-usec.htm.

Kromm, D.E. and White, S.E. (1992) *Groundwater Exploitation in the High Plains*, University Press of Kansas, Lawrence, Kansas.

Landry, C.J. (2000) A Free Market Solution to Groundwater Allocation in Texas: A Critical Assessment of the House Natural Resources Committee Interim Report on Groundwater. Available at: http://www.texaspolicy.com/pdf/2000-12-01-environ-water.pdf.

Llamas R. (2004) Water and Ethics: Use of Groundwater, UNESCO International Hydrological Programme, World Commission on the Ethics of Scientific Knowledge and Technology. Available at: http://racefyn.insde.es/academicos/descargas/Llamas/UWEG136322e.pdf

Llamas, R. and Custodio, E. (2002) Intensively exploited aquifers: Main concepts, relevant facts and some suggestions, UNESCO, IHP-VI, Series on Groundwater No. 4, UNESCO, IAH, Marcelino Botin Foundation. Available at: http://www.iah.org/News/2002/inexaqcd.pdf.

McKenzie, W. (2004) Forget gasoline prices, Texans need to talk water, *The Dallas Morning News*. Available at: http://www.mesawater.com/ReferenceResources/ref_materials_detail.asp?id=20.

Mesa Water (2005) Working to improve the future of Texas water. Available at: http://www.mesawater.com/default2.asp.

National Audubon Society v. Superior Court of Alpine County, 33 Cal.3d 419, 658 P.2d 709, 189 Cal.Rptr. 346 (1983).

Nebraska Association of Resource Districts (2005) Available at: http://www.nrdnet.org/index.htm.

Nebraska Revised Statutes (1978) Nebraska Legislature. Available at: http://srvwww.unicam.state.ne.us/Laws2005.html.

Nebraska Revised Statutes (2005) Nebraska Legislature. Available at: http://srvwww.unicam.state.ne.us/Laws2005.html.

Nebraska v. Wyoming (No. 108, Original, U.S. Supreme Court), Final Report of the Special Master (2001), p. 31. Available at : http://www.dnr.state.ne.us/PDF/FinalReport1.pdf.

Nebraska Water Policy Task Force (2004) Nebraska Department of Natural Resources. Available at: http://www.dnr.state.ne.us/watertaskforce/docs/TFResourcematerials.html.

100 Years of Rule of Capture: From East to Groundwater Management (2004) Texas Water Development Board, Report 361. Available at: http://www.twdb.state.tx.us/publications/reports/GroundWaterReports/GWReports/Report%20361/361_ROCindex.htm.

Ostdick, J.H. (2004) A Natural River, *Texas Parks and Wildlife*. Available at: http://www.tpwmagazine.com/archive/2004/jul/ed_7/.

Peck, J.C. (2003) Property rights in groundwater – some lessons from the Kansas experience, *Kansas Journal of Law and Public Policy,* XII, p. 493. Available at: http://64.233.167.104/search?q=cache:kTtQD9ipNPEJ:www.ku.edu/~kulaw/jrnl/v12n3/peck.pdf+%E2%80%9CProperty+Rights+in+Groundwater%E2%80%93Some+Lessons+from+the+Kansas+Experience&hl=en; http://www.ku.edu/~kulaw/jrnl/v12n3/peck.pdf.

Peck, J.C. (2004) Protecting the Ogallala aquifer in Kansas from depletion: the teaching perspective, *Journal of Land, Resources, and Environmental Law,* 24, p. 349, University of Utah S.J. Quinney College of Law, Salt Lake City, Utah.

Peck, J.C. and Nagel, D.K. (1989) Legal aspects of Kansas water resources planning, *University of Kansas Law Review*, 37(2), 199–318.

Pennsylvania v. West Virginia, 262 U.S. 553, 43 S.Ct. 658, 67 L.Ed. 1117 (1923).

Pike v. Bruce Church, Incorporated, 397 U.S. 137, 142, 90 S.Ct. 844, 847, 25 L.Ed.2d 174, 178 (1970).

Ponderosa Ridge LLC v. Banner County, 250 Neb. 944, 554 N.W.2d 151 (1996).

Potter, H.G., III, History and Evolution of the Rule of Capture (2004) *100 Years of Rule of Capture: From East to Groundwater Management,* Texas Water Development Board, Report 361. Available at: http://www.twdb.state.tx.us/publications/reports/GroundWaterReports/GWReports/Report%20361/1%20CH%20Potter.pdf.

Shah, T., Giordano, M., and Wang, J. (2004) Irrigation institutions in a dynamic economy: what is China doing differently from India?, *Economic and Political Weekly*. Available at: http://www.iwmi.cgiar.org/iwmi-tata/index.asp?id=1270.

Singh, K. (2002) *Co-operative Property Rights as an Instrument of Managing Groundwater*, World Water Council 3rd Word Water Forum, Kyoto, Japan.

Sipriano v. Great Springs Waters of America, Incorporated, 42 Tex.Sup.Ct.J. 629, 1 S.W.3d 75 (1999)

Sophocleous, M. (2005) Groundwater recharge and sustainability in the High Plains aquifer in Kansas, USA, *Hydrology Journal*, 13(2), 351–365.

South Dakota Codified Laws (2005) Available at: http://legis.state.sd.us/statutes/index.aspx.

Spear T Ranch v. Knaub, 269 Neb. 177, 691 N.W.2d 116 (2005).

Sporhase v. Nebraska, 458 U.S. 941, 102 S.Ct. 3456, 73 L.Ed.2d 1254 (1982).

State of Kansas v. State of Nebraska and State of Colorado 2002 (No. 126, Original, U.S. Supreme Court) Special Master's First Report, p. 45, and Case Management Order 6 para. 3.

Testimony of David L. Pope, Chief Engineer, before the Senate Natural Resources Committee on the Republican River Compact Litigation, January 23, 2003. Available at: http://www.ksda.gov/Default.aspx?tabid=334&mid=2356&ctl=Download&method=attachment&EntryId=229.

Texas Constitution (2005) Available at: http://www.capitol.state.tx.us/txconst/toc.html.

Texas Water Code (2005) Available at: http://www.capitol.state.tx.us/statutes/wa.toc.htm.

The Bimonthly Newsletter of the Oklahoma Water Resources Board (March–April 2001) Legislature Mulls First Interstate Groundwater Compact, Oklahoma Water Resources Board. Available at: http://www.owrb.state.ok.us/news/news2/pdf_news2/wnews/wn3_2001.pdf.

The Utton Center, Transboundary Resources (2004) The Model Interstate Surface and Groundwater Compacts Project, The University of New Mexico School of Law. Available at: http://uttoncenter.unm.edu/pdfs/MC_Project_Description.pdf.

Thorson, N.W. (1991 and 2004 Cumulative Supplement) State Surveys, Nebraska, *Waters and Water Rights,* Vol. 6, The Michie Company, Charlottesville, Virginia.

Water and the Future of Rural Texas (2001) Conference Proceedings, Texas Center for Policy Studies, Austin, Texas. Available at: http://www.texaswatermatters.org/pdfs/articles/waterconf.pdf.

Water Ranching in the Lone Star State: Texas Water Policy Update (2002) *Texas Center for Policy Studies*. Available at: http://www.texaswatermatters.org/pdfs/txwaterpolicydec.pdf; http://www.texaswatermatters.org/pdfs/tgxwaterpolicydec.pdf.

Waters and Water Rights (1991 and 2004 Cumulative Supplement) The Michie Company, Charlottesville, Virginia.

West v. Kansas Natural Gas Company, 221 U.S. 229, 31 S.Ct. 564, 55 L.Ed. 716 (1911).

Williams v. City of Wichita, 190 Kan. 317, 374 P.2d 578 (1962).

Wyoming Statutes (2005) Wyoming State Legislature, Wyoming Statutes. Available at: http://legisweb.state.wy.us/.

15 Institutional Directions in Groundwater Management in Australia

Hugh Turral[1] and Imogen Fullagar[2]

[1]*International Water Management Institute,127, Sunil Mawatha, Pelawatte, Battaramulla, Sri Lanka;* [2]*Charles Sturt University, C/o CSIRO Land and Water, GPO Box 1666, Canberra ACT 2601, Australia*

Now the stock have started dying, for the Lord has sent a drought;
But we're sick of prayers and Providence – we're going to do without;...
As the drill is plugging downward at a thousand feet of level,
If the Lord won't send us water, oh, we'll get it from the devil;
Yes, we'll get it from the devil deeper down.

From 'Song of the Artesi an Water' (1896)
by A.B. Banjo Paterson, Australian bush poet

Introduction

Australia is a large country, covering 7.69 million square kilometres, with a relatively small population of 20 million (Australian Bureau of Statistics, 2005). Agriculture accounts for a paltry 3% of gross domestic product (GDP) and only 6.88% of the land surface is under arable cropping, with 0.03% under permanent crops. In 1997, 5% of the labour force was directly engaged in agriculture (http://worldfacts.us/Australia.htm) compared with 22% in industry and 73% in services. These statistics set it a world apart from densely populated agrarian countries such as India and China.

Although agriculture has declined from being a major contributor to national wealth (>27% of gross national product (GNP) in the late 1980s), it still has a strong export focus. There have been considerable structural adjustments in agriculture in the last 20 years in response to Australia's commitment to free trade, removal of input and output subsidies and widespread application of 'user-pays' principles in service sectors. The recent strength of the mining sector with strong global demand for iron and aluminium has contributed to the relatively small contribution of agriculture to GNP.

Governance and Natural Resources Management

Australia is a democratic federation of six states and two territories, united by the Commonwealth government (federal government). Cohesion within this structure is cemented by centralization of income tax collection, the revenue of which is redistributed to the nine (central, state and territory) governments. There is a third layer of local government at the municipal (urban) and shire (country) levels. The state, territory and local governments can also contribute by raising some local revenue (e.g. states via petrol levies and local government via service levies).

Water is the responsibility of the state and territory governments (henceforth referred to as 'states' or jurisdictions) under the Australian Constitution, each having independent water laws and distinct policies. However, international issues, common jurisdictional concerns and Commonwealth leverage of Section 96 of the Australian Constitution (which allows the Commonwealth to grant financial assistance to any state on terms determined by the Commonwealth) have accelerated the development of a federal role in the national water policy (McKay, 2002).

Issues of national significance that concern the Commonwealth and all state governments are dealt with by the Council of Australian Governments (COAG). The COAG deals with a wide raft of issues through a number of ministerial councils. These councils facilitate development and implementation of national plans and proposals that would otherwise be impinged by the division of constitutional powers between the federal and state governments.

The Natural Resources Management Ministerial Council (NRMMC) was formed in 2001 'to promote the conservation and sustainable use of Australia's natural resources'. All Australian and New Zealand government ministers responsible for natural resource management issues are members. Decisions of the Council require consensus of the members. The reorganization through which the NRMMC evolved saw this Council absorb roles and responsibilities previously held by the Agricultural Resource Management Council of Australia and New Zealand (ARMCANZ) and the Australian and New Zealand Environment and Conservation Committee (ANZECC). Many current national water policies were therefore developed through the ARMCANZ and ANZECC.

Within this structure, the National Groundwater Committee (NGC) is a senior intergovernmental network that shares information and provides insight into the national groundwater policies and resource management, research directions, priorities and programmes. It also provides advice on groundwater issues, including those pertaining to surface water–groundwater interactions. The subject of groundwater is dominated by two issues:

1. salinity management;
2. extractive use.

Salinity management is perhaps more important, given the significance of irrigation-induced salinity problems, particularly in the state of Victoria, and of the parallel but slower development of dryland salinity in the state of western

Australia. Irrigation-induced salinity largely occurs because of the rise in water table, due to progressive accessions from irrigated fields and water supply infrastructure, where the groundwater is naturally saline or intersects naturally saline soils and rock formations.

Dryland salinity is emerging as a widespread and serious problem in catchments that have been cleared for dryland agriculture and pasture (National Land and Water Audit (NLWA), 2001): shallow-rooted crops and grasses transpire less water each year than the native scrub and forest, resulting in small net annual accessions, which, over 50–100 years, have also contributed to the rise in water table and attendant local salinization, particularly near streams and inland water bodies. The most alarming estimates of potentially affected areas for 2030 run to approximately 20 million hectares.

Since the main focus of this chapter is on the use of groundwater in agriculture, with only a passing reference to other sectors, it is instructive to set the context of irrigation development and water resources management in Australia.

A Brief History of Irrigation Development in Australia

In the first decade of the 19th century, Australia's agriculture dealt mainly with sheep, wool, beef and wheat production. The comparatively slow development of the irrigation sector compared with that of dryland reflects the high river flow variability characteristic of Australia, and the associated prerequisite of securing water supply through dam development. Despite the greater water resources in northern Australia, the history of urban and market access have largely dictated the geography of the irrigation industry, which is today dominated by development in the southern half of the Murray–Darling Basin (MDB) (see case study 1).

From 1901 (since the Federation) to the early 1990s, Australian governments were determined to 'drought-proof' the continent through river development (Tisdell *et al.*, 2002). Development of water courses and provision of security of supply were seen as a public good, necessary for the development of the nation. As agents for water resources, the states were the primary developers of surface water infrastructure. This resulted in extensive dam building and associated engineering works, which are represented by the Snowy Mountains hydroelectric scheme, a huge engineering feat that captures and redirects 5044.5 l (1121 gallons) of water from its natural course down the Snowy River into the MDB system. The dam's construction (which spanned 25 years and involved more than 100,000 workers) illustrates the level of cooperation between governments, which was fuelled by the general optimism of the development era.

Irrigation development was associated with concentrated efforts to settle high potential areas following the two World Wars – normally discussed as 'soldier settlement'. Private irrigation trusts were also established in the late 19th century, but were relatively small scale compared to state-sponsored developments. Public investments for dam construction, channel infrastructure and promotion of expansion within the industry made for an industry dominated by the heavily subsidized surface water irrigation industry. In contrast, groundwater development has been sparse, privately financed and localized, with a greater emphasis on non-agricultural use, partly due to the extensive development of surface water resources.

The goodwill generated by shared development ambitions and successful collaborative social and engineering projects have served intergovernment communication through similar jargon and shared responsibilities. This history underpins the relatively cooperative endeavour of water reforms today.

The development agenda of the early and mid 20th century declined since the 1970s, with the realization that the resource base in key areas (notably the MDB) was being jeopardized. It was realized, in the late 1970s, that the licensed volumes within the MDB exceeded the available supply on which interstate water-sharing arrangements were based (Turral, 1998). It was clear that the value of building additional dams would therefore come at a cost of filling existing storages. Quite clearly, in the developed irrigation sites, there was no surface water to harvest and any further development would reduce the security of supply to existing users.

Furthermore, the irrigation industry was increasingly aware of the environmental overheads of their own practices. The expense and political sensitivities relating to the riverine impact of salinity were expanding, and there was growing pressure for irrigators to internalize these costs by improving farm management. Rivers and inland wetlands were impaired by reductions in in-stream flows and, in some cases, through inversion of flow patterns, as irrigation water is released from dams in summer, whereas natural flows are mostly concentrated in winter in south and east Australia.

In the early 1980s, issues surrounding the proposal to dam the Franklin River resulted in an explosion of public debate and environmental awareness relating to river development. The Franklin Dam was a Tasmanian proposal, which was successfully halted by the Commonwealth Government on the basis of the World Heritage listing.

In terms of water development, the significance of the Franklin Dam was twofold:

1. It confirmed (by precedent) a Commonwealth power to intervene in activities previously held to be state responsibilities.
2. It clearly demonstrated, through polls, the public priority of environmental sustainability.

Coincidentally, the state governments were also reluctant to continue to subsidize the operation, maintenance and replacement of irrigation systems. As a result they were corporatized (Victoria) or fully privatized (New South Wales, NSW) in the 1990s, and are now run by professional managers responsible for farmer-dominated management boards. The governments also recognized the full environmental and financial costs of water diversion and transmission, and maintenance overheads of the existing infrastructure. Tisdell *et al.* (2002) clearly summarize:

> [A] singular construct of water capture and reticulation, which traditionally reflected the primacy of national development, was increasingly seen as failing to capture the multiplicity of water outputs, ecosystem functions and the changing societal objectives of maintaining in-stream values and water quality.

Water Property Rights

Surface and groundwaters are both licensed by, or on behalf of, the state governments, under state-specific water legislation and policy; licence details therefore vary considerably across the states.

A level of security is normally applied to water licences. This is traditionally based on the purpose for which the licence was originally issued. The accepted priorities of water supplies (from highest to lowest) are: town supply, stock and domestic, perennial crop (e.g. vineyards and orchards) and annual crops (e.g. grains).

Most water licences are specified in volumetric terms as an entitlement, based on a certain level of historical security of supply (exceeding availability in 99% of years, in the case of Victoria). Volumetric measurement and charging for surface irrigation water have been the norm throughout most of the MDB since the 1960s and date back much longer in Victoria. The actual amount a licence-holder can obtain in 1 year is determined *pro rata* by the announced allocation, which is reviewed every month, based on different formulas that incorporate available storage, plus minimum (1:100 year) expected rainfall volume, less the volume required by high-priority uses. The precise formulation of the allocation and entitlement rules varies from state to state, particularly in relation to environmental reserve, environmental flow rules governing dam operations and the ability to carry over unused allocations from one year to the next.

To some extent, this 'share' approach was the result of an explicit rejection of the 'prior appropriation' doctrine practised by the western states in America (Tisdell, 2002). It could nevertheless be contended that environmental and some native water titles can claim priority at least partially by virtue of history. The capacity of a share approach to entirely avoid prior appropriation issues also rests heavily on sound definition and hydraulic understanding of the water resource being licensed, implicitly assuming that these licensing frameworks account for any hydraulic connectivity between institutionally independent resources (e.g. surface water and groundwater).

In the MDB, interstate water shares were agreed in 1915 and those limits were not tested by water resources developments until it was realized (in the late 1970s) that the licensed volume exceeded the available resource, notably in NSW (Turral, 1998).

Subsequently, it was realized that the existing licensed volume already exceeded the sustainable water resource and that, at the prevailing rates of irrigation expansion, the actual diversion would exceed sustainable limits by 2020 (MDBC, 1996) and possibly approach the volume of annual runoff to the sea. A Cap on diversions of surface water within the MDB was agreed in 1995, set not to exceed the volume diverted at the extent of agricultural development in 1994. It was left to each state to work out how to implement the Cap and it has been independently audited annually since then. The idea of a rolling cap was implemented de facto, which allows states to overrun the Cap in low allocation years provided they balance this in subsequent above-average years. Since 2000, 3–4 years of consecutive drought, with less than the previous 1:100 year water availability, have put some strain on this arrangement. The largest volume of unused licences is in NSW, due to the existence of sleeper and dozer[1] users and relatively conservative withdrawals by many farmers in response to the lower security of supply in NSW, where there is considerably less interannual storage volume than in Victoria.

Water trading has been activated through private, state and central initiatives since the mid-1980s, although temporary trading has a long and informal history. The liberalization of water trading since the mid-1990s has activated some of this unused volume, putting further strain on the security of supply to existing users (Panta *et al.*, 1999). The market is dominated by temporary transfers of unused allocation within a season and activity reflects the general drought cycle and water resources availability, whilst permanent trades account for less than 1% of the licensed volume (Turral *et al.*, 2005). Most of the water trade is between irrigators within a particular state, and interstate trading is currently limited by questions of exchange rate between upstream and downstream transfers (Etchells *et al.*, 2004).

Institutional Reform in Water Resources Management in Australia

Reforms in water resources management in Australia have proceeded along three main lines, which have complementary origins in the community and in government: (i) state-driven water accounting and allocation reforms, pricing, cost recovery, removal of subsidies and administrative reform; (ii) a community-initiated movement for better land and water management – now commonly lumped under the banner 'LandCare'; and (iii) the development and specification of environmental flows, river flow rules and strategies to mitigate in-stream salinity and algal blooms. Land and water have been considered complementary factors in this process.

The Brundtland Report (World Commission of Environment and Development, 1987) highlighted the international importance of co-dependency between environmental and economic policy in achieving sustainability. Australia responded through the National Strategy for Ecologically Sustainable Development (NSESD, COAG, 1992) (http://www.deh.gov.au/esd/national/nsesd/strategy/index.html), which adopted the 'precautionary principle' as a guiding philosophy. This strategy had three broad objectives:

1. to enhance individual and community well-being by following a path of economic development that safeguards the welfare of future generations;
2. to provide equity between generations;
3. to protect biological diversity and maintain essential ecological processes and life-support systems.

Management of surface water has been high on the agenda and the laboratory has often been the MDB, due to the extent of irrigation and surface water development. Regulation of the river has allowed increased reliability in agricultural production through a combined dam storage volume equivalent to 2.8–3 years of mean annual flow, but this security has occurred at the expense of river health (MDMBC, 1996).

As a component of the NSESD, reform of the water industry was tied to microeconomic reforms via the National Competition Policy reform package (1995). Within this package, financial benefits of microeconomic reforms were distributed on the basis of performance against specific reform agendas, including that of water (Tisdell *et al.*, 2002), through 'tranche' payments of central tax revenue to individual states on compliance with agreed targets. In tandem with reforms focused on the MDB, the COAG began a process of reform aimed at removing subsidies and ensuring competition and economic efficiency. COAG (1994) water reforms were intended to allow water to move to its most productive use by enabling water markets and full cost recovery of the operation and maintenance of irrigation systems. This incentive initiated rapid institutional changes, including significant legislative amendment. By the necessity of relevant time frames, these were implicitly driven by the dominant surface water issues.

Recognizing continuing resource issues and the need for further reform to fully develop and deliver full cost pricing policies (van Bueren and Hatton MacDonald, 2004), the COAG agreed to the National Water Initiative (NWI) in 2004 (COAG, 2004). The NWI objective explicitly identifies its application both to surface and groundwaters, and more specifically the issue of surface water–groundwater connectivity. This implies an inconsistent implementation of earlier COA OPG, 2004 (1994) reforms across surface and groundwaters, which is generally acknowledged as a lag between implementation of surface and groundwater reforms in most states.

In agreement with the NWI, the COAG delegated the following responsibilities to the NRMMC:

1. overseeing implementation of the NWI, in consultation with other minister ial councils as necessary and with reference to advice from the COAG;
2. addressing ongoing implementation issues as they arise;
3. providing annual reports to the COAG on the progress with actions being taken by jurisdictions in implementing the NWI;
4. developing a comprehensive national set of performance indicators for the NWI in consultation with the National Water Commission (NWC) (set up to implement the NWI).

Somewhat contentiously, the NWI proposed a fund to buy back surface water for the environment based on financial contributions from the state and central governments. This was derailed for some time due to the Commonwealth's intention to fund NWI from the Natural Heritage Trust, at the expense of previously agreed initiatives and payments to the states.

The story of LandCare is rich, varied and interesting and is well documented elsewhere (Ewing, 1996). It began with genuine, community-based initiatives in irrigated salinity management in Victoria and catchment management for dryland salinity in western Australia in the mid-1980s. LandCare became a national programme in 1992, following the historic joint initiative of the Australian Conservation Foundation (an environmental NGO) and the National

Farmers Federation. By 2001, there were 4500 LandCare groups incorporating 50% of farmers and 35% of land administrators. This phenomenal growth in community-based management of natural resources and associated investment went largely unevaluated until 2000, when the National Land and Water Audit (1999–2002) was established to set a benchmark on resource availability, use and condition, and allow future evaluation of the impacts of community and other initiatives on the resource base. Simultaneously, many felt that there were too many voices from the plethora of LandCare groups, sometimes working at too local a scale. This resulted in the creation of umbrella groups for coordinated community-based management, now well established, such as the Catchment Management Authorities (CMAs) in Victoria. Despite fears of a creeping bureaucratization of grass-roots initiative, CMAs have emerged as a central force in natural resources management, where essentially the community decides and partly self-funds management plans and their implementation, using state agencies and commercial companies as advisers and consultants.

In conjunction with the NWI, there are other state-level initiatives, such as the 2004 Victorian White Paper 'Securing our Water Future Together' (DSE, 2004), that move the focus of land and water management to be framed more tightly within the concepts of environmentally sustainable development. With that broad introduction to the setting and recent institutional reform in the water sector as a whole, we now turn to the specifics of groundwater.

National Groundwater Resources and Use

A series of water resources assessments were conducted in Australia, with a primary focus on surface water. These assessments included:

- a review of Australia's Water Resources (1975), Australian Water Resources Council (1976), resulting among other outputs in 'Australian Rainfall and Runoff', a key work on hydrological data and methods in the continent;
- first national survey of water use in Australia (1981), Department of National Development;
- a review of Australia's Water Resources (1985), Australian Water Resources Council (1987);
- Water and the Australian Economy (AATSE) (1999);
- Water Account for Australia, Australian Bureau of Statistics (2000);
- National Land and Water Resources Audit (NLWRA) (2000), update of AWRC (1985).

Most of these studies were complemented by detailed hydrogeological and water resources assessments in the states, but had historically focused on resource development, and there was little information on actual groundwater use. The NLWR. A provides the most comprehensive national overview of groundwater availability and use in Australia to date. It estimated that the national groundwater availability amounts to 25.78 billion cubic metres per year on average, of which 21 billion cubic metres is of potable quality (NLWRA, 2002). Total abstraction in 1996/97 amounted to less than 10% of this at 2.49 billion cubic metres. On the face of it, this does not look to be a problem. However, poor distribution of groundwater use across available resources has resulted in overallocation of many good-quality and readily accessible groundwater stores – often the alluvial plains of prior and existing riverbeds within which surface water irrigation districts lie. The sustainable yield of groundwater in each state is shown in Table 15.1 and is disaggregated by salinity status, showing that about 63% is of high quality. It shows that salinity concerns are greatest in Victoria, western Australia and south Australia (SA). Salinity problems are on the rise in specific localities in NSW.

Nationally, about 50% of total groundwater abstraction is for irrigated agriculture (Table 15.2), but this figure rises to 65% in Victoria and NSW and is highest in SA at 80%. Groundwater allocation is a little over one-fourth of the total national water resources availability and less than one-third of the surface water allocation. The available resource in Northern Territories (NT) (Table 15.2) is enormous compared with actual allocation, so the fact that actual use exceeds allocation is not necessarily significant in resource management terms. The same story is broadly true for Tasmania. Rural water use includes stock and domestic water provision, and the majority of water abstracted from the Great Artesian Basin (covering large parts of NSW, Queensland and NT) is for pastoral use. A detailed breakdown of groundwater use is available for 286 out of 538 groundwater management units across the nation, and summary data are available for 377 of them. Groundwater is the sole source of water for many rural towns, mines and associated settlements.

Many surface-irrigated properties in northern Victoria and throughout NSW also have bores as drought insurance and for supplementing surface water supplies. Generally they abstract from deeper, higher-quality aquifers, which are separated from saline layers by an aquitard. Nevertheless, some provide water of suboptimal quality which is mixed with surface water before being applied to the crop (known colloquially as 'shandying').

In many areas, actual use is significantly less than allocation (Table 15.3). However, the local balance of use and conservation can be highly variable between years.

Table 15.1. Sustainable yield, by salinity status, of groundwater in Australia. (From National Land and Water Audit, 2002.)

	<500	500–1,000	1,000–1,500	1,500–3,000	3,000–5,000	5,000–14,000	>14,000	Total
NSW	554	4,237	129	790	480	–	–	6,189
VIC	302	422	244	367	207	1,377	797	3,717
QLD	1,422	1,030	113	160	35	23	–	2,784
WA	514	1,162	1,150	1,500	766	841	371	6,304
SA	–	290	709	102	21	25	–	1,146
TAS	1,585	767	–	178	–	–	–	2,531
NT	5,785	186	324	141	5	–	–	6,441
ACT	103	–	–	–	–	–	–	103
Total	10,264	8,094	2,670	3,238	1,515	2,266	1,168	29,215
%	35	28	9	11	5	8	4	100

Table 15.2. Mean annual groundwater extraction by category of use (million). (From National Land and Water Audit, 2002.)

	Irrigation	Urban/industrial	Rural	*In situ*	Total
NSW	643	160	205	0	1008
VIC	431	127	54	10	622
QLD	816	265	541	0	1622
WA	280	821	37	0	1138
SA	354	23	42	24	430
TAS	9	7	4	0	20
NT	47	48	33	0	128
ACT	2	0	3	0	5
Total	2003	1370	788	34	4171

Table 15.3. Total annual water allocations in Australia, in MCM. (From National Land and Water Audit, 2002.)

	Surface water allocation	Ground water allocation	Total allocation	Total water use	Δ allocation – use	% difference allocation–use
NSW	9,825	2,665	12,490	10,004	2,486	25
VIC	5,469	780	6,249	5,788	–461	7
QLD	3,202	983	4,185	4,591	406	–9
WA	855	1,138	1,993	1,796	197	10
SA	740	630	1,370	1,266	104	8
TAS	403	20	423	471	–48	–11
NT	53	73	126	179	–53	–42
ACT	76	7	83	73	10	12
Total	20,623	6,296	26,919	23,280	3,639	16

While groundwater development in western Australia, NT and the Australian Capital Territory is dominated by priority (town supply, stock and domestic) uses, intermittent surface flows have resulted in the agricultural development of groundwater as a primary agricultural source in many parts of SA. SA is also distinguished by the security of its surface water supply via the Murray River, which is a volume secured in agreement with Victoria, NSW and the Commonwealth (case study 1). This allows SA much tighter accounting mechanisms than can be accommodated by the less certain water budgets of other states.

Characteristics of Groundwater Irrigation Development in Australia

Groundwater development for irrigation has not received the significant subsidies characteristic of surface water irrigation. The process for irrigation development of groundwater has evolved directly from policies put in place to ensure that groundwater development processes could readily accommodate the high priority of remote town, as well as stock and domestic, supplies. The typical process has been for an irrigator to nominate preferred bore sites on a property, and apply for a groundwater licence. Assessments of nominated sites are made, and licences issued according to bore yield and need. The full cost of infrastructure (installation, operation and maintenance) is borne by the irrigator. In practice, bore owners have generally been fairly free to go about their business. Lack of a linear supply system (river or channel) limits natural centralization, which encourages communication between groundwater stakeholders.

By the very nature of this decentralized development, groundwater users are characterized as being highly independent, autonomous and protected by:

1. ownership of infrastructure located on private land;
2. limited detail of scientific understanding of cause-and-effect relationships between resource availability and resource use.

The private investment and operation of infrastructure make changes to groundwater management difficult and highly dependent on social willingness to comply (see case study 2).

Australian groundwater irrigation development is a natural response to surface water availability, markets and the expanding politics and compliance overheads of surface water development. In many established irrigation areas, groundwater development has been characterized by a tangible trade-off between poorer water quality and enhanced supply security. Table 15.4 summarizes resource and institutional differences between surface and groundwater irrigation.

Table 15.4. Characteristics distinguishing surface water and groundwater.

Characteristic	Surface water	Groundwater
Primary nature of development	Centralized	Decentralized
Infrastructure funding	(Historically) publicly subsidized	Private
Management of flow	Linearly regulated	Unregulated
Public awareness	High	Low
Security of supply	Low	High
Water quality	High (managed)	Variable
Physical extraction limit	Volume in storage	Bore capacity, draw-down
Capacity to enforce legal limits	High (linearly regulated)	Variable (private infrastructure on private land)
Monitoring and reporting	Regulatory and centralized	Variable, generally less than surface water
Primary financial costs of water use and entitlement	Levies	Infrastructure installation, maintenance and operation
Markets	Well established and widely available	Wide range. Generally developing
Ease of monitoring and building resource data	Relatively high	Low

Salinity management and extractive use have dominated public awareness of groundwater. Salinity management has been the dominant issue to date, given the significance of irrigation-induced salinity problems, particularly in the state of Victoria. Irrigation-induced salinity largely occurs because of rise in water table due to progressive accessions from irrigated fields and water supply infrastructure, where the groundwater is naturally saline or intersects naturally saline soils and rock formations.

Drivers for Change in Groundwater Management

There has been an increasing realization that surface and groundwater resources are inextricably linked – which is obvious at one level and yet quietly under-recognized, perhaps due to the relatively low historical use of groundwater.

Groundwater exploitation has risen in tandem with competition for surface water resources. The development of groundwater as a 'back-up' supply for irrigation properties is additionally increasing the demand for groundwater development in existing irrigation areas, in NSW and Victoria. Table 15.5 shows that groundwater use has tripled between 1983/84 and 1996/97 in NSW, Victoria and western Australia. Abstraction in Queensland actually declined, largely as a result of a programme to cap all the bores in the Great Artesian Basin, many of which had been flowing freely for years, gradually reducing artesian pressure and causing concern about 'senseless' wastage.

Although western Australia and the NT have the greatest reliance on groundwater, the primary users in these jurisdictions are urban, rural (town, stock and domestic) and mining. The capital of western Australia, Perth (population 1.5 million), is the largest groundwater-dependent city in Australia.

Despite various earlier initiatives to quantify water resources in Australia, it was progressively realized that, as a lot of groundwater use was neither licensed nor measured, steps would have to be taken to bring this in line with surface water management. Historically, the British riparian tradition of landowner access to groundwater had continued long after surface water had been declared

Table 15.5. Changes in mean annual groundwater use, 1983/84 to 1996/97. (From National Land and Water Audit, 2002.)

	Total use 1983/84 MCM	Total use 1996/97 MCM	% change in groundwater use
NSW	318	1008	217
VIC	206	622	202
QLD	1121	831	–26
WA	373	1138	205
SA	542	419	–22
TAS	9	20	122
NT	65	128	97
ACT	n/a	5	–
Total	2634	4171	58

Table 15.6. The extent of metering of groundwater use in 2000 in Australia. (From National Land and Water Audit, 2002.)

	Not known	No	Yes	Total
NSW	–	39	11	50
VIC	8	66	5	79
QLD	28	57	22	107
WA	40	134	–	174
SA	15	27	11	53
TAS	–	17	–	17
NT	2	26	27	55
ACT	–	3	–	3
Total	93	369	76	538

a state (and peoples') resource to be allocated through licensing. Table 15.6 shows that only a small number of groundwater management units were metered before 2000, although it is important to note that the majority of large agricultural abstractors, especially those operating within the large surface irrigation schemes were licensed and metered by this time.

Coupled with the lack of detailed knowledge on abstraction, the rising trends in total groundwater use prompted the introduction of legislation and initiatives designed to respond to three major principles of ecologically sustainable development of groundwater:

- Water level and pressure should be maintained within agreed limits and should not diminish.
- There should be no degradation of water quality.
- Environmental water needs should be determined and sustained.

National Framework for Groundwater Management

The National Framework for Improved Groundwater Management in Australia in 1996 (ARMCANZ, 1996a) set in train subsidiary policies and legislation in the states. Core recommendations were to publicly identify sustainable yield, allocation and use of aquifers as well as limit allocations to sustainable yields. Others included the enablement of trading of groundwater licences; improved integration of surface and groundwaters; management and licensing of high-yielding wells and provision of all drilling data by contractors; provision of funding for investigation in high-priority areas; and the introduction of full recovery of the costs of managing groundwater.

This framework resulted in tangible outcomes in terms of the definition of 72 groundwater provinces, and 538 groundwater management units, with associated water resources assessments and the initiation of groundwater management plans. Preliminary definitions of groundwater provinces and some management units go back to definitions made in the Water Review (1985), but these had only been partially developed. Figure 15.1 shows a summary of the degree of abstraction relative to sustainable yield in the groundwater

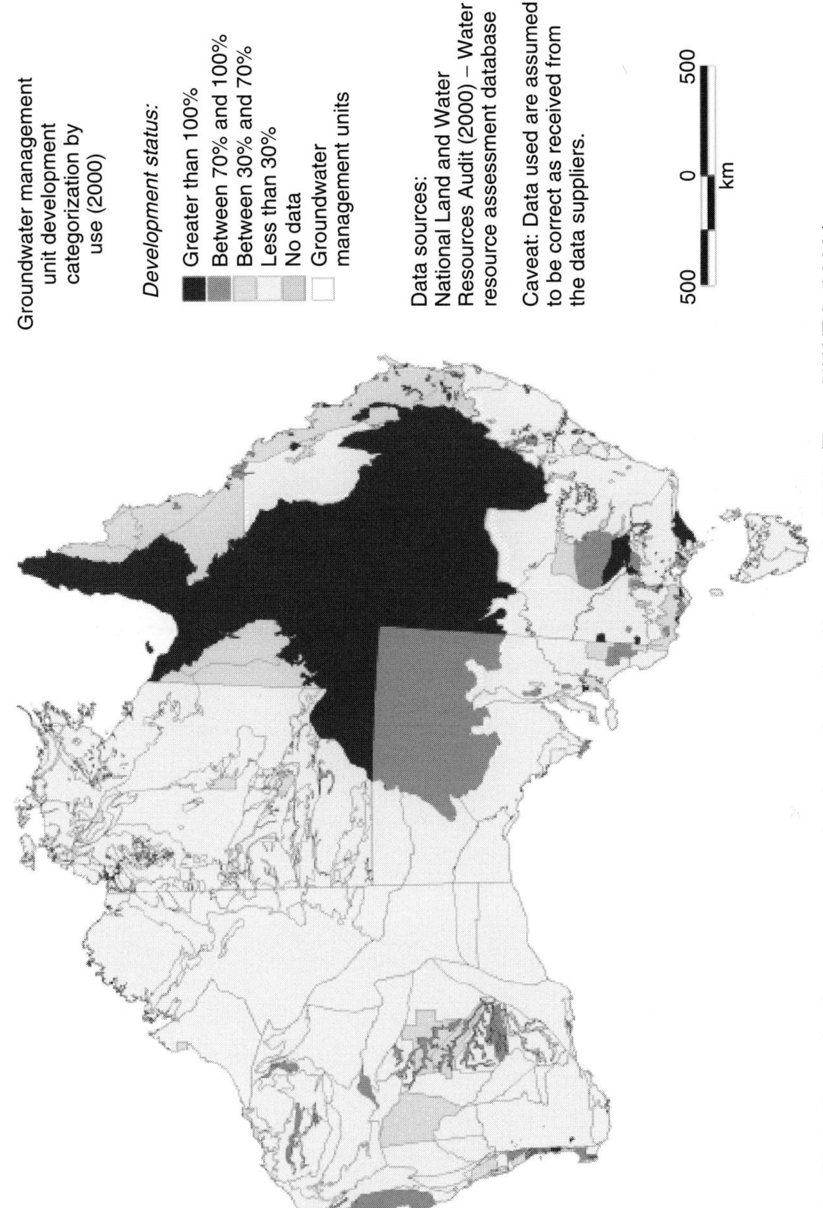

Fig. 15.1. Groundwater management units, categorized by use in 2000. (From NLWRA, 2002.)

management units of Australia. The management units are defined on the basis of water availability, water use and aquifer characteristics including depth, thickness and salinity. The NLWRA (2001) reported that more than 50% of the management units were extracting less than 30% of sustainable yield, with a further 19% between 70% and 100%, and 11% exceeding annual sustainable yield. Overall, 83 units (15%) were judged to be overallocated. Three management units, all in Victoria, had developed environmental allocation plans.

The framework is supported by two further national initiatives, and coordinated by the Department of Heritage and Environment – the National Principles for the Provision of Water for Ecosystems (ARMCANZ, 1996b) and the National Water Quality Management Strategy Guidelines for Groundwater Protection in Australia (ARMCANZ, 1995). A summary of groundwater-dependent ecosystems as envisaged in this and other work is given in Box 15.1.
There are two further supporting frameworks:

1. Overallocated Groundwater – A National Framework for Managing Overallocated Groundwater Systems has 13 recommendations designed to provide policy guidance for the states grappling with the serious issue of how to reduce the licensed volumes of overallocated groundwater aquifers. Associated with this policy paper is a Best Management Practice Manual, which suggests a broad range of approaches that are available to groundwater managers to reduce allocations and use (NRMMC, 2002a–c).
2. A National Framework for Promoting Groundwater Trading identifies the fundamental requirements for trading of groundwater as well as the impediments to groundwater trading.

The 13 recommendations address both the preconditions for trading and the requirement for a trading regime to operate. Methods to encourage trading are identified, as are the benefits of groundwater trading. The disadvantages of trading in overused systems are also identified. The document also asserts the following:

- The current level of monitoring of groundwater use (through the metering of bores) was low and more comprehensive data were required to correctly estimate sustainable yields.
- Commonly agreed methods for estimating sustainable yields and defining environmental water allocations for groundwater-dependent ecosystems were yet to be developed.
- Some states and territories have released new groundwater management policies; however, generally groundwater management reform was lagging behind those in surface water.

We will now turn to the central issue of *sustainable yield*: how this is defined, effected by groundwater-dependent ecosystems, and the characteristics of groundwater licensing and trade.

Box 15.1. Definitions of groundwater-dependent ecosystems in Australia. (Adapted from Hatton and Evans, 1998, 2003.)

- terrestrial ecosystems that show seasonal or episodic reliance on groundwater;
- river base flow systems, which are aquatic and riparian ecosystems in, or adjacent to, streams or rivers depending on the input of groundwater base flows, especially during dry seasons in seasonally dry climates or perennially in arid zones; hyporheic zones;
- aquifer and cave ecosystems, often containing diverse and unique fauna;
- wetlands dependent on groundwater influx for all or part of the year;
- estuarine and near-shore marine ecosystems that use groundwater discharge.

Sustainable Yield

Many countries (e.g. the USA and India) have widely adopted the concept of 'safe yield' (i.e. annual recharge) as a sustainable extraction limit. In many instances, this adoption is necessitated by high levels of groundwater development, but it has limited ability to account for hydraulic connectivity between water resources and environmental dependencies (Custodio, 2002; see also Llamas and Garrido, Chapter 13, this volume).

Australia's relatively recent development of groundwater allows a more conservative approach to sustainable yield. As any significant development of an aquifer will alter the water balance and have some impact, 'sustainability' must be interpreted as 'social acceptability of impacts' (Herczeg and Leaney, 2002). The central role of community in defining sustainable yield was noted by ARMCANZ (1996a):

> As any definition of sustainable yield embraces a range of technical as well as social, environmental and economic factors, it is necessary for considerable community input to make judgement of what is sustainable.

The NGC (2004) agreed the following definition of sustainable groundwater yield ('sustainable yield'):

> The groundwater extraction regime, measured over a specified planning timeframe, that allows acceptable levels of stress and protects dependent economic, social, and environmental values.

In adopting this definition, the NGC requested it be used with the explanatory notes provided, abridged in Box 15.2.

While implication within this definition and accompanying explanatory notes is to adopt a conservative approach to sustainable yield, the definition has been designed to allow for groundwater 'mining'. The willingness of the states to accept 'mining' – 'the exploitation of groundwater at a rate that is much greater than recharge' (Custodio, 2002) – as 'sustainable' has resulted in differences in the application of sustainable yield across the states. SA, in particular, accepts the notion of controlled depletion on the basis that (assuming no groundwater-dependent ecosystems) the groundwater is of no benefit if unused.

Box 15.2. Explanatory notes to accompany the nationally accepted definition of sustainable groundwater yield. (From NGC, 2004.)

Extraction regime

It is recognized that sustainable groundwater yield should be expressed in the form of an extraction regime, not just an extraction volume. The concept is that a regime is a set of management practices that are defined within a specified time (or planning period) and space. Extraction limits may be expressed in volumetric quantity terms and may further specify the extraction or withdrawal regime by way of accounting rules and/or rates of extraction over a given period and/or impact, water level or quality trigger rules. The limits may be probabilistic and/or conditional.

An oft-used means of defining the extraction regime has been by way of a maximum volume that may be taken in any single year. In some cases, where drawdown beyond the rate of recharge may be acceptable, it may be only for a specified period, after which time the rate may be less than the rate of recharge to compensate. In some cases and under specific circumstances (e.g. high or low rainfall years), the amount of water that may be taken may be greater or lesser than the longer-term value, and the conditions for this can be specified.

Acceptable levels of stress

The approach recognizes that any extraction of groundwater will result in some level of stress or impact on the total system, including groundwater-dependent ecosystems. The concept of acceptable levels of stress as the determining factor for sustainable yield embodies recognition of the need for trade-offs to determine what is acceptable. How trade-offs are made is a case- and site-specific issue and a matter for the individual states to administer.

The definition should be applied in recognition of the total system. That is, it should recognize the interactions between aquifers and between surface and groundwater systems and associated water dependent ecosystems.

In calculating sustainable yield, a precautionary approach must be taken with estimates being lower where there is limited knowledge. Application of the calculated sustainable yield as a limit on extractions must be applied through a process of adaptive management involving monitoring impacts of extraction. Sustainable yields should be regularly reassessed and may be adjusted in accordance with a specified planning framework to take account of any new information, including improved valuations of dependent ecosystems.

Storage depletion

The approach recognizes that extraction of groundwater over any time frame will result in some depletion of groundwater storage (reflected in a lowering of water levels or potentiometric head). It also recognizes that extracting groundwater in a way that results in any *unacceptable* depletion of storage lies outside the definition of sustainable groundwater yield.

Where depletion is expected to continue beyond the specified planning time frame, an assessment needs to be made of the likely acceptability of that continuation and whether intervention action might be necessary to reduce extraction. If intervention is likely to be necessary, planning for that action should be undertaken so that it can be implemented at the end of the specified time frame.

Major considerations in determining the acceptability of any specific level of storage depletion should be 'intergenerational equity', and a balance between

Box 15.2. *Continued*

environmental matters identified in the *National Principles for Provision of Water for Ecosystems* and social and economic values.

Protecting dependent economic, social and environmental values
The definition recognizes that groundwater resources have multiple values, some of which are extractive while others are *in situ* (e.g. associated water-dependent ecosystems) and all have a legitimate claim on the water resource.

The national definition of sustainable yield does not identify a standard planning time frame. The cumulative nature of extraction impacts and temporal response of aquifers can make the planning time frame a critical component of groundwater planning. These attributes of groundwater make sustainable yield estimations particularly subject to changes in social values and technical knowledge (see case study 2). Community understanding of groundwater availability can be difficult to progress with regard to the differences between the amount of water stored in an aquifer and the rate of recharge of that storage.

Groundwater Licensing, Management and Trade

Before the identification of groundwater management units and adoption of sustainable yield philosophies, it was not uncommon for water licences to provide access to a volume of water that could be taken as either surface water or groundwater ('conjunctive licences'). As a result of the COAG (1994) agreements to establish accounting mechanisms able to facilitate trade, conjunctive licences are progressively being separated into surface water and groundwater licences and this separation is considered complete in most states.

The identification of groundwater management units and adoption of sustainable yield practices within these management boundaries has allowed issue and management of groundwater licences to reflect that of surface water licensing. Thus groundwater licences comprise a share (still considered a volume in many areas) and an allocation. The introduction of groundwater management plans in overallocated areas alters the previously assumed 1:1 relationship between share and volume.

Where groundwater licences have been translated from volume to share through introduction of groundwater management plans, forecast of allocation is provided across the lifespan of the plan (typically 5–10 years). Thus, groundwater users have forewarning and can adapt if the plan requires a decrease in allocation. This is fundamentally different to the security offered by surface water and is of great importance to regional economies during drought where the storage/share ratio is low (e.g. NSW).

In locations where groundwater mining is not advocated, groundwater sharing plans typically address overallocation through an adjustment period by successively reducing the value of groundwater shares each year over the duration of the plan. In some areas where significant reduction was required (e.g. the Namoi, see case study 2), governments have provided financial support to assist regional communities to adjust to lower water availability. The other common practice to assist economic viability of communities in such instances is to develop carry-over capacities. As with surface water, this capacity allows unused (volumetric) allocation from 1 year to be transferred into the following year. Two primary constraints affect the capacity for such carry-over: (i) the physical limitation of bore yield; and (ii) the institutional limitations identified in the relevant groundwater management plan. While carry-over does not increase the net volume of available water over duration of a plan, it does allow for individuals to 'save' groundwater entitlements for drought years when surface water is not available.

The implementation of sustainable yield as an extraction regime rather than just a volume has generally been facilitated by the subdivision of groundwater management units into zones. They may be subject to different management constraints and practices (including trade) depending on zone-specific characteristics such as aquifer dynamics, level of development, water quality objectives, water level objectives and/or water pressure objectives. Case study 2 provides some insight into the manner in which zones can be used.

In accordance with COAG (1994) water reforms, groundwater management trade is progressively being enabled. Groundwater trade typically develops in fully allocated systems once enabled through institutional arrangements dictated via groundwater management plans.

Groundwater markets are geographically defined by groundwater management plans, and often restricted by institutional, technical and practical constraints applicable to zones subject to those plans. Generally speaking, groundwater trade in overallocated systems is considered a problem, and limited until overallocation has been addressed. Thus (nationally), groundwater trade is somewhat influenced by the priority development of groundwater management plans for overallocated resources and therefore tends to be localized (and can be restricted to zones within management areas).

The isolated nature of groundwater infrastructure and high costs of bore construction provide for narrow water market. Groundwater trade involves accessing more water from a bore rather than supplying more water via a channel. In practice, the high private overhead and risk of stranded assets associated with groundwater development for irrigation have limited the practical separation of groundwater property rights and land property rights

A national overview of groundwater markets was compiled in 2003 (Fullagar and Evans, 2003). This overview found that established rural groundwater trade markets existed only in SA and southern Victoria, for both temporary and permanent transfers. Prices for temporary trade ranged from AUS$0/m^3 to AUS$2.80/m^3. Prices for permanent trade ranged from AUS$0.325/m^3 to AUS$21.50/m^3. The broad range in prices is a direct reflection of the nature of markets within different groundwater management units: the niche wine markets in SA (notably McLaren Vale) allow far greater prices than do dominant crops in other states. Although only about 150 groundwater trades were estimated to occur annually in Australia (more than 50% of these in SA), expansion of groundwater trade is anticipated (Boyd and Brumley, 2003; Fullagar and Evans, 2003) as it is progressively enabled through implementation of the water reform process.

Integrating the Management of Surface and Groundwaters

As observed earlier, the focus of Australian public discussion and political interest in groundwater has now progressed from salinity management per se to recognizing the need for improved management across the surface and groundwater components of flow systems as well as the impact that limited surface water availability is having on groundwater development of adjacent aquifers. Thus, while saline base flows have been increasing, there is a risk that good-quality base flows will decrease.

Institutional (planning and management) separation of surface and groundwaters has allowed potential double allocation across a flow system (i.e. allocating the same yield once as surface water and again as groundwater).

In response to this issue, a national workshop addressing the management of hydraulically connected surface and groundwaters (Fullagar, 2004) recommended the adoption of five principles (see Box 15.3), the first of which was subsequently adapted and adopted as a component of the NWI objective. These principles are consistent with the issues and knowledge gaps that are handled by the NGC (2004).

Behind this work is the general belief that the sustainable productive capacity across a flow system (surface and/or groundwater) can be maximized by taking the 'right water, from the right place, at the right time' – this is the essence of the Australian interpretation of conjunctive water management.[2]

Managed aquifer recharge (including artificial groundwater recharge) is one aspect of surface and groundwater integration that has an interesting, if particular, history in Australia. There is increasing interest in capturing storm water, flood water and reclaimed or recycled water and diverting it to an aquifer either to recover lost storage or to enhance aquifer yield.

Before the 1960s, excessive private groundwater development for irrigation in the Burdekin delta, Queensland, led to sea water intrusion. In the mid-1960s, management of the Burdekin River was revised to provide for the replenishment of the delta aquifer through artificial recharge. The Burdekin became the largest groundwater-dependent irrigation scheme in Australia, with more than 35,000 ha of sugarcane and vegetables, adjacent to a surface-irrigated scheme of roughly the same area. Groundwater levels and yield have been systematically managed

Box 15.3. Recommended principles for managing hydraulically connected surface water and groundwater.

1. Where physically connected, surface water (including overland flows) and groundwater should be managed as one resource.
2. Allocation regimes should assume connectivity between surface water (including overland flows) and groundwater unless proven otherwise.
3. Overallocation of systems comprising connected surface water, groundwater and/or overland flows should be identified and eliminated by 2014.
4. Water users (surface water and groundwaters) should be treated equally.
5. Jurisdictional boundaries should not prevent management actions.

through artifical recharge from the Burdekin Falls dam since then. Recent economic analysis indicates that effective recharge may be adequately provided from irrigation return flows alone, with better benefits from the primary use of the irrigation water compared to direct recharge (see e.g. Hafi, 2003).

This example illustrates an unusual Australian development of surface water to respond to groundwater depletion, which contrasts with the more common problem of surface water depletion and increasing reliance on groundwater for drought management, whilst at the same time groundwater faces increasing degradation through salinity.

It is primarily economic costs of aquifer storage and recovery that have to date restricted practical interest to the high-value niche markets of SA. Noting the water values in McLaren Vale (see previous section on groundwater trade), it is not surprising that artificial recharge has created some interest. Water management in McLaren Vale involves the (privately initiated and funded) relocation and use of reclaimed water from an off-site treatment system (Grasbury, 2004). Interest in recharge has largely related to the need to secure winter storage in order to optimize use of this alternative water supply (10,000 million litres per year). In this instance, artificial recharge is economically viable and funding is not a primary issue. Trials have shown it to be a technically viable option (Hook *et al.*, 2002); however, obtaining necessary regulatory approvals have proven to be difficult: there are few precedents to build on, and obtaining approval thus requires a significant degree of government commitment.

Addressing surface water–groundwater interaction requires an understanding of the geographic distribution and volumes involved. Braaten and Gates (2003) made a statewide assessment of river systems in NSW, overlaying major streams with groundwater depth data and the locations of irrigation bores. The results demonstrated that river losses and/or gains are most closely correlated to groundwater levels in the mid-sections of the major rivers where alluvial systems are well developed, narrow and constricted, and groundwater depths are shallow.

Case study 1: managing groundwater in the Murray–Darling Basin

The profile of issues associated with surface water–groundwater interactions is perhaps best represented by the interjurisdictional activities that are in progress in context of the MDB (the catchment for the Murray and Darling rivers; Fig. 15.2). The MDB covers 1,061,469 km^2, and includes almost three-quarters of Australia's total irrigated land. About 70% of water used for agriculture in Australia is for irrigation in the MDB. The MDB extends over three-quarters of NSW, more than half of Victoria, significant portions of Queensland and SA and includes the whole of the Australian Capital Territory.

States retain responsibilities for natural resource management. The Murray–Darling Basin Commission (MDBC) is an interjursidictional institution established 'to promote and coordinate effective planning and management for the equitable, efficient and sustainable use of the water, land and other environmental resources of the Murray–Darling Basin' (MDBMC, 1992). The Commission reports to a ministerial council comprising ministers from each of the jurisdictional governments (including the Commonwealth) and a representative of the MDB community. Resolutions of the council require a unanimous vote.

The story of surface water allocation, the Cap on surface water in the MDB, has been presented earlier in this chapter; these policies were based specifically on river management and as such took no account of groundwater (MDBMC, 1996). Concerns relating to irrigation-induced salinity had been registered as early as 1911 within the MDB (Wilkinson and Barr, 1993). Accordingly, initial MDBC interest in groundwater was associated with water quality management and the impact of salinity to in-stream water quality – this interest subsequently expanded to encompass concerns regarding the mobilization of salts from dryland farming areas.

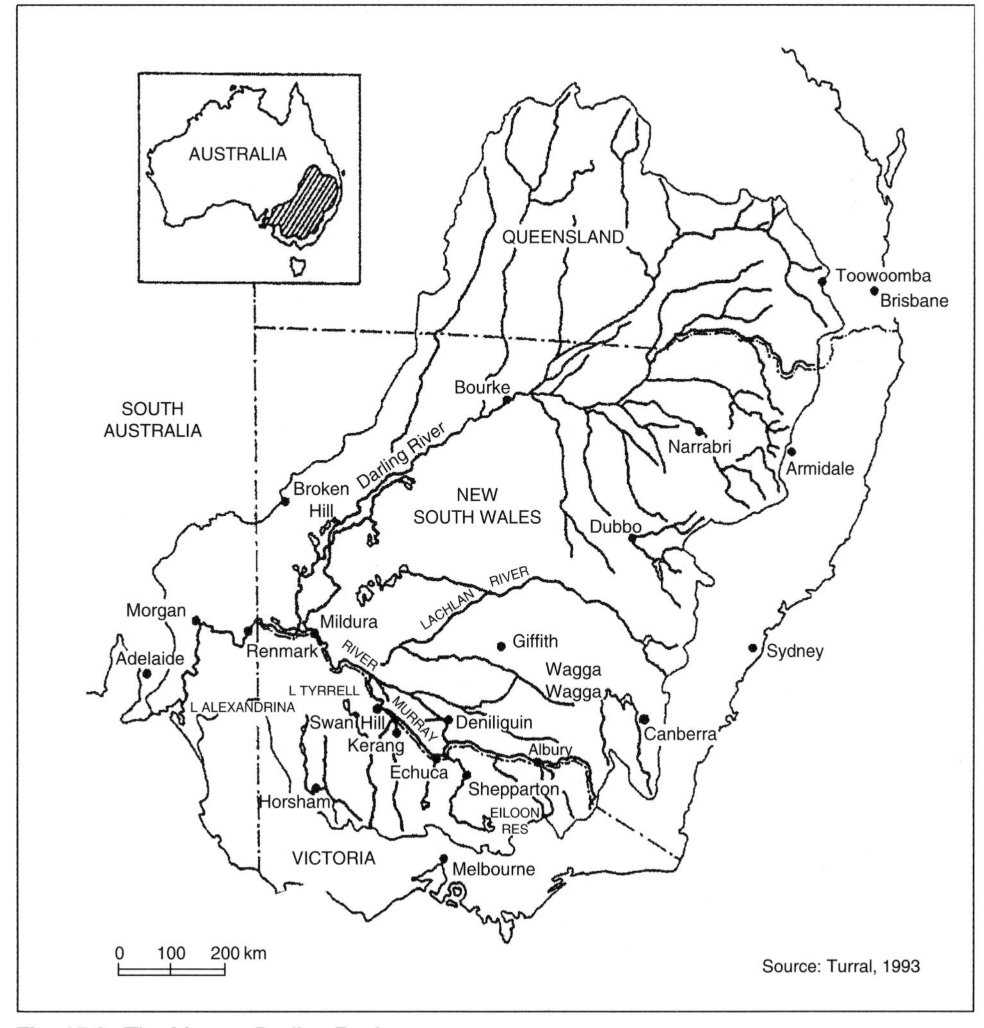

Fig. 15.2. The Murray–Darling Basin.

In 1996, a technical report (MDBC Groundwater Working Group, 1996) was released with the aim of 'progressing the setting of policy and programs to achieve a higher utilization of groundwater within the Basin's water resource allocation'. This report was followed in 1998 by another, which specifically outlined the impact that limited access to surface water would have on demand for groundwater, and the need to manage potential hydraulic impacts between surface and groundwaters (MDBC Groundwater Working Group, 1998). This report fed into general concerns that groundwater development could threaten river base flows – an impact with potential to thrust groundwater management into the central quantitative concerns of Cap agreements. The three means by which groundwater management may threaten the integrity of the Cap are (Fullagar, 2001):

1. reduced quantity of base flows through interception;
2. reduced quality of base flows through poor salinity management;
3. reduced capacity for governments to remain committed to the Cap in the event where viable alternative water supplies are lost.

A number of strategic studies were initiated to assess associated risks. These studies looked at: (i) the projection of groundwater extraction rates and implications for surface water; (ii) estimation of base flow in unregulated catchments of the MDB; and (iii) a review of groundwater property rights in Australia.

To provide a more comprehensive picture of water consumption within the MDB, the annual MDB Water Audit Monitoring Report (1999) began including groundwater consumption statistics in 1999/2000. Subsequent records (see Table 15.7) show a general increasing trend in groundwater consumption within the MDB, which peaked in response to the critical drought conditions of 2002/03.

The Review of the Operation of Cap (MDBCMC, 2000) found that the Cap had been a critical 'first step' in sustainable management of river resources in the MDB. The report included a recommendation to develop a groundwater management strategy for the MDB based on:

- jurisdictional management of sustainable yields;
- investigations clarifying how groundwater management practices may impact upon the integrity of the Cap in the future.

MDBC (2003) publicly released a report estimating an average reduction in surface water flow of 600 million litres for every 1000 million litres of groundwater use (Sinclair Knight Merz, 2003). Under groundwater development of the time, this amounted to a 2% undermining of the Cap, which was projected to increase to 7% in 50 years.

While the geological history of alluvial aquifer development implies some hydraulic relationship between groundwater and surface water, quantifying the potential for 'double allocation' is complicated by management and planning time frames, and time lags between groundwater flows and streams. Perhaps the most significant aspect of this work is the proactive manner in which the multiple jurisdictions have acknowledged and agreed to progress with a highly technical and political issue. This cooperation highlights the political importance

Table 15.7. Reported water use (GL) in the MDB 1999–2004.

Year	1999/ 2000	2000/01	Annual growth	2001/02	Annual growth	2002/03	Annual growth	2003/04	Annual growth	Growth 1999–2004
Ground water	1,103	1,240	12.4%	1,329	7.2%	1,632	22.8%	1,476	–9.6%	33.8%
Surface water	8,973	11,369	(capped)	10,960	(capped)	7,445	(capped)	8,780	(capped)	(capped)
Total	10,076	12,609	NA	12,289	NA	9,077	NA	10,256	NA	NA
Ground water % total	10.9	9.8	NA	10.8	NA	18	NA	14.39	NA	NA

given to ensure the long-term viability of existing surface water agreements underpinning management of the MDB. More broadly, groundwater interest within the MDBC structure is indicative of a wider interest in recognizing and realizing any potential environmental and/or productive opportunities associated with conjunctive water management.

Investigations associated with the development of an MDB groundwater management strategy continue. Consistent with broader water reforms, the primary focus of this research is to:

- establish consistent approaches to calculating sustainable yields for aquifers within the basin;
- build a framework for managing the combined use of surface and groundwaters;
- develop tools to help manage external groundwater impacts from irrigated areas;
- develop an approach to manage groundwater systems that have been overallocated;
- establish an evaluation process to help monitor and report progress against benchmarks and targets for managing groundwater resources.

In the following section, we focus on the development of groundwater policy at state level, with the case of NSW, and then take a more detailed look at an interesting example of efforts to bring an overexploited aquifer system back to sustainable levels in the Namoi Valley, which lies in NSW to the north of the MDB.

New South Wales: An Example of Integrating State and National Groundwater Policy

The total NSW groundwater resource is estimated at 5110 billion cubic metres, which is an enormous quantity of water, approximately 200 times the storage capacity of all dams in the state (DLWC, 2003). However, it has highly variable characteristics in terms of depth, yield, quality and spatial and temporal recharge. The sustainable yield is a tiny fraction of this (0.12%) at 6.19 billion cubic metres, of which 15% is too saline to use for most purposes. It is however a large resource and has been thought of as an effective buffer in drought.

In 1990, there were 70,000 licensed bores operating in the state of NSW, extracting 530 million cubic metres per year for irrigation, 15 million cubic metres per year for industry, commerce, mining and recreation and 60 million cubic metres per year for rural towns. Through the 1990s there has been increasing emphasis on high-value agriculture, with vegetables and fruits (grapes) leading the value table, and attracting higher-technology irrigation inputs (micro-sprinkler and drip irrigation) and accounting for a significant proportion of groundwater use. There has also been rapid development of groundwater since the early 1980s for conjunctive use on cotton and other commercial crops in the northern part of the state. There are few large dams in the northern river valleys and river flows are directly diverted, or harvested and stored in large on-farm dams known as 'ring-tanks' or 'turkey's nests'. Although cotton prices fluctuate considerably, the values shown in Table 15.8 indicate the price drivers for higher-value and intensified agriculture and the corresponding irrigated areas for each major crop.

The NSW Water Administration Act (1986) gave the minister of water resources the right to control, manage and use groundwater via the Department of Land and Water Conservation (DLWC), principally through licensing of use. Land use planning has been seen as crucial to the maintenance of groundwater quality and has been administered by the Department of Urban Affairs and Planning, working in cooperation with local government authorities under the remit of the Environmental Planning and Assessment Act of 1979. The protection of surface and groundwaters is governed by the Clean Waters Act of 1970 and the Environmental Offences and Penalties Act of 1989, both of which are administered by the Environmental Protection Authority (EPA).

In 1997, the government of NSW released its State Groundwater Policy Framework (DLWC, 1997), which was then supported by three subsidiary policies on: (i) groundwater quality protection (1999); (ii) groundwater quantity management (2000); and (iii) groundwater-dependent ecosystems (2000). The guiding principles of the policy framework are given in Box 15.4.

In 1998, a risk assessment was conducted for 98 aquifers across the state by the DLWC and 36 were found to be at high levels of risk. Of these, 4 aquifers suffered from water quality degradation and 32 from overallocation, and consequently 14 were embargoed from further development. The remaining potential for further groundwater development was judged to be limited to aquifers in some of the smaller inland river tributaries and valleys, some of the coastal sand and alluvial aquifer systems and 'unincorporated areas' (those within a groundwater province, but outside a designated groundwater management unit).

Implementation was also to be guided by risk assessment, so that increased focus and levels of management would be applied to more stressed aquifers on a priority basis. The management tools envisaged in the framework document included (DLWC, 1997):

Table 15.8. Value of water use in agriculture in Australia. (From National Land and Water Audit, 2002.)

	Gross value (million $)	Net water use (million m^3)	Irrigated Area (ha)	Value/ha ($/ha)	Value/million m^3 (million $/million m^3)
Livestock, pasture, grains, etc.	2,540	8,795	1,174,687	2,162	0.3
Vegetables	1,119	635	88,782	12,604	1.8
Sugar	517	1,236	173,224	2,985	0.4
Fruit	1,027	704	82,316	12,476	1.5
Grapes	613	649	70,248	8,726	0.9
Cotton	1,128	1,841	314,957	3,581	0.6
Rice	310	1,643	152,367	2,035	0.3
Total	7,254	15,503	2,056,581		

Box 15.4. Principles of the NSW Groundwater Policy Framework. (From DLWC, 1997.)

- An ethos for the sustainable management of groundwater resources should be encouraged in all agencies, communities and individuals who own, manage or use these resources, and its practical application facilitated.
- Non-sustainable resource uses should be phased out.
- Significant environmental and/or social values dependent on groundwater should be accorded special protection.
- Environmentally degrading processes and practices should be replaced with more efficient and ecologically sustainable alternatives.
- Where possible, environmentally degraded areas should be rehabilitated and their ecosystem support functions restored.
- Where appropriate, the management of surface and groundwater resources should be integrated.
- Groundwater management should be adaptive, to account for both increasing understanding of resource dynamics and changing community attitudes and needs.
- Groundwater management should be integrated with the wider environmental and resource management framework, and also with other policies dealing with human activities and land use, such as urban development, agriculture, industry, mining, energy, transport and tourism.

- groundwater management plans where necessary;
- supporting guidelines for local government and industry;
- creation of aquifer resources and vulnerability maps;
- an education strategy;
- legislative mechanisms for groundwater management;
- licensing tools and conditions for users that better reflect resource protection objectives;
- economic instruments applicable to groundwater management.

At the time the framework was released, there were already 13 groundwater management plans in existence, and a further 5 in preparation, and the experience gained thereby was effectively incorporated into the policy. Groundwater management plans are to be reviewed on a 5-year basis and reporting is undertaken by community-staffed Groundwater Management Committees, supported where necessary by state funds. Reporting is biennial, and requires comparison of measurable indicators against the plan's targets.

Much of the ensuing debate in NSW has hinged on the definition of sustainable yield, and within this, determination of volumes available for development. Statewide this has been defined as 100% of the long-term average recharge with further reductions advised to reserve water for groundwater-dependent ecosystems. A pilot process was undertaken in the Namoi Valley to reduce abstractions to sustainable levels – through consultative processes and committees – and defining sustainable yield was at the core of the negotiations.

Since different formulas are used in different states and territories, and in many the amount of data is increasing and the reliability of assessment is improving, there are some cases where the estimate of sustainable extraction has actually risen since 2000.

Case study 2: the Namoi River

The Namoi River catchment lies in north-east–central NSW and covers approximately 42,000 km^2, as shown in Fig. 15.3. The river flows 350 km from east to west and there are three major storages on the main stem and its tributaries: Keepit, Chaffey and Split Rock dams. The catchment includes part of the Liverpool Plains that has been subject to long-term investigations of fertilizer and agrochemical pollution of groundwater. Rain generally occurs in summer but is highly variable between years and seasons, from as high as 1100 mm/year over the Great Dividing Range in the east (upper catchment) to as little as 470 mm/year in the downstream area in the west. As in the rest of south-eastern Australia, potential evaporation generally exceeds rainfall rising from 1000 mm/year in the east to more than 1750 mm/year in the west.

Groundwater is generally sourced from quaternary alluvial aquifers running along the major stream lines, but there are also two low-yielding sandstone aquifers. The total volume of groundwater storage is estimated to be 285 billion cubic metres, of which 89% is of low salinity (less than

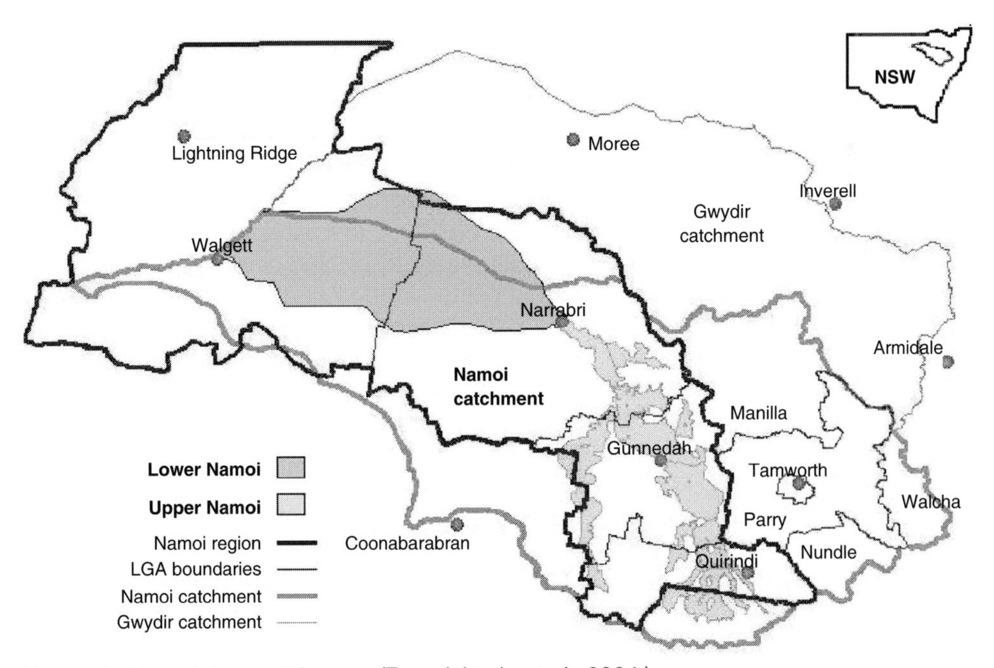

Fig. 15.3. Namoi river catchment. (From Ivkovic *et al.*, 2004.)

1000 mg/l total dissolved solids (TDS)). In 1988, there were 1639 high-yielding tube wells, mainly in the paleo-channels or alluvium adjacent to the river, with maximum yields as high as 200 l/s. Average groundwater use at this time was 200 million cubic metres per year, which was equivalent to recharge these aquifers. Small volumes are also sourced from porous sandstone aquifers of the Great Artesian Basin, which stores the bulk of the groundwater in the Namoi Valley (243 billion cubic metres) at depths of 520–810 m below ground level. Although TDS are generally less than 1200 mg/l, the water has high sodium content and is not suitable for irrigation, but is used for stock watering and for town and rural drinking water. Groundwater in fractured rocks (basalt) is sometimes sourced for stock and domestic supplies, but yields are low and success in drilling is variable.

The Namoi accounts for about 40% of NSW's total groundwater use and is one of the most intensively developed irrigation areas in the state, with largely private investment through agri-business (e.g. Auscott and Twynhams) and large landholders who have moved into intensive irrigation development. Cotton is essentially a 'young industry', with highly mechanized large-scale layouts, mainly using furrow and bed irrigation. Substantial research has been undertaken into tightly scheduled irrigation and irrigation agronomy, coupled with trials on micro-irrigation and drip tape, but the consensus is that furrow and bed irrigation is best suited to the vertisol soils and has cheaper capital and operational costs, which attract less risk with volatile cotton prices.

Groundwater has been extensively monitored since the 1970s, with 560 piezometers at 240 sites in 1995, and a further 470 licensed bores monitored on 175 properties (Johnson, 2004). This data allowed the completion and calibration of a groundwater model of the Lower Namoi in 1989 and its subsequent refinement.

The chart in Fig. 15.4 shows the rapid development of groundwater, principally to irrigate cotton, lucerne and wheat since the late 1970s, increasing from less than 15,000 ha to around 35,000 ha. The surface-irrigated area in 1988 was marginally larger at 36,544 ha. The Commonwealth Scientific and Industrial Research Organization (CSIRO) undertook the first assessment of groundwater use in 1991, and recharge was estimated to be just over 200 million cubic metres per year. After community consultation, a contentious agreement was brokered to implement a policy of 'controlled depletion' of 220 million cubic metres per year on average, in the full knowledge that the economic life of the aquifer would then be only 30 years (i.e. till 2020). The idea of controlled depletion meant that an annual average recharge plus a further 10% or so annual depletion would be allowed.

However, it was not long before many people in the community as well as in public administration decided that a more sustainable long-term solution would be preferable, and that mining the aquifer was in very few peoples' interest. Further assessment and modelling studies indicated that average usage in the Namoi was below recharge, but at the same time it was overallocated with sleepers and dozer licences and punctuated by periodic overuse, corresponding to low surface water allocation years (Fig. 15.5). However, at this stage, the

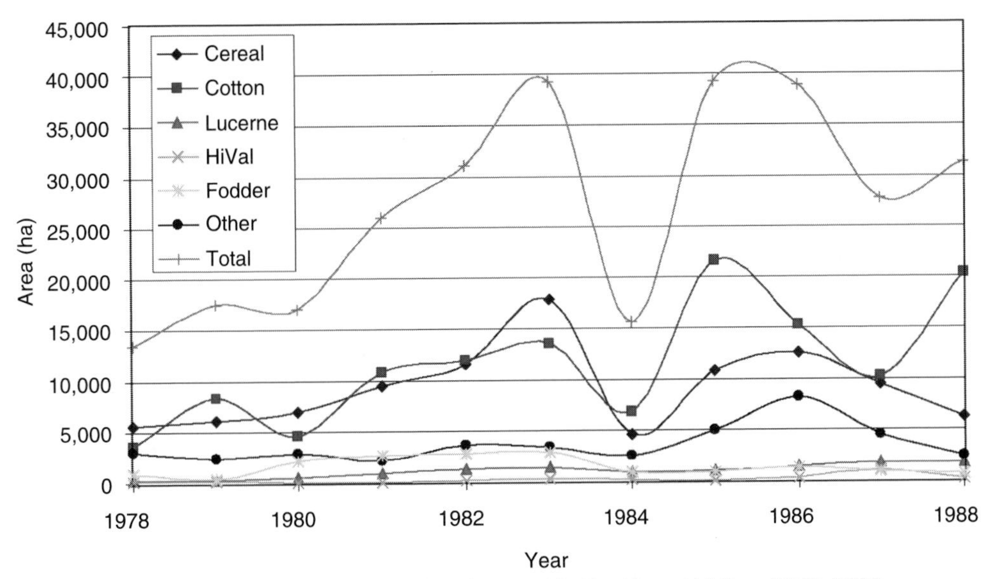

Fig. 15.4. Groundwater-irrigated area development in the Namoi Valley, 1978–1988.

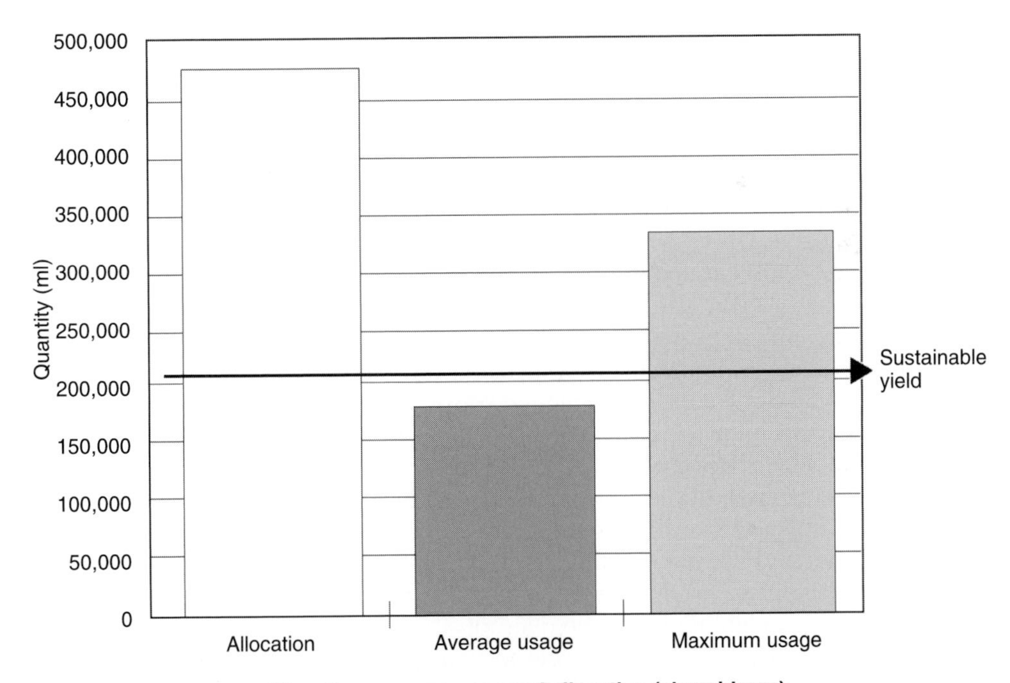

Fig. 15.5. Comparison of allocation and use in the Upper and Lower Namoi groundwater systems, 1998 (DLWC).

assumption was still that sustainable extraction equated to 100% long-term average annual recharge and that there were no groundwater-dependent ecosystems in the valley.

Even at this most optimistic formulation of sustainable extraction, some subsystems were 3–4 times overallocated, and no further development was allowed in all zones (see Table 15.9) except zone 6 where the water table continued to rise. There was an in-principle agreement to phased reductions in allocations to 35% of existing values during 1996–1998, but in practice this proved very difficult to agree and implement. Following national and state initiatives for groundwater management reforms, a series of modelling assessments were undertaken and then supported by a Social Impact Study, conducted by CSIRO with, and on behalf of, the community. Initial stakeholder assessments on fair reallocation and the definition of environmental flows were made to develop the full process (Nancarrow *et al.*, 1998a,b). The main focus of the assessment was to understand differential treatment of active and unused licences, as well as the likely impacts on the community and their expressed priorities. The consultation was conducted in 1999, and the main characteristics are summarized in Box 15.5.

The results were incorporated into the Water Sharing Plan, seeded in 1999, which was expected to be formalized in 2000 and followed by swift implementation. The Namoi study was effectively a pilot for other groundwater management units in NSW, but in the end, the final plan was not agreed and published until 2003, and began implementation only in 2004, having progressed through one of the most severe and extended droughts on record (2000–2004).

The principle source of contention concerned the definition of sustainable extraction and the preference of many in the community to maintain this at 100% annual average recharge. In 1999, DLWC supported the continued abstraction of 100% annual average recharge and proposed a 10- to 15-year period to determine and implement a transition to incorporating an environmental share of the resource. In response to the paper 'Perspectives on the Sustainable Development of Groundwater in the Barwon Region', presented to the Namoi Groundwater Management Committee, the Nature Conservation Council of NSW drafted a hard-hitting response (http://www.nccnsw.org.au/water), and suggested the immediate and precautionary implementation of 70% as an environmentally sustainable yield in underused zones, with a 10-year transition for the overexploited zones. They proposed formulas to cut back allocations to sustainable limits for each zone, which were eventually adopted in the water sharing plan after some modification (2003).

Simultaneously, combined surface and groundwater assessment studies were undertaken with hydraulic and social impact models linked together (Letcher and Jakeman, 2002). The investigators noted that many such studies require approximately 3 years for model completion, by which time the initial key issues might no longer be relevant. They commended the development of models to be sufficiently flexible for reapplication to other problems, and to emphasize the difference between outcomes and policy developed from models, compared to accurate prediction. Despite considerable community involvement, they note that great effort is required to explain model outputs and accept

Box 15.5. Community consultation and participation in development of the water sharing plan in the Namoi Valley: the NSW socio-economic assessment. (Adapted from Nancarrow *et al.*, 1998a.)

1. Understanding the catchment:
   (a) community water profile – socio-economic characteristics and history, water use profile;
   (b) identifying change processes in the catchment;
   (c) identifying key issues.

2. Goal setting (principle of balance of benefits and costs):
   (a) understanding government goals and objectives;
   (b) understanding community expectations of the water reform process;
   (c) communities' aspirations and concerns for the future.

3. Generating management options:
   (a) development of a range of appropriate options.

4. Identifying effects:
   (a) on different uses, population groups, industry sectors, communities and over time;
   (b) extractive and non-extractive uses – matrix of sectoral uses and options;
   (c) socio-economic effects – checklist of financial effects inside and outside catchment, socio-demographic structure, community institutions and vitality, heritage values, environment.

5. Assess effects:
   (a) preliminary and detailed;
   (b) extent, likelihood, intensity, timing and duration;
   (c) impacts of no-change – development of a common reference scenario;
   (d) detailed studies – clear statement of assumptions; quality assurance principles – focus, long-term horizon and equity; targeted sensitivity analysis; identification of appropriate methods and techniques; identification of data sources.

6. Determining preferred option:
   (a) impact display table;
   (b) trade-offs – weighting strategies to cope with differential benefits;
   (c) risk and uncertainty analysis.

7. Developing impact management strategy.

8. Reporting:
   (a) required reporting on important steps (i.e. all the foregoing).

9. Monitoring:
   (a) monitoring for management, feedback and adjustment;
   (b) review of objectives and actions;
   (c) generation of monitoring questions;
   (d) identification of key factors or variables to be monitored.

10. Evaluating and adjusting.

uncertainty and iterative solutions. They also noted that stakeholders must be encouraged and assisted to have more realistic expectations on the appropriate and inappropriate uses of models and their outputs, and implied that gaining feedback through public seminars and discussions was insufficient.

The final Water Sharing Plan (MLWC, 2003) documents the agreed reductions in allocation and the rules associated with allocation and monitoring. The plan defines 13 separate groundwater management zones within the Namoi Valley and determines the long-term average recharge for each one (Table 15.9). The largest zone, in terms of geographic area, water resources and use, is the Lower Namoi, a contiguous near-stream alluvial aquifer. However, wells that were drilled deeper through unconsolidated sediments of the Lower Namoi and into the Great Artesian Basin were not included in the plan. The crux of the matter is the process by which allocations will be reduced to address current overallocation (Box 15.6).

Previous drafts of the plan were consistently opposed by the Nature Conservation Council of NSW (see Report Card on Water Sharing Plans), which recommended against gazetting the Namoi Water Sharing Plan in 2003 on the grounds of insufficient allocation for the environment.

The final assessment of extractable water for agriculture was undertaken on the basis of environmental health requirements (taken at approximately 30% of annual recharge) and other high-priority uses (utility licences and native title use), and considers the long-term aspects of climate variability. The domestic and stock rights were calculated separately, and then the actual agricultural demand was also determined (Table 15.9). In fact native title rights in the Namoi amounted to zero and so had no impact in this case, and it can be seen that the stock and domestic and utility licence volumes are generally modest. A simple formula that pro-rated new licensed volume in proportion to available resources, reserved licence and prior licence volumes resulted in the revised figures and the percentage reductions in zonal allocations summarized in Table 15.9. It can be seen that there are no reductions in zones 6, 9

Table 15.9. Summary table of groundwater allocations by zone in Namoi, water sharing agreement 2003 (in thousands m^3). (From Water Sharing Plan 2003.)

Zone	Estimated annual average recharge	Stock and domestic rights	Estimated water requirement	Local utility access licence	Reductions in access licence (% volume)
1	2,100	39	8,510	1,716	87
2	7,200	359	23,810	59	70
3	17,300	470	56,017	199	69
4	25,700	667	82,590	4,660	73
5	16,000	262	36,042	56	45
6	14,000	274	11,448	97	0
7	3,700	89	6,321	4,407	41
8	16,000	166	48,204	–	67
9	11,400	187	11,342	–	0
10	4,500	36	1,420	–	0
11	2,200	210	8,740	–	75
12	2,000	73	7,487	–	73
Lower Namoi	86,000	3,304	172,187	–	51
Total	208,100	6,136	474,118	11,194	

Box 15.6. Objectives and performance indicators for the groundwater sharing plan, Namoi.

The objectives of the plan include:

- protection maintenance and enhancement of ecosystems dependent on groundwater;
- protection of the structural integrity of the aquifers and of their water quality;
- management of extraction so that there is no long-term decline in water levels;
- preservation of basic landholder rights access to the groundwater sources and assurance of fair, reliable and equitable access through management of local impacts and interference effects;
- contribution to the protection, maintenance and enhancement of the economic viability of groundwater users and communities;
- assurance of sufficient flexibility in account management to encourage efficient use of groundwater resources and to account for the effects of climate variations.

The performance indicators selected to monitor the objectives include:

- change in groundwater level and climate adjusted levels;
- change in groundwater level adjacent to dependent ecosystems;
- change in groundwater quality;
- change in economic benefits derived;
- extent to which domestic, stock, water utility and native title rights have been met;
- change in structural integrity of the aquifer.

and 10, and water levels have been rising in zone 6 due to recharge from surface irrigation and other surface water–groundwater interactions.

The plan makes allowance for future revision of estimates of sustainable extraction volumes and sets limits on the maximum (over)abstraction within 1 accounting year, compared to the longer-term (3 years) average extraction as reported to the minister. Typically, the maximum 1-year overabstraction limit is 25% greater than the nominal long-term value given in Table 15.9.

Water availability is determined by continuous monitoring and compares the average abstraction with the extraction limit over the current and preceding 2 years, with some upper limits set on water availability in some zones. Water accounting is conducted annually over a water year that runs from 1 July to 30 June. To minimize interference between adjacent bores, no new agricultural bores can be approved within 100 m of an existing well or 200 m from an existing property boundary, and are subject to further expert hydrogeological findings as appropriate. They must also be more than 400 m from an existing monitoring well and 500 m from an existing domestic water supply well. Finally the plan was scored by the DLWC on how well it met the 38 targets of the State Water Management Outcomes Plan. The transition period allowed for the full implementation of environmental allocation was finalized at 10 years. It will be implemented through re-specification of licences, such that the sustainable licence volume is now formally allocated with supplementary water allocations that will be gradually reduced to zero over the transition period.

A socio-economic evaluation of the plan was conducted by the University of New England, Armidale, NSW, in late 2003 (Institute for Rural Futures (IRF), 2003). This was preceded by a number of studies undertaken generally for water sharing plans in NSW by Australian Consultants International Limited (ACIL) in 2002, and by the DLWC in conjunction with the CSIRO-conducted exercise. A number of expert commentaries were also written by other observers, including the Australian Bureau for Agriculture and Resource Economics (Topp, 2000), which illustrates not only the importance and pioneering nature of the Namoi case, but also the pluralistic and broader interests and perspectives brought into play by the state, the water users and the environmental lobby groups. It also shows that different studies are employed at different times for different purposes, even if they seem to cover the same territory – for example, dealing with public or users' perception and priorities in the evolution of a plan and a more dispassionate, objective assessment of the impacts of that plan after it has been declared.

The IRF study looked in detail at the economic impacts by commodity and zone, using primary data, secondary data and a farm modelling analysis. The farm analysis was extended to regions, and complemented by industry and social impacts. It was conducted at a time when an earlier version of the Water Sharing Plan was deferred for 6 months, and simplified water allocation reductions, similar to those voluntarily agreed by the user community, had been reinstated.

Some farmers indicated that they would acquire, or try to acquire, increased surface water supplies to substitute for 'lost' groundwater allocation, and set their future farming strategies accordingly; hence various scenarios of future water use were investigated, including the impact of trading. However, the authors lamented the lack of reliable information on the interaction between streams, irrigated fields and aquifers and the extent to which surface water could be substituted for groundwater.

Groundwater-irrigated farms were estimated to contribute AUS$384 million or 56% of the gross value of agricultural production in 2000–2001. The analysis of all zones indicated a future loss of production of AUS$26.7 million in 0–9 years (under the plan) and a further AUS$42.3 million in 10–20 years (post plan), considerably more than the structural adjustment compensation of AUS$18 million proposed by the NSW government. An alternative plan of AUS$120 million compensation had also been proposed, but cut back to this value, amounting to an average of about AUS$70,000 per affected property. As a result, it was felt that some owner-operators would be forced to amalgamate and expand or to cease operation due to reduced net income of reduced water allocation. The mitigating impacts of new enterprises, new technology and possible higher-price regimes in the future were all positive. Overall, it was expected that irrigated production would contract, with cereals reducing far more than irrigated cotton, and would be partially compensated by an increase in rain-fed wheat and sorghum. Lucerne production would decline and there would be an increase in feedlot cattle production and a corresponding reduction in open grazing. Little change in high-value cropping was anticipated.

At a regional scale, it was estimated that gross regional product would decline by 2% in 0–9 years and by 4% thereafter (10–20 years), with corresponding reductions in household income of 2% and reductions in employment of 2%. Social impacts were not quantified, but explained in qualitative terms, such as loss of employment, reduction in school population, reduced local spending and knock-on effects on service industries. The report identified the town of Gunnedah as the focal point of declining cotton production, which was expected to concentrate closer to existing service centres in Narrabri.

Although the Water Sharing Plan was developed in close consultation with the community over a long period, individual property owners are reported to have spent as much as AUS$ 250,000 in trying to challenge the plan in court (Rural Reporter, 30 August 2003).

The story continues to unfold with the same pressures from users, environmental groups and resource managers coming into play. The plan was due to be implemented towards the end of 2004, but was delayed and is now scheduled for implementation in 2006. In the current iterations, research continues on surface water–groundwater interactions and the resulting effects on water allocation policy (Ivkovic *et al.*, 2004). Preliminary conclusions indicate localized reductions in stream base flow, likely to be attributed to groundwater use. More extensive investigation continues, but it is likely that there will be further pressure on limiting both surface and groundwater abstraction, until a balance that is acceptable to the community has been achieved. This will no doubt continue to be a robust and noisy process.

Lessons for Groundwater Management in Other Countries

Although there are obvious structural differences between Australia and developing countries such as those in South Asia and China using groundwater, there are still useful insights to be gained. The contextual differences include population, particularly the farm population (20 million well users in India vs. 70,000 in NSW) and corresponding farm size, where Australian holdings range from hundreds to thousands of hectares. As a result, the number of wells in Australia is relatively modest and licensing and metering are *not* the daunting tasks presented in, for example, the Indian subcontinent. The lessons for Australia itself can be briefly summarized as follows:

1. Ensure that groundwater and/or surface water reforms happen in tandem to avoid lags in policy development and implementation.
2. Recognize groundwater–surface water interactions and aim to use these proactively rather than reactively.
3. Ensure that sustainable yield takes into account the temporal and geographic distribution of water use as well as the sustainable volumes available for development.
4. Zonal approaches can be used to fine-tune sustainable yield management.

5. Ownership of policies is critical to compliance, especially where overallocation or isolated infrastructure is involved.
6. Interindustry and interjurisdictional issues relating to aquifer development should be pre-empted – economic inequities between industries can complicate resolution.
7. Regular monitoring and reporting underpin management, understanding and compliance – groundwater issues can only be managed if they are recognized or addressed early enough.

However, Australia shares a common heritage of a philosophy of state-sponsored development of agriculture and irrigation in particular. This has been focused particularly on the development of a commercial agricultural economy with a major focus on exports. As the world market has become more competitive and rural sector's economic share of GNP has declined, the state has been less inclined to support the agriculture and has plunged it into global free trade with an enthusiasm and commitment seen in few other parts of the world. Coupled with the rising conviction that environmental management is of crucial importance, the federal government, through the COAG, has pursued reform objectives, based on clearly defined economic, environmental and social principles. These have been developed by state leaders and explained, sold and forced on the states' populations through combinations of incentives and penalties.

This has occurred against a background of genuine (if expensive) public participation in natural resources management, which has been transformed from disparate local initiatives into a national movement, and then rationalized to some extent through catchment-based management organizations that retain a strong community ownership and membership. Politically, the environmentally conscious urban electorate has become significantly more powerful than the rural lobby, whilst at the same time the true guardians of the rural environment are those who live and work there – predominantly farmers. In contrast, there is probably no broad-based consensus on economic reform and national competitiveness (masked at the ballot box by other issues), and this has allowed the central government to take the lead on potentially unpopular reforms with much less public participation and discussion.

Public participation involves genuine dialogue and often rancorous discussion supported by publicly available information. Although some information is recognized to be still far from perfect, there is a good general understanding of the resource base and its constraints, if less than perfect knowledge of actual groundwater use. With respect to groundwater, there has been a big step forward in the understanding of allocation in relation to sustainable resource use and this has led to hard-to-negotiate adjustment programmes to reduce overallocation, which is intrinsically easier than dealing with overconsumption, which in India is the real problem in absentia of rational energy pricing and any form of allocation system (see Shah, Chapters 2 and 11, this volume). On the inside, the debate is noisy and fragmented, giving a very different impression to the 'contestants' on the ground compared with observers trying to synthesize experience and progress from the outside. However, noise and dispute are welcome signs of a dynamic and healthy process, and in the end contribute to more balanced sets of outcomes than administration by fiat, whether it is honoured in practice or in the breach.

Public availability of data, commitment to find more when it is insufficient and access to modelling and other impact assessments, commissioned by the community, by the state or in collaboration, all contribute to a more transparent and better-argued politics in natural resources management.

There is an increasing tendency to look at structural differences between developed and developing countries and then say 'obviously this cannot be done' or 'that does not apply'. There is an increasing body of literature questioning integrated water resources management, especially its more prescriptive formulations (Biswas *et al.*, 2005). However, sound principles and practices need to be applied if we wish to achieve sustainable development of water resources and not overdevelop or degrade the resources for future generations.

This chapter shows that groundwater management is a complex, multifaceted process that is dynamic and has continually changing contexts, problems and challenges, just as with surface water. It also illustrates clearly that surface and groundwater management needs to be integrated in many cases, although this adds further complexity, more stakeholders, greater need for data and so on.

However, structural differences between Australia and, say, India mask differences in the size and importance of groundwater as a sector. In India, it is a much more significant contributor to both the economy and the individual welfare and, as such, should be accorded serious attention concerning its future sustainability. The recognition of this importance has either escaped the government's notice (by now, unlikely) or has been submerged by other conflicting short-term agendas and solutions. The Australian experience shows that initiative and active involvement by different interest groups working at different levels and for different ends can move towards a longer-term agenda, for broadly similar reasons of welfare and stability that confront developing countries.

An important point is that an effective process, based on a combination of policy, economics, science and participation can be, and has been, established. Attention to detail has been a fundamental plank in groundwater reforms, considering resource availability, use, environmental consequences, economic benefits and losses, and accounting for the range of stakeholders' perspectives and views. This does not mean that all stakeholders' needs and concerns are satisfied – far from it – but they are ultimately negotiated and cajoled towards what is believed to be a better position. A commitment to monitoring should ensure that results can be evaluated and the effectiveness of different policies and positions determined in a continuing and dynamic cycle of 'adaptive management'.

Countries such as India can learn from broader federal mechanisms of carrot and stick policies, applied to their own contexts. The Australian case shows how a strong and purposive government can rub shoulders with true public participation. There are positive lessons in the detailed development of process, interagency cooperation and genuine participation at state level. In India, managing approximately 20 million tube well owners looks like an impossible issue even though they represent only 1/50th of the whole population. At state level this number may reduce to a million tube well owners, amongst other millions of citizens – and becomes immediately more tractable, although daunting. Guiding and resourcing local authorities to manage jointly and locally with the community require commitment, clear direction and professional and service-oriented public agencies.

None of these reforms have happened overnight in developed countries and have a backdrop of a long history of changes in technology, management, ideology and public institutions. Change does not happen rapidly and cannot be expected to do so, miraculously, in developing countries. Solutions adapt to problems through the simple and pragmatic business of trying them out and gaining experience, confidence and trust. Exact models of management cannot be expected to be transplanted and made to work in different contexts, but different components offer potential to provide solutions if there is the broad policy and incentive structure to maintain commitment to learn and adapt on the ground.

To find solutions it is necessary to define problems, and there is great potential to do this more effectively, thoroughly and in more detail from a range of stakeholders' perspectives. How to do this with large numbers of stakeholders remains a challenge, which is only partly solved by increasing education and awareness.

Notes

1 Sleepers and dozers are licence-holders who pay for their entitlement annually, but use little or none of it. Typically they run mixed farms with rain-fed crops and substantial livestock holdings, for which they keep water entitlement as insurance in drought years, either for fodder production or direct stock watering. There are no 'use-it or lose-it' provisions (as in the US prior appropriation doctrine) for water licences in Australia.

2 'Conjunctive water management' encompasses both productive and environmental objectives, and some account of any hydraulic interdependency between surface and groundwaters is generally implicit to Australian use of the term. Consistent with the notion of sustainable yield as a 'regime' rather than volume and a rejection of 'prior rights', Australian terminology assumes fairly specific 'flow system' connotations and thus may be distinguished from aggregation of conjunctive use (e.g. according to Raju and Brewer, 2000) and activating use of aquifer storage services (e.g. characteristic of conjunctive water management in the USA, Blomquist *et al.*, 2004).

References

ARMCANZ (Agriculture and Resource Management Council of Australia and New Zealand) (1995) *National Water Quality Management Strategy*. Guidelines for groundwater protection in Australia. ARMCANZ, Canberra, Australia.

ARMCANZ (Agriculture and Resource Management Council of Australia and New Zealand)

(1996a) *Allocation and Use of Groundwater: A National Framework for Improved Groundwater Management in Australia.* Occasional Paper No. 2 Commonwealth of Australia 1997.

ARMCANZ (Agriculture and Resource Management Council of Australia and New Zealand) (1996b) *National Principles for the Provision of Water for Ecosystems.* ARMCANZ, Canberra, Australia.

Australian Bureau of Statistics (2005) http://www.abs.gov.au/websitedbs/d3310114.nsf/Home/.

Biswas, A.K, Varis, O. and Tortajada, C. (eds) (2005) *Integrated Water Resources Management in South and South-East Asia.* Oxford University Press, New Delhi, India.

Blomquist, W., Schlager, E. and Heikkila, T. (2004) *Common Waters, Diverging Streams.* Resources for the Future, Washington, DC.

Boyd, T. and Brumley, J. (2003) Optimising groundwater trade. *Symposium Proceedings of About Water,* 28th International Hydrology and Water Resources Symposium, Vol. 2, pp. 273–279.

Braaten, R. and Gates, G. (2003) Groundwater-surface water interaction in inland New South Wales: a scoping study, *Water Science Technology,* Vol. 48(7), pp. 215–224.

Council of Australian Government (COAG) (1994) *Water Resources Policy.* Report of the working group on water resources policy. Canberra, Australia.

Council of Australian Governments (COAG) (2004) *National Water Initiative.* Full documentation available at: www.coag.gov.au.

Council of Australian Governments (1992) *National Strategy for Ecologically Sustainable Development,* Australian Government Printing Service, Canberra, Australia.

Custodio, E. (2002) Aquifer over-exploitation: what does it mean? *Hydrogeological Journal* 10(2); 254–277.

Department of Land and Water Conservation, NSW (DLWC) (1997) *The NSW State Groundwater Policy Framework Document.* DLWC, Sydney, NSW.

Department of Land and Water Conservation, NSW (DLWC) (1998) *Water Sharing, the Way Forward.* DLWC, Sydney, Australia.

Department of Sustainability and Environment (2004) *Victorian Government White Paper: Securing Our Water Future Together.* Melbourne, Australia.

DLWC, NSW (2003) *NSW State Groundwater Dependent Ecosystems Policy. Parammata,* Sydney, Australia.

Etchells, T., Malano, H., McMahon, T.A. and James, B (2004) Calculating exchange rates for water trading in the Murray–Darling Basin, Australia, *Water Resources Research,* Vol. 40(12), W12505.

Ewing, S. (1996) Whose landcare? observations on the role of 'community' in the Australian Landcare programme. *Local Environment,* 1(3), pp. 259–276. Journals Oxford, Oxford University Press, Oxford.

Fullagar, I. (2001) *Two Hands Clapping in the MDB.* Groundwater Workshop proceedings.

Fullagar, I. (2004) *Rivers and Aquifers: Towards Conjunctive Water Management (*Workshop proceedings*).* Bureau of Rural Sciences, Canberra, Australia.

Fullagar, I.M. and Evans, R. (2003) Trading groundwater concepts', *Symposium Proceedings of About Water,* 28th International Hydrology and Water Resources Symposium, 2, pp. 281–287.

Grasbury, J. (2004) *The Willunga Basin Pipeline.* Irrigation Australia 2004 (Conference proceedings).

Hafi, A. (2003) *Conjuctive Use of Groundwater and Surface Water in the Burdekin Delta Area.* The Economic Record, June 2003, 79 (special issue), pp. 52–62(11). Blackwell, Oxford.

Herczeg, A.L. and Leaney, F.W. (2002) *Groundwater Flow Systems and Recharge in the mcLaren Vale Prescribed Wells Area.* Report to Onkaparinga Catchment Water Management Board. CSIRO technical report No. 10/02. CSIRO Land and Water, Canberra, Australia.

Institute for Rural Futures (2003) *The Impact of the Namoi Groundwater Sharing Plan. A Socio-economic Analysis of the Impact of the Reduction in Groundwater Allocations in the Namoi Valley.* University of New England, Armidale, NSW, Australia.

Ivkovic, K.M., Letcher, R.A. and Croke, B.F.W. (2004) *Groundwater-river Interactions in the Namoi Catchment,* NSW and their implications for water allocation. 9th Murray–Darling Basin Worskhop.

Johnson, R.C. (2004) *Spatial Analysis of the Groundwater Resources for the Lower Namoi Valley*. The Regional Insitute, NSW, Australia.

Letcher, R.A. and Jakeman, A.J. (2002) *Experiences in an Integrated Assessment of Water Allocation Issues in the Namoi River Catchment, Australia*. In: iEMSs 2002. The International Environmental Modelling and Software Society, Lugano, Switzerland. pp. 85–90.

McKay, J. (2002) *Encountering the South Australian Landscape: Early European Misconceptions and Our Rresent Water Problems,* Hawke Institute working papers series No. 21, Hawke Institute, University of South Australia.

MDBC Groundwater Working Group (1996) *Groundwater Development Potential in the Murray Basin*. Technical Report 1. Murray–Darling Basin Commission, Canberra, Australia.

MDBC Groundwater Working Group (1998) *MDB Groundwater: A Resource for the Future*. Murray–Darling Basin Commission, Canberra, Australia.

MDBC (Murray–Darling Basin Commission) (1999) *Water Audit Monitoring Report 1997/98*. Murray–Darling Basin Commission, Canberra, Australia.

MDBC (Murray–Darling Basin Commission) (2000–2005) *Water Audit Monitoring Report* (series), Murray–Darling Basin Commission, Canberra, Australia.

MDBC (Murray–Darling Basin Commission) (2003) *Projections of Groundwater Extraction Rates and Implications for Future Demand and Competitions for Surface Water*. (Author Sinclair Knight Merz.)

MDBCMC (Murray–Darling Basin Ministerial Council) (1992) *Murray–Darling Basin Agreement*. Murray–Darling Basin Ministerial Council, Canberra, Australia.

MDBCMC (Murray–Darling Basin Ministerial Council) (1996) *Setting of the Cap*. Murray–Darling Basin Commission, Canberra, Australia.

MDBCMC (Murray–Darling Basin Ministerial Council) (2000) *Review of the Operation of the Cap*. Murray–Darling Basin Commission, Canberra, Australia.

MLWC (Ministry for Land and Water Conservation), NSW (2003) *Water Sharing Plan for the Upper and Lower Namoi Groundwater Sources, 2003 Order*. Department of Land and Water Conservation, Sydney, Australia.

Nancarrow, B.E., Syme, G.J. and McCreddin, J.A. (1998a) *The Development of Stakeholder-Based Principles for Defining Environmental Flows in Modified Rivers*. Australian Research Centre for Water in Society, CSIRO Land and Water. Consultancy report No. 98-25. Canberra, Australia.

Nancarrow, B.E., McCreddin, J.A. and Syme, G.J. (1998b) *Developing Fair Processes for the Re-allocation of Groundwater for Long Term Sustainability in the Namoi Valley*. Australian Research Centre for Water in Society, CSIRO Land and Water. Consultancy report No. 98-40. Canberra, Australia.

NGC (National Groundwater Committee) (2004) *Knowledge Gaps for Groundwater Reform*. Proceedings from workshop 12–13 November 2003, Canberra, Australia.

NLWRA (National Land and Water Resources Audit) (2001) *Australian Water Resources Assessment: Surface Water and Groundwater –Availability and Quality, 2000*. Land and Water Australia, on behalf of the Commonwealth of Australia. ACT.

National Land and Water Resources Audit (2002) *Australian Catchment, River and Estuary Assessment 2002*. Australian Government, Canberra, Australia.

NRMMC (2002a) *Groundwater Quality Protection*. Commonwealth of Australia.

NRMMC (2002b) *Groundwater Trading*. Commonwealth of Australia.

NRMMC (2002c) *Managing Overallocated Groundwater Systems*. Commonwealth of Australia.

Panta, K.P., McMahon, T.A., Turral, H., Malano, H.M., Malcolm, W. and Lightfoot, C. (1999) *Water Allocation Strategies for the Lachlan River Valley*. Centre for Environmental Applied Hydrology, University of Melbourne, Australia.

Raju, K.V. and Brewer, J.D. (2000) *Conjunctive Water Management in Bihar*. Working

paper series of the Institute for Social and Economic Change, Bangalore, India.
Sinclair Knight Merz (2003) *Projections of Groundwater Extraction Rates and Implications for Future Demand and Competition for Surface Water*. Murray–Darling Basin Commission, Canberra, Australia.
Tisdell, J. (2002) *English and Dutch Water Auction Experiments*. In: Catchword. Newsletter of the Cooperative Research Centre for Catchment Hydrology. No. 104.
Tisdell, J., Ward, J. and Grudzinski, T. (2002) *The Development of Water Reform in Australia*. CRC for Catchment Hydrology, Technical report 02/5.
Turral, H.N. (1998) *HYDRO-LOGIC? – Reform in Water Resources Management in Developed Countries with Major Agricultural Water Use: Lessons for Developing Countries*. ODI research study. Overseas Development Institute, London.
Turral, H.N., Etchells, T., Malano, H.M.M., Wijedasa, H.A., Taylor, P., McMahon, T.A.M. and Austin, N. (2005) Water trading at the margin: the evolution of water markets in the Murray–Darling Basin, *Water Resources Research*, 41(7).
Van Bueren, M. and Hatton MacDonald, D. (2004) *Addressing Water-Related Externalities: Issues for Consideration*. Paper presented at the Australian agricultural and resource economics society water policy workshop, 10 February, Melbourne, Australia.
Wilkinson, R. and Barr, N.F. (1993) *Community Involvement in Catchment Management: An Valuation of Community Planning and Consultation in the Victorian Salinity Program*. Victorian Department of Agriculture, Melbourne, Australia.
World Commission of Environment and Development (1987) *Our Common Future (The Brundtland Report)*, Oxford University Press, Oxford. Available at: www.brundtlandnet.com.

16 Sharing Groundwater Information, Knowledge and Experience on a Worldwide Scale

Jac A.M. van der Gun

International Groundwater Resources Assessment Centre (IGRAC), PO Box 80015, 3508 TA Utrecht, The Netherlands

Importance of Sharing Groundwater-Related Information, Knowledge and Experience on a Worldwide Scale

Rationale behind sharing information

Sharing information is at least as old as mankind, maybe even as old as primitive life on earth. Parents bring up their children by teaching them all they consider necessary or useful for their proper development into independent and happy human beings. When children grow up, the initially predominantly unidirectional flow of information gradually changes into a balanced bidirectional process. However, sharing information is not restricted to the parent–child relationship, but can be observed between all people who have something in common: partners in marriage, relatives, friends, neighbours, members of the same community or a nation, colleagues, business partners, etc.

Why do we share knowledge and experience? By instinct, we know that knowledge and experience are important for survival and for feeding into the learning processes that help making us more successful in daily life. Sharing knowledge and experience has the potential to accelerate these learning processes significantly. Sharing information simply improves efficiency, stimulates development and reduces the probability of making wrong decisions. It is evident that good relations between people stimulate their preparedness for sharing information.

Importance of a knowledge base related to groundwater

It is commonly agreed that major water issues of our time include meeting basic human needs for water, securing food supply, protecting ecosystems and avoiding loss of life and other damages due to flooding and droughts. Over time, it has been recognized that appropriate action on these major issues is impossible without having access to sufficient and relevant information on the systems considered. As stated by the South African Minister of Water Resources Ronny Kasrils (2003): 'No sustainable development of a scarce natural resource is possible without understanding the resource and managing it wisely according to this growing understanding.' Therefore, the knowledge base on our water resources has to be continuously developed and updated.

Where a knowledge base is needed on water resources in general, it is even more indispensable in relation to groundwater. Groundwater is an invisible resource: it is hidden underground and after appearing at the surface it is formally not groundwater any more. This makes groundwater a component of the water cycle that is comparatively difficult to understand for most people. Unlike rain and surface water, groundwater is veiled in mystery. Only few people have a conceptually correct idea of aspects such as: how groundwater in their region is stored underground; how voluminous it is; how it moves and at what speed; what is its quality and how this quality may change; how groundwater is linked with surface water, local ecosystems and the environment; how to develop the groundwater resources efficiently and protect them against pollution and other problems. Even more difficult is it to imagine how the state and functions of a groundwater system may respond over time to intensified rates of groundwater exploitation and other changing boundary conditions.

It goes without saying that sufficient information and knowledge is needed if we want to make optimum use of such a natural resource and if we want to manage and protect it properly. However, even the nature of the information and knowledge needed varies widely. On the one hand, area-specific (i.e. georeferenced) information is required to:

- unveil at least some of the mysteries of this hidden resource in order to make groundwater systems under standable to water resources planners and decision makers;
- contribute to the proper identification of groundwater-related potentials and problems;
- facilitate the prediction over time of the groundwater system evolution in response to changing natural and anthropogenic boundary conditions.

On the other hand, professionals engaged in groundwater cannot fulfil their jobs adequately without also having access to some more generic information on groundwater, especially on:

- relevant scientific and technological principles, such as geology, hydraulics, drilling, pumping, hydrology, groundwater quality, eco-hydrology, water economics, water law and behaviour sciences;
- methods and technology for assessment, development and management of groundwater.

Consequently, knowledge bases that combine both local information with broader principles related to groundwater are required for guiding towards proper development and management of groundwater resources.

Added value of taking a global or regional perspective

It is evident that the success of groundwater development and management depends to a large extent on local efforts in collecting relevant information and on the local expertise available. If the local dimension is so important, what are the potential benefits of taking a global or regional perspective, by sharing knowledge and experience across and far beyond national boundaries?

In this context it is useful to make a distinction between generic knowledge and area-specific information. Generic knowledge is in principle universally applicable; hence sharing this knowledge on a worldwide scale produces efficiency, as pointed out at the beginning of this section. Important actors involved as well as programmes and activities they have embarked upon for globally disseminating this generic knowledge are reviewed in the following sections.

The value of sharing area-specific information internationally may seem less obvious at first glance. However, it becomes more evident in the context of specific challenges:

- *Managing transboundary groundwater resources*: This is an emerging issue. Many countries share aquifers with neighbouring countries and there is a growing need to jointly manage them (or at least coordinate management efforts) because of steadily increasing pressures such as scarcity, pollution and environmental impacts. International exchange of information on shared groundwater systems is among the first and most indispensable steps to put the process towards transboundary aquifer management into motion.
- *Understanding global or regional patterns and processes*: This is obviously of first importance for those studying groundwater phenomena at a global, continental or other supranational scale (e.g. world water balance, world climatic processes, occurrence of different types of water-related problems). However, it is useful as well to those focusing on spatially more restricted areas, by providing an overall context and reference, and by making them more aware of patterns and how these are produced.
- *Recognizing potentials, problems and trends related to groundwater*: For groundwater investigators and planners alike, analogies between different areas of the globe may be of great help in the preliminary estimation of groundwater potentials and in the early diagnosis of trends and problems. Global patterns of relevant variables may reveal similarities that provide a basis for tentative diagnosis. The basic principle is that information collected for more intensely investigated, monitored or pressured areas may give hints on likely conditions or trends for analogous areas that have been studied less and/or have been exposed thus far to a lower external pressure.
- *Benefiting from experiences gained under similar or analogous conditions*: This is related to the previous comment. Observed similarities or analogies in groundwater conditions are a support in developing ideas on appropriate action for groundwater development, use and management. Knowing the actions implemented elsewhere under similar or analogous conditions enriches the overview of possible actions to be considered. Knowing

whether (and why) they have been a success or a failure helps in making optimal decisions and avoiding less effective measures.
- *Contributing to standardization of variables, methods and observational practices*: It is clear that the potential benefits outlined above, but also the validity of research outcomes based on data from different areas, depend on the consistency between all data-sets used. Lack of standardization in definitions, observational practices and methods for processing may lead to serious errors in interpretation. Sharing area-specific information internationally will undoubtedly contribute to international standardization, which will raise the quality of any analysis based on international data-sets.

Important International Actors and Programmes

Bilateral and multilateral cooperation projects related to groundwater

Especially since the 1960s, numerous international projects have been carried out in the context of development cooperation or cooperation between 'befriended nations'. The general idea behind these projects is that the development of a country or a part of it may be accelerated by international cooperation, either in a bilateral (country-to-country) or in a multilateral setting (country and international organization). The projects are operating on the basis of a mix of national and foreign or international inputs, with a formal project agreement as a certain guarantee for having these inputs available when they are needed. Inputs from the donor countries or international organizations tend to include financial support, vehicles, equipment and other materials, as well as personnel for supplying professional capacity and/or transfer of knowledge and technology.

A substantial part of all these development cooperation projects was and still is related to water resources, and many of these focus on groundwater. Since the 1960s, several donors and recipient countries have become aware that available groundwater resources were being underused in many areas, due to limited knowledge on these subsurface resources or due to insufficient access to the technology or the funds needed to exploit them. Unlocking this natural resource to its full extent was therefore seen as a promising strategy to improve economic and social development. This idea has triggered a large number of groundwater projects all over the world. Many projects focused on regional exploration and assessment of groundwater resources to provide a basis for implementing groundwater development initiatives. Even more projects have been given the explicit objectives to drill large numbers of wells and to assist the local population in making efficient use of the tapped groundwater resources. The impacts of these projects are enormous. Many areas in the world where groundwater abstraction used to be insignificant have changed into areas of intensive groundwater exploitation within a few tens of years, yielding immense social gains by securing water supplies and providing water for economic activities such as irrigated farming. The aquifers of these areas are tapped now by large numbers of wells, many of them much deeper than the traditional ones.

Important multilateral donor agencies involved from the onset in these international groundwater-related projects are the United Nations Development Programme (UNDP), the United Nations Office for Technical Co-operation (UNOTC, later renamed to UN/DTCD and successively UN/DESA), and other organizations within the UN system such as the Food and Agricultural Organization (FAO), the World Health Organization (WHO) and the World Bank (IBRD). On a more regional level, the European Union (EU; formerly the European Economic Community or EEC) and the Organization of American States (OAS) should be mentioned. Bilateral donors with respect to the above-mentioned groundwater-related projects are also numerous. They include donor agencies such as the British ODA (now DFID), the French ORSTOM, the German GTZ, the Japanese JICA, the American USAID, the Dutch DGIS, the Danish DANIDA, the Swedish SIDA, the Norwegian NORAD, the former USSR government and many others. Examples of this category of international cooperation projects on groundwater are presented in Boxes 16.1 and 16.2.

While groundwater exploitation quickly intensified in many parts of the world and produced enormous benefits, problems related to groundwater

Box 16.1. Groundwater Resources Development in the Altiplano, Bolivia, 1969–1973. (From Naciones Unidas, 1973.)

- Cooperation between government of Bolivia and UNDP;
- UN-OTC support (10 foreign experts) having the lead in all operations;
- Focus on regional development of the Altiplano by exploration, exploitation and use of groundwater;
- Approximately 120 wells drilled and provided with motorized pumps;
- Demonstration of the use of groundwater for irrigation;
- Only limited attention for institutional development and transfer of knowledge;
- Main project output: wells, pumps and increased irrigated area.

Box 16.2. Water Resources Assessment Programme Yemen (WRAY), 1982–1995. (From Negenman, 1995.)

- Cooperation between governments of Yemen and the Netherlands (DGIS);
- Technical support by TNO experts, limited in number (2–4 residents, on average) and in an advisory role;
- Focus on institutional development and transfer of knowledge;
- Part of the institutional development was the development of a structural national water resources assessment programme as a basis for water resources development and management;
- Main output: a national organization competent regarding water resources matters and provided with advanced technical knowledge and tools.

gradually surfaced. Some of these were a direct consequence of intensified groundwater development (groundwater depletion, activated sea water intrusion, modified ecosystems, land subsidence, etc.); others – such as groundwater pollution – were largely caused by external factors that have become more prominent recently. Growing awareness on these problems and their implications for sustainable development has reshaped the groundwater-related international cooperation programmes. The focus has been gradually shifted from groundwater development to groundwater resources management. The related projects now often address groundwater in an integrated water resources management (IWRM) context and/or they implement measures for protecting and augmenting the groundwater resources. This shift in approach has been adopted by virtually all recipient countries and donor organizations mentioned above.

Early projects were generally very much focused on physical outputs and often paid insufficient attention to the national counterparts involved. Transfer of knowledge was often unintentional and occurred as a 'spin-off'. Over time, awareness has grown on how crucial the national institutions and their staff are for sustainable effects of project efforts. This progressive awareness has modified the approach in most programmes and projects. Progressively, more attention is being paid to structural transfer of knowledge to local counterparts and to institutional development.

The objectives of the international groundwater projects as described above are not limited to giving support to agriculture. Many of the projects have been designed primarily for other purposes, such as rural domestic water supply, urban water supply or water resources management. Nevertheless, international cooperation projects have played an important role in the agricultural groundwater revolution. They have mobilized know-how from all over the world that facilitated tapping large quantities of groundwater for agricultural purposes, and the examples provided by the international projects have stimulated private initiative and investment in groundwater development enormously. The main challenge today is to make the fruits of this development sustainable, by properly managing and protecting the groundwater resources.

Global associations, organizations, programmes, projects and working groups on groundwater

Apart from the bilateral and multilateral projects and programmes already described, there are numerous entities that in one way or another contribute to worldwide exchange of information, knowledge and experience on groundwater. While there are too many to mention all, a number of the key international actors and programmes are reviewed here.

International Association of Hydrogeologists

The International Association of Hydrogeologists (IAH) is a professional association that was founded in 1956 at the International Geological Congress in Mexico City, after lengthy discussions dating from as early as 1948 (Day, 1992; IAH, 1994, 2003). The first IAH congress took place in Paris in 1957.

The initial aims of the association were getting to know each other, sharing professional expertise and furthering hydrogeological science. In IAH's annual report of 2003 the objectives are formulated as follows:

> 'to advance public education and promote research (and disseminate the useful results of such research) in the study and knowledge of hydrogeological science. The Association seeks to achieve these objectives by:
>
> - Publishing journals, book series, newsletters and other occasional publications in both hard copy and electronic format for the benefit of members and the wider community interested in the objectives of the Association.
> - Promoting international cooperation among hydrogeologists and others with an interest in groundwater through commissions, working groups and joint projects.
> - Encouraging the worldwide application of hydrogeological skills through education and technology transfer programmes, and sponsoring international, regional and national meetings open to all.
> - Cooperating with national and international scientific organizations, to promote understanding of groundwater in the international management of water resources and the environment.'

IAH has been affiliated with the International Union of Geological Sciences since 1964.

IAH is a very active and influential worldwide professional association, governed by the IAH Council (11 members) and a General Assembly. By the end of 2003, the association counted more than 3700 members in more than 130 countries, scattered over all continents. There is a special fund (Burdon Fund) for sponsoring members in developing countries. Commissions (formerly Working Groups), National Chapters (in 40 countries) and regional groups or committees facilitate the efficient implementation of activities.

The main activities of the IAH include: (i) organizing or coorganizing international meetings and sponsoring national meetings; (ii) publishing a journal, newsletter and publications; (iii) organizing and participating in international projects on hydrogeological subjects. IAH's website address is: www.iah.org.

Foremost among the international meetings and conferences organized by IAH are the IAH congresses, which are the most important international meetings for hydrogeologists. The October 2004 congress in Zacatacas, Mexico, was the 33th IAH Congress since the association was established; thus the average frequency is about one in 18 months.

IAH's scientific journal started in 1992, initially under the name *Applied Hydrogeology*, but at the beginning of 1995 it was renamed *Hydrogeology Journal*. It appears bimonthly and favours papers with an applied hydrogeological and/or area-specific focus.

To date, IAH has produced two series of publications: *International contributions to Hydrogeology* and *Hydrogeology: Selected Papers*. An important early co-production (IASH, IAH, UNESCO/IHD, IGS) was the *International Legend for Hydrogeological Maps* (1970, in four languages).

IAH currently has 12 commissions (permanent technical groups) on the following subjects: Hydrogeological Maps (the oldest one, established in 1959);

Hydrogeology of Karst; Mineral and Thermal Waters; Groundwater Protection; Hydrogeology in Developing Nations (Burdon Commission); Education and Training; Hydrogeology in Urban Areas; Transboundary Aquifers; Hydrogeology of Hard Rocks; Managing Aquifer Recharge; Groundwater Dependent Eco-Systems; Aquifer Dynamics; and Coastal Zone Management. In addition, a Working Group on Groundwater and Climate Change is in formation.

The association closely cooperates with the United Nations Educational, Scientific and Cultural Organization (UNESCO), for example, in the project to develop the *Hydrogeological Map of Europe* '1:1.5 M' (30 sheets; started in 1965 and now nearly completed), in the WHYMAP project (Groundwater Resources of the World) and in the ISARM initiative (Internationally Shared Aquifer Resources Management). More information on these projects will be provided later.

International Association of Hydrological Sciences

The International Association of Hydrological Sciences (IAHS), an association involving all components of the hydrological cycle, was founded in 1922 in Rome as the International Section of Hydrology of the International Union of Geodesy and Geophysics (IUGG). It is now one of the seven autonomous bodies that together constitute the IUGG.

IAHS is the oldest learning society in the field of water. It aims to serve the needs of humanity through the promotion of hydrological science and the stimulation of its applications. It has approximately 3700 members, from 129 countries. Websites of IAHS are: www.wlu.ca/~wwwiahs/index.html and www.cig.ensemp.fr/~iahs.

The activities of IAHS include organizing scientific meetings (assemblies, symposia and workshops), producing and disseminating publications and carrying out research projects. Among the publications are the *Hydrological Sciences Journal*, a Newsletter and the *Red Books* series started in 1924 in which almost 300 titles have appeared, several of them related to groundwater.

IAHS has nine scientific commissions, including the International Commission on Groundwater (ICGW). The latter is active in:

- organizing international symposia such as ModelCARE and Groundwater Quality (GQ);
- collaboration with the American Geophysical Union (AGU), the International Association of Hydrogeologists (IAH) and the International Groundwater Modeling Centre (IGWMC);
- organizing working groups, often under UNESCO's IHP.

The relative importance of IAHS with regard to groundwater has undoubtedly declined somewhat after IAH came into being and achieved a strong position among hydrogeologists.

Association of Geoscientists for International Development

The Association of Geoscientists for International Development (AGID) is a global association uniting more than 1400 geologists, hydrogeologists, engineers and other professionals involved in international development activities with an earth-scientific scope. Notwithstanding a number of very useful activities – such as the *Preliminary Bibliography on Groundwater in Developing Countries* (Stow *et al.*, 1976) – regarding groundwater, it cannot compete in importance and activity level with associations such as IAH and IAHS. AGID's website can be found at: http://agid.igc.usp.br.

United Nations Educational, Scientific and Cultural Organization

Through its Water Science Division, UNESCO has developed a strong profile in water since the International Water Decade (1965–1974). This decade was followed by the International Hydrological Programme, carried out in phases, with IHP-1 starting in 1975. IHP is UNESCO's intergovernmental multi-disciplinary scientific programme in hydrology and water resources; the programme of each phase reflects the needs and/or priorities of the member states and is formally approved by these states. IHP is currently in its sixth phase (IHP-6, 2002–2007, see Box 2.3).

UNESCO/IHP Headquarters in Paris is cooperating with IHP National Committees and with IHP regional and cluster offices (Apia, Brasilia, Montevideo, Kingston, Port au Prince, Venice, Moscow, Cairo, Nairobi, New Delhi, Tehran, Jakarta, etc.). Forging cooperation is the main mechanism for laying foundations. As of early 2003, the IHP network included 13 UNESCO centres and 11 UNESCO university chairs.

Groundwater-related objectives of UNESCO's Water Science Division are: development of scientific knowledge in hydrogeology (especially quantity, quality, salt water, arid hydrology) and related training. Research is promoted by working groups, supported centres, publications, participation in projects, etc. Training is promoted by supporting various international training centres (e.g. Barcelona groundwater course; UNESCO-IHE).

Over the years, UNESCO has produced a considerable quantity of training materials, manuals, guides, proceedings and publications on groundwater, has participated in the Hydrogeological Map of South America and is still involved in the preparation of the Hydrogeological Map of Europe. Recent important projects and programmes on groundwater are: World Groundwater Resources Map (WHYMAP), Internationally Shared Aquifer Resources Management (ISARM) and the Working Groups on Groundwater Indicators and on Non-renewable Groundwater Resources. Well-known series of publications are UNESCO's *Technical Papers in Hydrology, Studies and Reports in Hydrology*

Box 16.3. Themes of UNESCO's IHP-6 (2002–2007) 'Water interactions: systems at risk and social challenges'

1. Global changes and water resources;
2. Integrated Watershed and Aquifer Dynamics;
3. Land Habitat Hydrology;
4. Water and Society;
5. Water Education and Training.

Box 16.4. Recent issues in UNESCO's 'Series on Groundwater'

Produced in the framework of IHP (available in hard copy and on CD):

1. Internationally Shared (Transboundary) Aquifer Resources Management (2001);
2. Groundwater contamination inventory (2002);
3. Groundwater studies (2004);
4. Intensively exploited aquifers (2002);
5. Submarine groundwater discharge (2004);
6. Groundwater resources of the world and their use (2004);
7. Groundwater and fractured rocks (2003).

and *Technical Documents in Hydrology*. Some of the recent publications related to groundwater are listed in Box 16.4.

United Nations Food and Agricultural Organization

Although FAO (website: www.fao.org) has been involved in groundwater assessments since the 1960s, followed by groundwater modelling for regional assessment of Africa and by extensive regional groundwater programmes in the Near East and Africa during the 1970s and 1980s, FAO's role in generating and disseminating new knowledge on groundwater is less pronounced than that of UNESCO.

FAO's activities related to groundwater usually reflect an irrigation or drainage perspective. Nevertheless, the organization has contributed to developments in groundwater quality studies and in groundwater management. Furthermore, the long-standing expertise on legal aspects of water has resulted in active involvement in legal and institutional aspects of internationally shared groundwater management.

Many FAO publications are relevant for those interested in groundwater, in particular the publications of the series of FAO Technical Papers/Water Reports (25 reports over the period 1993 through 2003).

A very interesting service to the international community is *AQUASTAT*, FAO's web-based worldwide water database (www.fao.org/ag/AGL/aglw/Aquastatweb/Main/html/aquastat.htm). It presents key data on water variables at the country level, such as total internally renewable water resources and their breakdown into groundwater, surface water and 'overlap'; national breakdown of irrigated lands in groundwater-dependent irrigation and surface water–dependent irrigation; dependency on internally generated water resources; and per capita indicators.

United Nations Environmental Programme

The United Nations Environmental Programme (UNEP) covers a very broad field, of which groundwater represents only a minor part. Nevertheless, several of UNEP's activities are contributing to a better understanding and characterization of the world's groundwater resources. Examples are its recent global assessment of problems and options for groundwater management (UNEP, 2003), and in particular the periodically published *Global Environmental Outlook*. The last version of this *Global Environmental Outlook* – GEO-3 – includes a global and regional synopsis of groundwater conditions (UNEP, 2002). The associated Internet-based GEO-3 database (http://geocompendium.grid.unep.ch/index.htm) includes – among other things – useful groundwater information at a country level, for all countries of the world.

International Atomic Energy Agency

The Isotope Hydrology Section of the International Atomic Energy Agency (IAEA) (Vienna; website: www.iaea.org) develops activities on isotopes and related geochemistry in the water cycle. It does so by conducting research, training and workshops, and by supplying isotope laboratory services to those interested in the application of isotope techniques. Major ongoing projects of the section regarding groundwater are:

- mapping palaeo-waters of the world (non-renewable or fossil groundwater);
- Dead Sea (including study on the origin of salt groundwater);
- groundwater of eastern and southern Africa (main subjects: recharge, surface water–groundwater interrelations in conjunctive management, pollution transport).

The section is cooperating with relevant international organizations such as UNESCO and IAH by participating in their projects (e.g. WHYMAP, UNESCO Working Groups on Non-renewable Groundwater and on Groundwater Indicators) or in the form of joint programmes (e.g. with UNESCO in the Joint International Isotopes in Hydrology Programme). A biannual training course on isotopes in hydrology is presented in Graz, Austria. In addition, IAEA is organizing many short courses in different parts of the world.

The World Bank

The World Bank (IBRD) finances development programmes and projects in many countries of the world. Within its programmes, the role of groundwater is most pronounced in the water and sanitation sector and the irrigation sector programmes. Apart from financing and taking care of the technical work involved in the project cycle, the World Bank is also contributing to the international exchange of experience in groundwater. This is done in particular by studies on the impact of groundwater-related projects on human welfare and by dissemination of publications on 'good practice' or 'lessons learned'. A rather technical example of the latter is the publication *Community Water Supply: The Handpump Option* (Arlosoroff *et al.*, 1987), which presents results of testing and monitoring 2700 pumps of 70 different models in 17 countries. A recent World Bank activity focusing on the promotion of 'good practice' in groundwater is the Groundwater Management Advisory Team (GW·MATE) programme.

GW·MATE was designed to support the thrust 'from vision to action' of the World Water Forum of March 2000 and has the overall objective to give worldwide support to groundwater resources management and protection. It does so by supporting and strengthening the groundwater components of Bank-financed projects and Global Water Partnership (GWP) actions, by harvesting global experience and by disseminating 'best-practice elements' internationally. GW·MATE issues a *Briefing Note Series* with the following characteristics:

- the notes are intended to give a concise introduction to the theory and practice of groundwater resources management and protection, in a convenient and readily accessible format.
- primary target audiences include water resources managers of limited groundwater experience and groundwater specialists with limited exposure to water resources management.

So far, 13 titles are available in the Briefing Note Series: 9 on Core Series Topics, 4 on Supplementary Series Topics. Contribuitions to guides and books related to groundwater management are another output of GW · MATE.

GW·MATE is a component of the Bank-Netherlands Water Partnership Programme (BNWPP), using trust funds from the Dutch and British governments. More information can be found on the website at: www.worldbank.org.gwmate.

World Meteorological Organization

The World Meteorological Organization (WMO) (http://www.wmo.ch/index-en.html) has a long-standing activity record in the promotion of hydrological/meteorological monitoring and the analysis of observed meteorological and hydrological variations in time. To this end, networks of national organizations collaborating by exchanging information are operational for approximately 50 years. Although the organization's activities mainly focus on meteorology and on surface water, attention to groundwater is increasing and is being supported, inter alia, by the WMO Resolution 25 ('Exchanging hydrological data and information', Cg-XIII, 1999) and by the World Hydrological Cycle Observation System (WHYCOS), a decentralized global programme intended to improve regional monitoring networks.

WMO's famous *Guide to Hydrometeorological Practices* (2nd edition, 1970; 1st edition, 1969), renamed *Guide to Hydrological Practices* in later editions, pays only limited attention to groundwater (groundwater levels only), but has been, and still is, of great interest to many groundwater specialists all over the world.

World Water Assessment Programme

Many international conferences emphasize that water is at the heart of sustainable development, trigger debates on a global water crisis and call for immediate action. However, in spite of the many valuable water resources assessments in the past, until recently there has been no global system in place to produce a systematic, continuing and comprehensive global picture of water and its management. The World Water Assessment Programme (WWAP), a joint initiative of the water-related agencies under the UN, was established to fill this gap. The Internet address www.unesco.org/water/wwap provides information on the programme.

Certainly the most visible output of WWAP is the World Water Development Report (WWDR). According to the preface of its first report (WWDR-1), the report is;

> 'designed to give an authoritative picture of the state of the world's freshwater resources and our stewardship of them. The WWDR builds upon past assessments and will constitute a continuous series of assessments in the future. The WWDR is targeted to all those involved in the formulation and implementation of water-related policies and investments, and aims to influence strategies and practices at the local, national and international levels.'
>
> (WWAP, 2003)

After the presentation of WWDR-1 in 2003 at the Third World Water Forum in Japan, a second report was prepared for presentation at the Fourth World Water Forum in Mexico in March 2006. Numerous water-related agencies and specialists from all over the world contribute to this report. Groundwater is focused upon in the chapter on 'Water Resources', but the space allocated by WWAP for this subject is very limited.

Global Water Partnership

The Global Water Partnership (GWP), created in 1996 and based in Sweden, is a working partnership among all those involved in water management: government agencies, public institutions, private companies, professional organizations, multilateral development agencies and others committed to the Dublin–Rio principles. The mission of GWP is 'to support countries in the sustainable management of their water resources'.

One of the products of GWP is its 'ToolBox', a comprehensive set of guidelines for IWRM. The tools are organized in three groups:

- *Enabling environment*: policies, legislative framework, financing and incentive structures;
- *Institutions*: creating an organizational framework and building institutional capacity;
- *Management tools*: water resources assessment, planning for IWRM, efficiency in water use, social change instruments, conflict resolution and regulatory instruments.

GWP and its ToolBox do not (yet) have a pronounced focus on groundwater. GWP's website can be accessed at: www.gwpforum.org.

International Water Management Institute

The International Water Management Institute (IWMI), with its headquarters in Sri Lanka, is active in applied research and capacity building. It deals with issues related to water management and food security: water for agriculture; groundwater; poverty; rural development; policy and institutions; health and environment.

IWMI's groundwater research to date has focused on groundwater use and management in irrigation, primarily in south and south-east Asia. At present – and especially with the Comprehensive Assessment in Agriculture, of which this paper is a part – a more comprehensive approach to research and synthesis of knowledge within groundwater management is being developed on a wider scale.

Some of the information to be shared can be found at the website: www.cgiar.org/iwmi/. An interesting global product prepared by IWMI is the World Water and Climate Atlas, based on weather data collected over the period 1961–1990.

International Groundwater Resources Assessment Centre

The International Groundwater Resources Assessment Centre (IGRAC) was founded in early 2003 in Utrecht (the Netherlands) with funding from the government of the Netherlands made available after UNESCO and WMO formulated an initiative for such a centre. The need for IGRAC was motivated by the following main factors:

- Generally perceived *poor access to geo-referenced information on groundwater at a global scale*: Even if the information does exist, it is often so difficult to access that most of it in practice fails to contribute to the analysis and planning of groundwater at the global, regional and even national levels.
- *Inadequacy of groundwater data acquisition* in many countries: Many groundwater systems have not been explored and assessed sufficiently, while variations in time of the groundwater conditions – essential information for adequate management – are monitored only exceptionally.

There is no doubt that a global groundwater centre can make an important contribution to filling these gaps. Therefore, it has the potential to significantly reduce the currently widespread inefficiencies in groundwater-related activities and to help countries in defining their priorities regarding groundwater.

Under the general objective of contributing to adequate development and management of the world's groundwater resources, in conjunction with surface water resources, the fundamental objectives adopted by IGRAC are:

- enhancing worldwide knowledge on groundwater, by promoting related data and experiences to be shared and by making this information widely available on the basis of centralized retrieval services and targeted dissemination;
- contributing to the acquisition of more and better groundwater data, by means of guidelines and protocols for groundwater assessment and monitoring.

IGRAC pursues these objectives by the development of a Global Groundwater Information System (GGIS), by activities related to Guidelines and Protocols for groundwater data acquisition (G&P) and by participation in strategic global and regional projects with a strong groundwater component (WHYMAP, WWAP, UNESCO working groups, ISARM, etc.). Cooperation with UNESCO, WMO, IAH, national groundwater organizations and many other partners all over the world is among the key mechanisms to achieve IGRAC's goals.

The ambition of IGRAC is to become a central international platform for groundwater information, where parties from all parts of the world share their information, knowledge and experience on groundwater. Emphasis is on geo-referenced information. As such, it fits into the family of global centres collecting geo-referenced information on components of the water cycle, such as the Global Precipitation and Climate Centre (GPCC) at Offenbach, Germany, the Global Runoff Data Center (GRDC) at Koblenz, Germany, and the Global Environmental Monitoring System on Water (GEMS/Water) at Burlington, Canada. Most of IGRAC's products can be found on the website at: www.igrac.nl.

International Groundwater Modeling Centre
IGWMC was established in 1978 at the Butler University, Indianapolis, USA, and had a branch office in Delft, the Netherlands, during the 1980s. IGWMC originally had the profile of a clearing-house for groundwater modelling software, and presented groundwater-modelling courses as a secondary activity. The centre now operates from Colorado School of Mines in Golden, Colorado, USA. It sells software and it provides advisory services and training related to groundwater modelling. Its website address is: www.mines.edu/igwmc/.

Regional institutions, programmes and networks on groundwater

Regional institutions, programmes and networks on groundwater are numerous. It is virtually impossible to give a complete overview. Therefore, only a few of them (probably the ones most visible to the global community) are briefly mentioned here, region by region.

Europe
Interesting from the groundwater point of view are – among others – the European Union (EU), the United Nations Economic Commission for Europe (UN/ECE), ISARM-Balkans and the Association of European Geological Surveys (EGS).

The EU is important for groundwater for at least two reasons: (i) the Water Framework Programmes for financing innovative research on Europe's water resources; and (ii) the EU Framework Directive on Water, aiming to establish a framework for the protection of Europe's water systems (including groundwater), which forces all states of the EU to assess their water resources properly and to upgrade their water-monitoring systems according to a common standard.

In recent years, the UN/ECE Task Force on Monitoring and Assessment (now known as the UN/ECE Working Group on Monitoring and Assessment) has made very substantial efforts related to groundwater, which has resulted in Europe-wide reports on:

- inventory of transboundary groundwaters;
- problem-oriented approach and the use of indicators;
- application of models;
- state of the art on monitoring and assessment of groundwater.

ISARM-Balkans is a network in which representatives of the Balkan countries and Turkey are trying to forge the cooperation needed for transboundary aquifer management in their region.

North, Central and South America
A powerful regional organization for the Americas is the Organization of American States or Organización de Estados Americanos (AOS/OEA). In cooperation with the Montevideo Office of UNESCO/IHP, it has been, and still is, actively involved in many groundwater-related projects and efforts such as the Hydrogeological Map of South America and ISARM of the Americas.

As a regional equivalent of the global IAH, the Asociación Latinoamericana de Hidrología Subterránea para el Deasarrollo (ALHSUD) has been established as the professional association of hydrogeologists of Latin America.

Middle East and Northern Africa

For this region, several regional programmes on groundwater are carried out by the United Nations Economic and Social Commission for Western Asia (UN/ESCWA, Beirut), the Arab Center for the Study of Arid and Dry Lands (ACSAD, Damascus), l'Observatoire du Sahara et du Sahel (OSS, headquarters in Paris) and the Arab Networks on Groundwater, usually in cooperation with the Cairo office of UNESCO/IHP. More agriculturally oriented regional organizations with keen interest in water resources are the Centre for Environmental Development for the Arab Region and Europe (CEDARE, Cairo) and the International Centre for Research in the Dry Areas (ICARDA, Aleppo).

The OSS, with headquarters in Paris (UNESCO building), is working under the theme 'Fight against Desertification' (Earth Summit in Rio, Agenda 21) with its main objectives:

- consolidating and improving observational programmes (including equipment, standards and info systems);
- increasing knowledge on shared resources;
- optimizing local management of natural resources (e.g. by databases and exchange of experience and know-how);
- sustaining the fight against desertification in Africa.

Its membership includes five East African countries, nine West African countries, five North African countries, four European countries and several sub-regional, civil society and international organizations. Its website is: www.unesco.org/oss.

ACSAD's scope of activities includes water resources, soil science, plant studies, animal studies and economics. Its website is: www.acsad.org. A very relevant groundwater-related coproduction in this region is the Hydrogeological Map of the Arab Region and Adjacent Areas (UNESCO/ACSAD, 1988). Furthermore, attention is being paid in the region to shared aquifers and to non-renewable groundwater resources.

Central and Southern Africa

The most visible regional network is the Southern African Development Community (SADC). Among its many activities, it is carrying out groundwater-related activities in cooperation with the Gabarone office of UNESCO and with other multi- or bilateral parties. More information can be viewed on the website at: www.thewaterpage.com/sadcWSCU.htm.

Asia and the Pacific

An important organization in this region is the Coordinating Committee for Coastal and Offshore Geosciences Programmes in East and South-East Asia (CCOP), with its Technical Secretariat located at Bangkok, Thailand. It organizes courses, workshops and projects with a focus on exchanging knowledge and experience within the region. Its interests include integrated coastal zone management and groundwater resources. Member countries of CCOP are: Cambodia, China, Indonesia, Japan, Korea, Malaysia, Papua New Guinea, Philippines, Singapore, Thailand and Vietnam. CCOP's website address is: www.ccop.or.th.

Other regional organizations actively regarding groundwater in this region are the United Nations Economic and Social Commission for Asia and the Pacific (UNESCAP) and the Jakarta office of UNESCO.

Selected Themes

Structured interaction between groundwater professionals

This may be the oldest way of sharing information and experience on groundwater internationally. Since the time groundwater subjects were considered to have a scientific dimension, scientists have been communicating on their findings and ideas on groundwater. An example from the early days of groundwater science is the famous publication on groundwater flow by Dupuit (1863). Books, papers in journals with international circulation and international meetings related to groundwater have become important means to share knowledge and experience worldwide and to trigger international scientific debate. Recently emerged mechanisms and tools for structured professional interaction are international working groups, newsletters and electronic discussion or conferencing platforms.

The international interaction between groundwater professionals is now very well facilitated, in particular as far as knowledge on groundwater at a theoretical, methodological and generic level is concerned. There is ample opportunity to meet at the regular global IAH Congresses and IAHS Conferences, and at the regional and thematic conferences, workshops or symposia organized by national or regional organizations, often in cooperation with IAH, IAHS and/or water-related UN organizations. Proceedings of such meetings are a valuable support to reach a wider group of professionals – beyond those that attended the meetings – and they provide access to the papers for later consultation.

International commissions, working groups and research projects organized and/or funded by UNESCO/IHP, IAH, IAHS, EU and other foundations provide additional opportunities for structured interaction between groundwater professionals.

Several scientific publishers bring groundwater-related books on the market. In addition, there are important journals and publications series, contributing to international dissemination of groundwater knowledge. At the international level, of particular interest for groundwater are the following journals:

Hydrogeology Journal (IAH); Groundwater; Journal of Hydrology; Water Resources Research; Hydrological Sciences Journal (IAHS)

and publications series:

International Contributions to Hydrogeology (IAH); 'Red Books' (IAHS); UNESCO/IHP Series on Groundwater; FAO Technical Papers – Water Reports; etc.

Electronic versions of papers, reports and publications are becoming widely accessible on the Internet, among others via the websites of IAH, UNESCO/IHP, IGRAC, etc.

In conclusion, the exchange of theoretical, methodological and generic knowledge seems to be constrained more by the absorption capacity of the groundwater community than by a lack of structured means for sharing knowledge. Different professional backgrounds or specializations, different roles (scientists vs. practitioners) and lack of time constitute important bottlenecks in the effective exchange of knowledge.

Enhanced knowledge as a result of international interaction among groundwater professionals certainly had its impact on the agricultural groundwater revolution. It has disseminated scientific and technical knowledge on groundwater development and has made many water sector professionals more aware of the potentials offered by groundwater. The challenge now is to make the benefits of the revolution sustainable. No doubt structured international interaction between groundwater professionals may provide a valuable contribution to this endeavour by alerting them to side effects of increased groundwater exploitation and by facilitating joint development and enhancement of the knowledge and methodologies required for effective groundwater resources management.

Raising public awareness

The World Water Vision's slogan 'making water everybody's business' launched at the Second World Water Forum in 2000 (WWF2) correctly highlights the fact that everybody on the globe has a stake in water. Consequently, water is not a subject matter to be understood and handled by water specialists only, but rather a matter of concern to everybody. Politicians and other decision makers need to understand in general lines how to exploit and use the water resources properly, ensure sustainability, protect water quality and minimize negative impacts of exploitation. Water users need to know how to benefit optimally from water, both for domestic and productive uses. The general public, finally, needs to understand how individual behaviour – on a voluntary basis or enforced by regulations – contributes to conservation and protection of water resources and the related environment.

In general, public awareness on water is still rather limited. Water is often not yet sufficiently prominent on the national water agenda and in the national budgets, while decision makers lack vision to make proper decisions on water and the general public fails to adopt water-friendly behaviour. Do water professionals fail in raising public awareness on water? There is no doubt that significant efforts are being made already on this subject at local, national, regional and global levels. But, admittedly, these efforts do not have a very long history and the process of raising public awareness is rather time-consuming.

Milestones in global activities to raise awareness on fresh water are listed in Table 16.1. In spite of widespread criticism, it cannot be denied that the listed conferences have produced a great impact on the awareness of politicians, decision makers and professionals on water and related matters. Publicity around these international events is a mechanism to reach the general public and to raise its awareness on the main issues on water, but opinions differ on how effective this is and how public awareness may be further enhanced.

Apart from these large international conferences, international water professionals and their organizations are exploring and using other methods for raising

Table 16.1. Fresh water milestones over the period 1972–2003. (From Stockholm to Kyoto after UN, 2003: Freshwater Future – www.wateryear2003.org.)

Place and year	Event	Remarks and citations
Stockholm, 1972	UN Conference on the Human Environment	'We must shape our actions throughout the world with a more prudent care for their environmental consequences.'
Mar del Plata, 1977	UN Conference on Water	One of the recommendations of the Mar del Plata Action Plan is: assessment of water resources. ('. . . . . . relatively little importance has been attached to water resources systematic assessment. The processing and compilation of data have also been seriously neglected.')
New Delhi, 1990	Global Consultation on Safe Water and Sanitation for the 1990s	New Delhi statement: 'Some for all rather than more for some.'
Dublin, 1992	International Conference on Water and the Environment	Dublin statement on water and sustainable development (water is finite and vulnerable resource; participation needed; central role of women; water is an economic good).
Rio de Janeiro, 1992	UNECD Earth Summit	Rio Declaration ('establishing a new and equitable partnership' . . .) and Agenda 21 ('The holistic management of freshwater . . . and the integration of sectoral water plans and programmes within the framework of national economic and social policy are of paramount importance for action in the 1990s and beyond.')
Noordwijk, 1994	Ministerial Conference on Drinking Water Supply and Environmental Sanitation	Action programme assigns highest priority to basic sanitation.
Copenhagen, 1995	World Summit for Social Development	Copenhagen declaration ('Alleviate poverty by providing water supply and sanitation').
Istanbul, 1996	UN Conference on Human Settlements (Habitat II)	
Rome, 1996	World Food Summit	

Table 16.1. *Continued*

Place and year	Event	Remarks and citations
Marrakech, 1997	First World Water Forum	Marrakech declaration (access to water and sanitation is a basic need; shared water to be effectively managed; support and preserve ecosystems; encourage efficient water use).
The Hague, 2000	Second World Water Forum	World Water Vision: 'making water everybody's business'.
Bonn, 2001	International Conference on Freshwater	Recommendations for action: governance; mobilizing financial resources; capacity building and sharing knowledge.
Johannesburg, 2002	World Summit on Sustainable Development (Rio + 10)	Millennium Development Goals.
Kyoto, Osaka, Shiga, 2003	Third World Water Forum, Japan	First edition of the World Water Development Report (WWDR-I).

awareness on water. These methods include CDs, documentary movies and in particular publications. Two important and authoritative publications of this nature are the WWAP (2003) and the Global Environment Outlook (UNEP, 2002).

The WWDR is produced by the WWAP, in which 23 UN agencies concerned with fresh water cooperate, assisted by numerous water specialists from all over the world. The first edition was published in 2003, at the occasion of the Third World Water Forum in Japan. The second edition is currently in preparation and will be presented at the Fourth World Water Forum in Mexico, 2006. The WWDR focuses on factual information and attempts to present an up-to-date picture of the world's freshwater resources, their use and their management. The Global Environment Outlook, produced by UNEP, is a rather similar type of publication, but covering a much wider field ('environment'), of which water is only a limited part. Nevertheless, it contains valuable information on water in a geographical context.

Groundwater is even less familiar to politicians, decision makers and the general public than other components of the water cycle such as surface water and precipitation. This is probably so because of the invisibility of groundwater, the complexity of its occurrence and the limited efforts put in by groundwater specialists so far to bring groundwater under public attention. International organizations involved in groundwater are aware of this and make efforts to elevate groundwater in the general perception. The steady increase of attention paid to groundwater successively in the First, Second and Third World Water Forums illustrates these efforts.

Mapping and assessment

Mapping and assessment have an explicit geographic dimension. The information focused upon is primarily geo-referenced information, meant to document conditions at specific locations or in specific areas or zones.

Mapping

Groundwater mapping has its origins at the level of individual groundwater exploration projects. Maps have emerged as a tool to understand the spatial variation of observed groundwater parameters, to interpolate between observational locations, to forge a common understanding of the groundwater systems concerned and to preserve information in a format convenient for later activities. Over time, many countries and states have developed groundwater mapping programmes for their entire territory or for areas in which their most important groundwater resources are located. Little uniformity existed initially between these programmes, both on the parameters mapped and on the way information was presented. The desire to improve methodologies and to 'speak a common cartographic language', especially in cases of groundwater bodies crossing international boundaries, triggered international cooperation to develop a uniform internationally applicable methodology for hydrogeological mapping. As a result, IAH, UNESCO and FAO jointly produced 'A Legend for Hydrogeological Maps' (*Red Books Series* IASH, 1962), which was followed by a coproduction by IGS, IAH, IAHS and UNESCO/IHD entitled 'International Legend for Hydrogeological Maps' (1970, in four languages). In 1974, UNESCO published a specialized quadrilingual supplement to the 1970 legend entitled 'Legend for geohydrochemical maps'. The 1970 and 1974 legends have been adopted and further elaborated by Struckmeier and Margat (1995). A related mapping guideline on groundwater vulnerability has been published in the same series (Vrba and Zaporozec, 1994).

The hydrogeological mapping methodology and standard legend were tested extensively in the project 'International Hydrogeological Map of Europe'. At its start in 1960, this project was a forerunner among regional hydrogeological mapping projects and in scale is still the most detailed. Several other regional mapping projects followed, most of them using the international legend mentioned earlier. Table 16.2 gives an overview of existing regional, continental and global groundwater maps.

Figure 16.1 is a simplified version of the world map on groundwater resources produced by WHYMAP. The map was produced at a scale of 1:50 million, but a more detailed version at a scale of 1:25 million is in preparation. This map is meant for educational purposes and for raising awareness rather than for assisting hydrogeologists in their daily activities. The legend was derived from the international legend for groundwater maps, but with some modifications and simplifications.

Regional and global assessment

Hydrogeological maps present an overview of the hydrogeological conditions in the area concerned, but largely at an exploratory level. Although this is very useful, many tasks at various levels – from local to global – demand more quantitative information, particularly on the quantity and quality of the groundwater resources, on how they relate to other factors and on how they are changing over time. Groundwater resources assessment aims to produce such additional information, usually in the form of extensive and consistent data-sets.

Table 16.2. Existing global, continental and regional hydrogeological maps. (Modified after an unpublished internal document prepared by UNESCO.)

Name	Scale (one to *x* million)	Number of sheets	Year	Authors
Groundwater in North and West Africa	1:20	2	1988	UN/DTDC
Hydrogeological Map of Asia	1:8	6	1997	Ed. Jiao Shuqin *et al.*
Hydrogeological Map of Australia	1:5	1	1987	Lau, J.E., Commander, D.P. and G. Jacobson
Hydrogeological Map of the Arab Region and Adjacent Areas	1:5	2	1988	ACSAD and UNESCO
Hydrogeology of the Great Artesian Basin	1:2.5	1	1997	Australian Geological Survey Organization (Habermehl and Lau)
Hydrogeology of North America	1:13.333	2	1988/1989	Heath, R.C.
International Hydrogeological Map of Africa	1:5	6	1992	OAU/AOCRS (editor Safar-Zitoun, M.)
International Hydrogeological Map of Europe	1:1.5	28	25 sheets produced by 2003	BGR and IAH
Les Eaux Souterraines dans la Communeauté Européenne	1:5		unpublished	BRGM (Margat, J.)
Mapa hidrogeológico de America del Sur	1:5	2	1996	UNESCO/PHI and Government of Brazil
Middle East Hydrogeology	1:8	1	1998	Tübinger Atlas des Vorderen Orients
The National Atlas of the United States of America, Principal Aquifers	1:5	1	1998	Miller, J.A.
World Map of Hydrogeological Conditions and Groundwater Flow	1:10	6	1999	Dzamalov, R.G. and Zektser, I.S.
Groundwater Resources of the World[a] (predecessor of a similar map at scale 1:25 million)	1:50	1	2004	UNESCO, BGR, CGMW, IAEA and IAH (WHYMAP, 2004)

[a]The map depicts not only the type of aquifer conditions (hydraulic potential), but also the hydrological potential (recharge class). The 1:50 million version marks in addition the zones in which only saline groundwater occurs.

The role and usefulness of assessment at the level of an aquifer (or part of it) is generally understood and recognized. Hence, assessment is widely practised as a basic step in the development of balanced programmes for groundwater resources development and management. It underpins local actions and enables a better prediction of effectiveness and possible side effects.

Water resources assessment at the national, regional and global levels addresses other needs and often focuses more on policy than on 'action on the ground'. National governments and international organizations want to know the opportunities and problems offered by water in the near future. It allows them to define priorities in their policies and programmes, to allocate budgets and other means accordingly and to decide on which aspects of the political agenda efforts are needed to raise public support. It will be clear that data-sets on selected cases in certain areas – although extremely useful to demonstrate typical concepts and processes – are not enough to satisfy these needs. Full coverage is needed of the territory concerned: a nation, region or even the entire globe. Global water resources assessment is strongly boosted by the WWAP.

In response to the needs for water resources assessment on a global level, FAO has developed its *AQUASTAT* database on water and agriculture, and UNEP has organized the environmental database of its *GEO* data portal. Both global databases aim for consistent sets of data on a large number of attributes, defined at the level of countries. FAO's original data are based on enquiries circulated to representatives in the different countries; UNEP's data are being collected by the World Resources Institute (WRI).

The recently established *GGIS* database of IGRAC (see: www.igrac.nl), intending to bring together all information relevant for groundwater, draws heavily on data from these and other global or regional databases. *GGIS* not only builds a global database on groundwater-related attributes, but it also provides possibilities for online visualization. One of the *GGIS* views is 'country-oriented', like the *AQUASTAT* and *GEO* databases, and contains 77 standardized attributes. In addition to these administratively defined spatial boundaries, IGRAC has developed a system of Global Groundwater Regions, in order to organize data according to more physically based units. For this second view, 43 uniform attributes have been defined. The Global Groundwater Regions are depicted in Fig. 16.2. These regions form the highest level in a hierarchical system to delineate physical groundwater units on earth: global groundwater regions – groundwater provinces – aquifers. The global groundwater regions can be subdivided into a number of groundwater provinces; each province encompasses a number of aquifers, which in turn can be subdivided into aquifer beds, and so on. A map of South America's groundwater provinces is shown in Fig. 16.3 for illustrative purposes. A similar map for Australia has been prepared by the Australian Water Resources Council (2004). The added value of *GGIS* goes beyond bringing together data from different sources and visualizing them; it actively incorporates new data as well.

Programmes like WWAP with its WWDR and UNEP's Global Environment Outlook are heavily depending on data from global databases like *AQUASTAT*, *GEO* and *GGIS*, together with those of the GRDC, GEMS/Water, the GPCC and other global data centres. The quantity, quality and consistency of all these data are still limited. Large efforts are needed to upgrade the databases and the underlying data acquisition programmes.

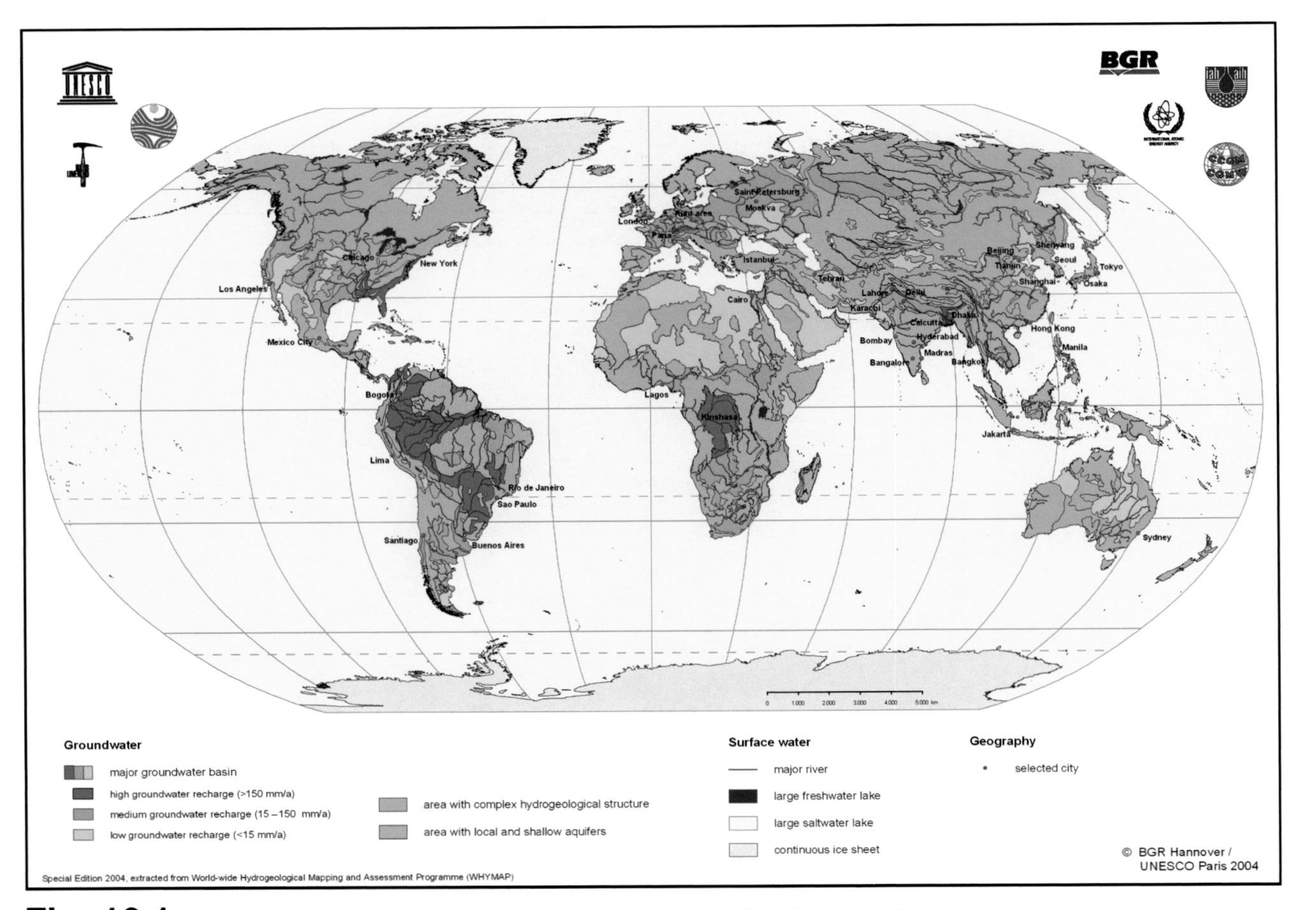

Fig. 16.1. Groundwater resources of the world. (From WHYMAP, 2004.)

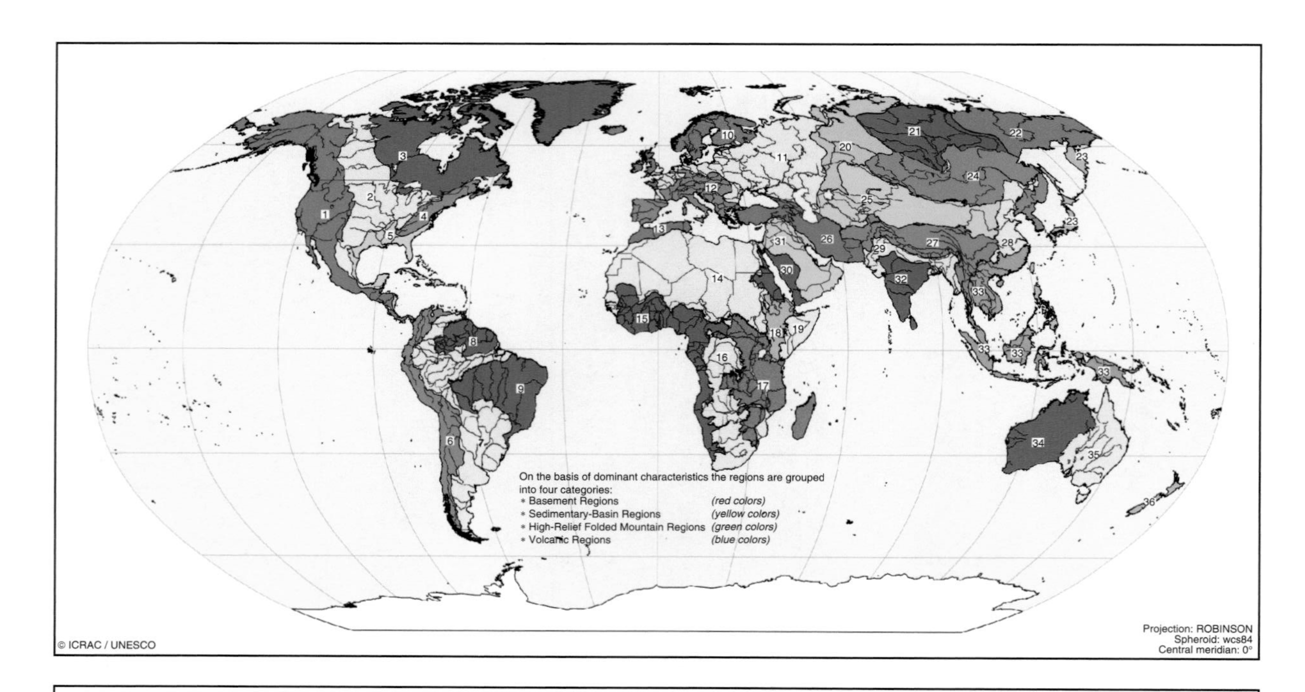

Fig. 16.2. Global groundwater regions. (From IGRAC, 2004.)

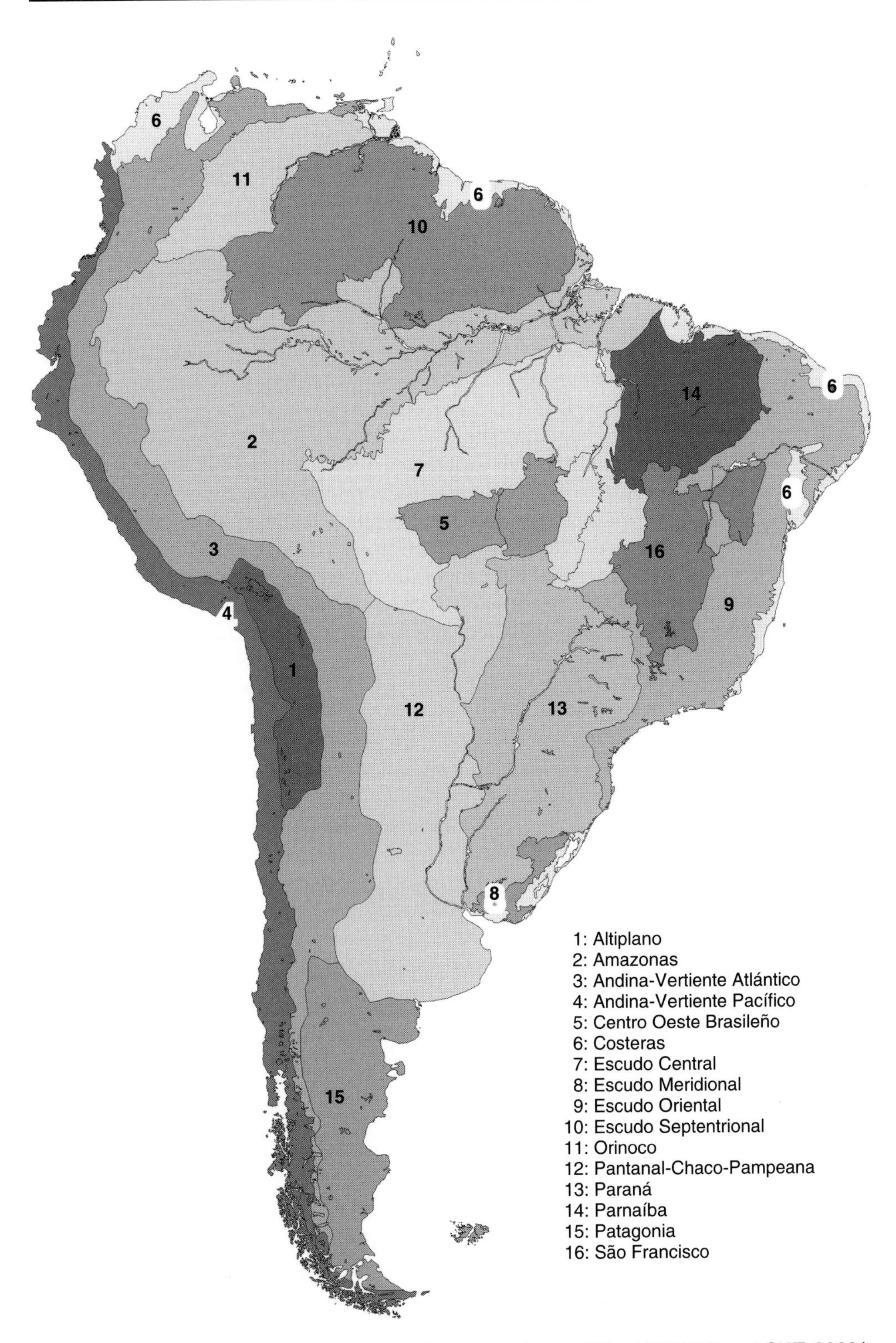

Fig. 16.3. A map of South America's groundwater provinces. (After UNESCO and CIAT, 2000.)

The WWDR and UNEP's Global Environment Outlook present groundwater information in a very broad interdisciplinary context. A recent UNESCO publication on the world's groundwater resources more specifically addresses the needs of groundwater specialists (Zektser and Everett, 2004).

Guidelines and 'lessons learned'

Groundwater assessment and monitoring

Guidelines may be of great benefit to those practically involved in groundwater exploration, groundwater resources assessment or groundwater monitoring. To carry out such activities efficiently and effectively, experience is at least as important as a professional background in groundwater. A guideline aims to absorb such experience and to disseminate it by suggesting some kind of 'best practice' under given circumstances. Guidelines speed up the learning process of those who have only limited experience, they may inspire more experienced colleagues to improve their daily practices and they may contribute to a larger degree of uniformity or even to standardization (WMO, 2001). Standardizing data acquisition is not only convenient; in some cases it is even essential for properly assessing the meaning and value of data-sets from different sources. Examples of generally implemented standardization tools are the laboratory protocols for the determination of water quality parameters and the field protocols for geophysical surveys and aquifer testing. Without a specification of the protocols used, the numerical results have only limited value.

An early guideline covering more or less the entire field of groundwater exploration, groundwater resources assessment or groundwater monitoring is UNESCO's publication *Groundwater Studies: An International Guide for Research and Practice* (1972). This guide, completely revised in 2004, is in reality very close to a textbook and is more useful as a general background than for providing step-by-step guidance on specific activities. The guidelines *Water Resources Assessment: Handbook for Review of National Capabilities* (WMO/UNESCO, 1997) pays ample attention to institutional capacities required for water resources assessment. Many other guidelines, however, focus on narrower, more specific subjects and usually offer more details. These guidelines have often been written to support activities inside a certain organization or country, but several of them may provide excellent support far beyond that region. A good example is the USGS series entitled *Techniques of Water Resources Investigations of the United States Geological Survey*, which currently consists of nine chapters with a total of more than 50 different topics, part of them on groundwater. Only a limited number of guidelines, primarily from IAH and UNESCO, have been developed explicitly for worldwide dissemination. Guides with the characteristics of protocols are produced and disseminated by organizations for standardization, both international such as the International Organization for Standardization (ISO) and national such as the American Standards for Testing Materials (ASTM) and the German Industrial Norms (DIN). Quite a number of them are related to groundwater.

Many hydrogeologists are not sufficiently aware of the many guidelines and protocols available in the public domain that may help them in their data acquisition activities. Therefore, IGRAC developed a database on such guidelines and protocols and made it publicly accessible from their website (www.igrac.nl) as a service to the international groundwater community. The database has been fed by an initial inventory of more than 420 available documents, and it is intended to be expanding continuously by additional references brought under IGRAC's attention by groundwater specialists from all over the world. On the basis of observed gaps and on interactions with groundwater professionals, IGRAC initiated a programme of developing new guidelines. Since 2004, two international working groups are active in this context, with the aim of developing guidelines on 'baseline groundwater monitoring' and 'exploitable groundwater resources', respectively.

Groundwater resources management

Recent guidelines concern groundwater resources management. While during the early 1990s worldwide attention was raised for integrated water resources management, several international organizations have been producing publications and guidelines in order to promote good practice in water resources management and to communicate 'lessons learned' (e.g. World Bank, 1994; Burke and Moench, 2000; FAO, 2003). Regarding groundwater, several relevant international sets of guidelines have appeared in recent years, among which the ToolBox of the Global Water Partnership (although addressing groundwater to a limited extent only) and the Briefing Notes Series of the World Bank's GW·MATE. Both products are accessible from the Internet at www.gwpforum.org and www.worldbank.org.gwmate, respectively. The titles of GW·MATE's Briefing Notes are listed in Table 16.3.

Transboundary groundwater resources management

Groundwater systems do not respect administrative boundaries: many groundwater systems therefore are shared between two, three or more countries. Whereas debates on international river basins have been taking place for many years, the attention for transboundary or 'internationally shared' aquifers is only very recent. The establishment of internationally shared aquifer resource management/transboundary aquifer resource management (ISARM/TARM) as a commission under IAH and UNESCO has been instrumental to raise awareness of the need for transboundary aquifer management arrangements between countries. ISARM/TARM is also active in forming – in cooperation with regional and national entities – regional networks for transboundary aquifer management. Examples are the networks in the Balkans and in the Americas. The scope of activities is multi-disciplinary and includes hydrogeological, legal, institutional, socio-economic and environmental aspects. In addition, there are transboundary aquifer management projects for large shared aquifers such as the Guaraní aquifer in South America and the Iullemeden aquifer in northern Africa. Organizations like the Global Environmental Facility (GEF), World Bank

Table 16.3. GW·MATE's Briefing Notes on Groundwater Management: key concepts and tools.

Number	Title	Year of publication
1	Groundwater resource management – an introduction to its scope and practice	2003
2	Characterization of groundwater systems – key concepts and frequent misconceptions	2003
3	Groundwater management strategies – facets of the integrated approach	2003
4	Groundwater legislation and regulatory provision – from customary rules to integrated catchment planning	2003
5	Groundwater abstraction rights – from theory to practice	2002
6	Stakeholder participation in groundwater management – mobilizing and sustaining aquifer management organizations	2003
7	Economic instruments for groundwater management – using incentives to improve sustainability	2003
8	Groundwater quality protection – defining strategy and setting priorities	2003
9	Groundwater monitoring requirements for managing aquifer response and quality threats	2004
10	Groundwater dimensions of national and river-basin planning – ensuring an integrated strategy	Not yet available
11	Utilization of non-renewable groundwater – a socially sustainable approach to resource management	2003
12	Urban waste water as groundwater recharge – evaluating and managing the risks and benefits	2003
13	Groundwater resources development in minor aquifers – management strategy for village and small town water supply	2005
14	Natural groundwater quality hazards – avoiding problems and formulating mitigation strategies	2005
15	Groundwater-dependent ecosystems	Not yet available

and UNESCO are involved in these projects, along with regional and national organizations.

If sharing information on groundwater between countries is at all needed, it is certainly needed in the case of internationally shared aquifers. One of the first steps in the activities of the regional networks for transboundary aquifer management therefore is an inventory. This inventory allows information to be shared on each country's part of the transboundary aquifers, thus putting pieces of the puzzle together. UN/ECE pioneered such an inventory for Europe and the inventory model has been used later by other regional groups. Findings so far are that this inventory step is already very difficult and time-consuming, but it breaks the ice for next steps in which not only information has to be shared, but a good degree of trust as well. IGRAC is cooperating with ISARM by organizing ISARM's website (including regional pages) and by assisting in the processing and visualization of inventory data.

It can be observed that at present the degree of sharing information is insufficient to allow shared aquifer management to be planned and implemented. It will still take many more years and dramatically increased efforts to achieve satisfactory levels of sharing information.

Conclusions and a Look into the Future

Exchange of information and the agricultural groundwater revolution

Worldwide exchange of information has undoubtedly been one of the key factors triggering the 'agricultural groundwater revolution' in quite a number of countries scattered over the world. It has been a mechanism to raise awareness on the potential benefits of the groundwater resources and to spread knowledge on how to explore, develop and use them properly.

However, once agriculture has been expanded significantly by tapping groundwater, a number of associated problems usually start to develop. This means that the main challenge shifts from expanding groundwater exploitation to keeping the resource and its beneficial use sustainable, while at the same time avoiding significant damage by groundwater abstraction to nature and the environment. Intensive sharing of information, knowledge and experience is likely to have a positive effect on the outcomes of these efforts. Given the mentioned shift in focus, it follows that the types of information and experience to be exchanged have to change accordingly.

Current state of sharing information on groundwater worldwide

Important contributions to the global sharing of information, knowledge and experience on groundwater are being made by many actors and programmes. Many of the endeavours are focusing on information, knowledge and experience of a generic nature, but others are explicitly related to area-specific information.

It can be observed that international exchange of generic information, knowledge and experience on groundwater has a much longer tradition and is much more advanced than that of area-specific information, knowledge and experience. Where the former may produce more than any groundwater specialist is able to absorb, the latter is still in its infancy and does not yet satisfy more than very elementary demands. This means – among other things – that existing global or regional pictures of groundwater conditions are still deficient and inaccurate, that many analogies regarding groundwater are not yet properly identified or understood, that highly relevant patterns and trends may still fail to be recognized, and that transboundary groundwater management does not take off because information on the aquifers to be managed jointly is lacking.

With so many organizations in one way or another active in the described field, it is difficult to ensure coordinated efforts and to avoid duplication. However, there is an increasing tendency to join forces in international programmes under the aegis of international organizations such as UNESCO and IAH. This creates synergy, reduces inconsistencies and increases efficiency.

A look into the future

Although the continuation of exchanging groundwater information on a generic level remains important, it follows from the earlier-described current state of affairs that priority needs to be given to the *enhancement of sharing area-specific (or geo-referenced) information, knowledge and experience*. Documentation of the world's groundwater systems on freely accessible platforms, such as those initiated by FAO (*AQUASTAT*), UNEP (*GEO*) and IGRAC (*GGIS*) needs to be intensified, improved and diversified. Present-day deficiencies of these systems include the scarcity of data in general; the lack of uniformity/synchronization of the data and thus an inherent poor quality of the processed information; and the lack of groundwater-monitoring data, which implies that changes in groundwater conditions are very poorly known, precluding rational and efficient actions to exploit, manage and protect the groundwater resources properly. Furthermore, the information is presented predominantly in the form of variables and indicators; analysis of the corresponding numbers and well-documented accounts of experiences on groundwater are rare.

In the endeavour to establish an enhanced GGIS to better service the international community, IGRAC is challenged in many ways. One of the challenges is to convert the currently ad hoc data inflow mechanisms into more structural ones; this will require strong and active international networks of motivated people, which can only be achieved in the longer term. Another challenge is diversification of the information inside the GGIS. Important categories of additional information to be exchanged and made centrally available are:

- monitored change in groundwater conditions;
- activities undertaken and results obtained in groundwater resources management (sharing experiences as an efficient mechanism for learning);
- national or local organizations and projects active in collecting and providing area-specific information (metadata);
- principal maps, reports and publications on groundwater (metadata).

Beyond the improvement of these platforms and the related databases, there are many more modalities for improving the process of sharing information and experience on groundwater. One of the evident options is building communities of practice and interactive networks on groundwater; expanding the awareness on groundwater effectively to non-technical target groups (politicians, the general public) is another important one.

Managing groundwater resources is a very complex activity and characterized by trial and error. Information is only one of the many factors at play, but an important one. Sufficient area-specific information will help us in choosing which road to embark upon and in keeping track of whether it brings us where we want to be. Local information is crucial in this respect, but – as explained in this paper – the international exchange of information and experience may produce significant added value and thus contribute to better results.

References

Arlosoroff, S. *et al.* (1987) Community water supply. *The Handpump Option: A Joint Contribution by the UNDP and the World Bank to the International Drinking Water Supply and Sanitation Decade*. The World Bank, Washington, DC.

Australian Water Resources Council (2004) Groundwater provinces map of Australia, published on the website of the Australian Department of the Environment and Heritage. Available at: http://www.deh.gov.au/soe/2001/inland/water01–2d.html

Burke, J. and Moench, M. (2000) *Groundwater and Society*. United Nations, New York.

Day, J.B.W. (1992) A brief account of the International Association of Hydrogeologist. In: *Applied Hydrogeology* (International Journal) Vol. 0/1992, IAH.

Dupuit, J. (1863) Etudes théoriques et pratiques sur le mouvement des eaux dans les canaux découverts et à travers les terrains perméables. Dunod, Paris.

FAO (2003) Groundwater Management: The Search for Practical Approaches. FAO Water Report no. 25, Rome.

IAH (1994) Membership Directory 1994.

IAH (2003) Annual Report 2003.

IAH, UNESCO and FAO (1962) A legend for hydrogeological maps. *Red Books Series*. IASH.

IGS, IAH, IASH and UNESCO/IHD (1970) International legend for hydrogeological maps (in four languages).

IGRAC (2004) Global Groundwater Information System. Available at: http://igrac.nitg.tno.nl/ggis_start.html

Kasrils, R. (2003) Preface. In: Xu, Yongxin and Hans E. Beekman (eds) *2003: Groundwater Recharge Estimation in Southern Africa*. UNESCO-IHP Series No. 64, UNESCO, Paris.

Naciones Unidas, PNUD (1973) Desarrollo de los Recursos de Aguas Subterráneas en el Altiplano. Resultados del Proyecto. Conclusiones y recommendaciones.

Negenman, A.H.J. (1995) Capacity building and water resources management in Yemen: the WRAY project experience. In: Romijn, E. and de Roon, J.C.S. (eds) *Netherlands Experiences with Integrated Management*. RIZA, Lelystad, The Netherlands, pp. 33–52.

Stow, D.A.V., Skidmore, J. and Berger, A.R. (1976) Preliminary Bibliography on Groundwater in Developing Countries. AGID Report No.4.

Struckmeier, W. and Margat, J. (1995) Hydrogeological Maps: A Guide and a Standard Legend. IAH Series *International Contributions to Hydrogeology* 17, Heise, Hannover.

UNEP (2002) *Global Environment Outlook 3: Past, Present and Future Perspectives*. Produced by UNEP GEO team, Division of Early Warning and Assessment (DEWA), UNEP, Nairobi, Kenya.

UNEP (2003) *Groundwater and Its Susceptibility to Degradation*. UNEP/DEWA, Nairobi, Kenya.

UNESCO (1972) *Groundwater Studies: An International Guide for Research and Practice*. UNESCO, Paris.

UNESCO (1974) Legend for geohydrochemical maps (quadrilingual supplement to the 1970 legend on hydrogeological maps). UNESCO, Paris.

UNESCO and ACSAD (1988) Hydrogeological map of the Arab region and adjacent areas. Two sheets, scale 1:5 million.

UNESCO, BGR, CGMW, IAEA and IAH (2004) Map of Groundwater Resources of the World, 1:50 000 000. Special edition for the 32nd International Geological Congress, Florence/ Italy, 20–28 August 2004.

UNESCO and CIAT (2000) *Mapa Hidrogeológico de América del Sur*. Digital coverages on CD, prepared for UNESCO/IHP.

UNESCO, IAH, UNECE and FAO (2001) *Internationally Shared (Transboundary) Aquifer Resources Management: Their Significance and Sustainable Management*. A framework document. IHP-VI, IHP Non Serial Publications in Hydrology, UNESCO, Paris.

USGS *Techniques of Water-Resources Investigations of the United States Geological Survey*. US Department of the Interior, Geological Survey. US Government Printing Office, Washington, DC.

Vrba, J. and Zaporozec, A. (eds) (1994) Guidebook on mapping groundwater vulnerability. *International Contributions to Hydrogeology*, Vol.16, IAH/IHP.

WHYMAP (2004) Identical to UNESCO et al. (2004).

WMO (2001) Exchanging hydrological data and information. WMO Policy and Practice. Geneva, Switzerland, WMO-No. 925, 2001.

WMO and UNESCO (1997) Water resources assessment. *Handbook for Review of National Capabilities*, USA, pp. 1–113.

World Bank (1994) *A Guide to the Formulation of Water Resources Strategy*. World Bank Technical Paper No. 263. The World bank, Washington, DC.

WWAP (2003) Water for People, Water for Life. The United Nations World Water Development Report. Published by UNESCO, Paris/Berghahn Books, Oxford, UK.

Zektser, I.S. and Everett, L.G. (2004) Groundwater resources of the world and their use. UNESCO, IHP-VI Series in Groundwater No 6.

17 Groundwater Use in a Global Perspective – Can It Be Managed?

KAREN G. VILLHOLTH AND MARK GIORDANO

International Water Management Institute, 127 Sunil Mawatha, Pelawatte, Battaramulla, Sri Lanka

Introduction

Groundwater has long been second to surface water in terms of its importance for human use and the attention devoted to it by the general public and water sector managers. However, this picture is quickly changing as groundwater increasingly supplants surface water in many areas of the world as the primary and preferred source of water for all types of use, i.e. domestic, agricultural (crop and livestock) and industrial. This change is being driven by groundwater's inherently beneficial properties in terms of both quality and quantity combined with easy access through better and cheaper drilling and pumping techniques. While its 'in-stream' values, as is the case with rivers, have not been widely acknowledged, the critical role groundwater plays in maintaining important surface water systems, riparian and other types of vegetation as well as vital ecosystems is also increasingly recognized. However, this recognition has unfortunately emerged in many cases in a retrospective manner, as the signs of overdraft and degradation gradually become manifest in the depletion and deterioration of the associated aquifers, rivers, lakes, wetlands and other water-related ecosystems. Groundwater is surfacing, so to speak, in people's awareness mostly as a result of the increasingly observable problems rather than as a reaction of gratitude for all the benefits that it is providing humankind. The saying: 'You never miss your water till your well runs dry' is very suitable in this context. However, the question then turns to whether the impending accruing groundwater-related problems can be countered and curbed based on this increased general awareness and appreciation of the resource. Can groundwater use in today's world be actively managed, and how?

This chapter highlights some salient characteristics of groundwater as a fundamental resource for human existence, the contemporary use of the resource, particularly in agriculture, and the present challenges associated with its management in a local and global context. The objective is to summarize, in a kaleidoscopic and more philosophical way, the chapters presented in this volume, *The Agricultural*

Groundwater Revolution: Opportunities and Threats to Development, and suggest answers to the above questions.

The Contemporary Story of Groundwater Use

Groundwater is generally a reliable and good quality water source, and with modern technology for drilling, electrification and pumping, it is widely accessible throughout most parts of the world today. In fact, these technological advances are primarily accountable for the recent, remarkable increase in global abstraction of groundwater. The history of global intensive groundwater use is less than 50 years old and much of the modern increase in global water use has been contributed by groundwater. Surface water use has remained constant or increased at a slower rate, simply because resources are running out or the feasibility of capturing and storing them is low. What is also remarkable about today's groundwater use is that the increase is continuing on a global scale, with only patches of declining or stagnating trends. Global aquifers hold an enormous water reserve that is several times greater than surface water resources (UN/WWAP, 2003). Groundwater could, in principle, be exploited at an aggregate level that is higher than it is today. However, the overriding limitations to further groundwater use in the future will continue to be environmental problems associated with the desiccation of aquifers and the socio-economic problems related to increasingly unequal access, especially in developing countries, to the resource as the groundwater levels decline and the aquifers become contaminated as a side effect of intensive use and generally increased pressure on natural resources.

In addition, the classical problem of uneven geographical distribution of surface water resources also applies to groundwater, at least at more regional scales, and the general mismatch between the location of high demand (high-population, potential-intensive agricultural areas) and groundwater availability is very real and relevant. Often, and logically, groundwater is developed and in further demand in dry and semiarid areas where surface water is scarce or seasonal. But such regions are typically underlain by non-replenishable or slowly replenishable aquifers unfit for intensive exploitation, putting a natural break on unlimited growth in use. Likewise, half of the world's population today lives in coastal areas (Post and Lundin, 1996) where groundwater traditionally provided secure and adequate water supply. However, these areas are increasingly threatened by deterioration of water quality due to salt water ingress from widespread and intensive groundwater extraction (Kaushal *et al.*, 2005).

It is essential to focus on the agriculture in the context of global groundwater use for the simple reason that volumes used in this sector significantly exceed other uses, e.g. industrial and domestic, at the global scale. Especially in many arid and semiarid regions of the world that coincide with nations in development, such as India, North China and Pakistan, groundwater use is critical for food security. Here the management challenges are manifold in the sense that a balance between securing groundwater-dependent livelihood and ensuring the long-term environmental sustainability is required. But even in more developed countries in less arid regions such as the USA, Australia and Mexico, as well as in Mediterranean countries like Spain, groundwater supplies significant water for agricultural use and its management presents great challenges. High dependence on groundwater also occurs in humid countries, but more for industrial and domestic uses (e.g. Japan, the former USSR and north European countries like Denmark and the Netherlands (Margat, 1994)). Here, the volumes drawn are generally not threatening the resource base from a quantity point of view – it is more water quality issues that present the major challenges.

Yet another category of countries includes those that potentially could benefit from an intensification of groundwater use for agriculture and associated development, such as parts of sub-Saharan Africa (see Masiyandima and Giordano, Chapter 5, this volume), Nepal and eastern India (see Shah, Chapter 2, this volume). In these cases, present limitations to such development seem to be associated with poor energy access, lack of infrastructure and market access, lack of credit possibilities and possible cultural or demographic barriers.

Groundwater is now surpassing surface water in importance in many regions of the world, in terms of water supply for irrigation. The 2005 FAO AQUASTAT database on irrigated area lists the countries Algeria, Bangladesh, India, Iran, Libya, Saudi Arabia, Syria and Yemen, as those depending more on groundwater than surface water for their irrigation. Expanded groundwater use in a global context can be seen as a second step in the continued and accelerated quest for water for human development. Basically, surface water was accessed, appropriated and allocated first, as this resource was more visible and readily available and most human settlements confluenced with rivers and streams where water was traditionally secured. As surface water resources are being exhausted and strained in terms of quality and the options to dam them have diminished, groundwater has become the second-generation resource to be captured and appropriated.

Taking this analysis further and linking it with the general hydrological cycle, there is now a trend towards focusing on rainwater as the 'new' water source to capture for direct use as well as for storing and optimizing. What is interesting in this scheme is that in a sense we are moving progressively backwards, or upstream, in the hydrological cycle to look for water, because rainwater feeds groundwater and groundwater feeds surface water (Fig. 17.1). More importantly, there is a tendency, amongst lay people but also water professionals, to look at water sources independently and consider them as isolated, new resources that can be explored without affecting the others. One example is the artificial recharge movement in India that attempts to capture rainfall and runoff for local replenishment of aquifers for the improvement of livelihood and to counteract groundwater declines (see Sakthivadivel, Chapter 10, this volume). However, referring to Fig. 17.1, it is obvious that these sources are intricately interlinked, and capturing one will diminish the availability of the downstream sources. Realizing this and respecting this simple, but fundamental, upstream–downstream and mass balance concept is crucial to any kind of water resources management. Following these arguments, it is also clear that groundwater cannot be managed in isolation and that integration of all water resources needs to be considered in overall assessment and planning.

Hence, the issue of groundwater use today is essentially not technical but managerial – how to balance the benefits of use with the associated negative

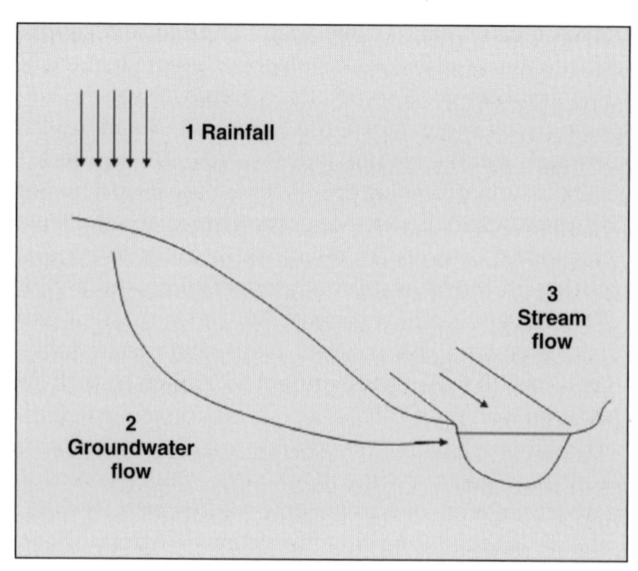

Fig. 17.1. Groundwater as a sequential component of the hydrological cycle.

impacts? Although this is increasingly being realized by water managers, practitioners and scientists, it is proving to be one of the most challenging tasks for humankind as we move into the 21st century as population increases, demand for higher living standards in the developing world and climate change with associated increases in extreme events all mix together to put higher pressures and threats on already strained resources, including groundwater.

While paradigms for groundwater management are slowly emerging and various models are investigated and tested (see Schlager, Chapter 7; Kemper, Chapter 8; and Moench, Chapter 9, this volume), it is also clear that from actual cases that management based on strict control of groundwater development and use (demand management) is generally difficult to implement and enforce, perhaps especially in developing countries (see Shah, Chapter 2; and Wang *et al.*, Chapter 3, this volume). This is to a large extent attributable to the fact that groundwater has many 'open-access' properties, leaving little incentive for users to curtail their use because they cannot fully capture the associated benefits (see Schlager, Chapter 7, this volume). The generally easy access to the resource for individuals combined with this fact actually presents the core dilemma in groundwater management, and as of today there seems to be very few examples of solutions addressing this dilemma.

Drivers of Groundwater Development

Two fundamental human drivers for groundwater development for irrigation may be summarized as survival and profit. These two drivers may in fact be considered as extremes of a continuum governed primarily by the stage of development of a certain region, exemplified by the small-scale farmer in central India trying to improve the outcome of his or her small rain-fed plot by supplemental irrigation from groundwater on one end, and a large-scale commercial mid-west farmer in the USA optimizing his maize yield and speculating on world food prices on the other end. The large-scale users, of course, exert the largest abstraction pressure on the resource per person. However, when many small-scale users are conglomerated within larger areas, like in parts of India and China, the aggregate effect may be similar (Fig. 17.2). Furthermore, when large numbers of users are involved, as is often the case in countries with limited resources for monitoring and governance, cooperative management becomes difficult. The ironic fact is that both of these extreme cases represent situations in which groundwater is considered 'overabstracted' today, and the national and local authorities are concerned about the sustainability of the present-day exploitation of the resource.

This spectrum view of groundwater use is a simplification of the reality but serves to show that intensive groundwater exploitation in agriculture today is a common global phenomenon across quite different socio-economic settings, resulting in essentially the same types of physical or environmental impacts. Although the scope and capability for addressing the problems is potentially more favourable for countries like Spain, the USA and Australia (see, respectively, Llamas and Garrido, Chapter 13; Peck, Chapter 14; and Turral and Fullagar, Chapter 15, this volume), it is clear that by no means is the curbing of groundwater intensive use easy in any setting.

Figure 17.2 gives a sketch of the spectrum of groundwater use as a function of development. 'Use' and 'development' here are broad illustrative terms, not put to any quantitative scale.

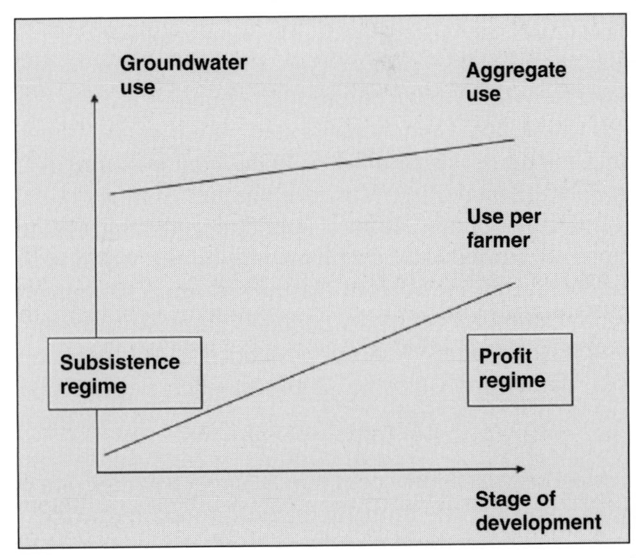

Fig. 17.2. Sketch of spectrum of groundwater use as a function of development.

Groundwater Overexploitation – What Is That?

With groundwater resources increasingly being utilized to fulfil human requirements and the demands seemingly insatiable, the question of sustainability naturally arises. Within the last decade or so, the debate around defining limits to sustainable groundwater use has intensified (Custodio, 2002; see also Llamas and Garrido, Chapter 13; and Turral and Fullagar, Chapter 15, this volume). Basically, there is no clear and unambiguous definition of such a limit, be it designated 'sustainable yield' or 'exploitable groundwater'. However, there is a growing consensus that such a concept is a valuable tool for legitimization, informed discussion and consensus building on management: defining areas where additional groundwater use should be curtailed or where investment in management, rather than development, should be made. There is also a growing recognition that such a concept is not restricted to an assessment of the physical availability of groundwater in a certain area of concern. Equally important is the assessment and reconciliation of the positive and negative impacts of increased utilization on society as well as the environment. Examples where such approaches are taken up more systematically as part of the national groundwater management approaches include countries such as Australia, India and South Africa. The effectiveness of such an approach hinges on its credibility to the stakeholders and decision makers. Furthermore, realizing the dynamics of society and the gradual improvements in information and data availability, the assessments should be ongoing (see Turral and Fullagar, Chapter 15, this volume).

Recognition of Groundwater Problems

There is a large gap in the level of information and documentation of the national and subnational state of groundwater between developed and developing countries (see van der Gun, Chapter 16, this volume). Furthermore, literature from developing or transition countries is more often in local languages as in China and Latin America (see Wang *et al.*, Chapter 3; and Ballestero *et al.*,Chapter 6, this volume) or in the grey literature as in sub-Saharan Africa and Central America (see Masiyandima and Giordano, Chapter 5; and Ballestero *et al.*, Chapter 6, this volume), impeding general access to it. Nevertheless, literature on groundwater problems around the world is increasingly reflecting the general upward trend in use and its impacts. Rather than repeating these here, a reference to summary papers is given which covers both developed and developing countries (Burke and Moench, 2000; Danielopol *et al.*, 2003; FAO, 2003; Llamas and Custodio, 2003; Moench *et al.*, 2003; Morris *et al.*, 2003; Moench and Dixit, 2004; Shah *et al.*, 2006; see also the five regional chapters in this volume).

Most of the scientific community agrees that there is a problem with present-day groundwater use in many regions around the world, basically because of the way the groundwater resource itself, the environment and poor segments of the societies, especially in developing countries, are adversely affected. The latter refers to the inability of poor or disadvantaged people, in general, to cope with degradation of natural resources, both because they are often more directly dependent on them for their livelihood and because they are less capable of adapting to the increased competition for the resources and are most often left with poor access to poor-quality water, even for their basic needs.

It is interesting to consider the fact that intensive groundwater use over just one generation, or essentially the last 3–4 decades, has drawn down underground water resources to an unprecedented level in human history, and there is no likelihood that water management in the future will make it possible to revert to earlier levels, or even maintain status quo. Basically, this blue underground treasure, which is only partially replenishable, is permanently lost and with it, valuable wet ecosystems as well as an important buffer capacity against droughts. Again this impact strikes harder in already marginal and resource-stressed areas and regions of the world where poor people already tend to accumulate.

Groundwater pumping most often occurs in an uncontrolled and indiscriminate manner, be it in developed or developing countries. Entitlement to groundwater is most often associated with access to land and financial resources (for drilling and pumping costs) more than formal rights and regulations to the resource itself. This can result in the classic 'tragedy of the commons' problem often associated with groundwater, but also misuse from other perspectives. For example, high-quality groundwater might be used for agriculture while poor people seek drinking water supplies from contaminated surface sources.

Despite the recognition of the problems associated with intensive groundwater use in many countries among the scientific community, there may not be the same consonance regarding the groundwater problems among decision makers and actual groundwater users, and even then it may be very difficult to reach agreement on primary problems, root causes and key issues responsible for the problems,[1] let alone the remediation measures to put in place. This clearly illustrates that the management of groundwater needs to consider the whole spectrum of users as well as direct and indirect stakeholders, including the 'silent' or subordinate users, namely the environment and the disadvantaged groups of society (often represented only through international environmental organizations, e.g. IUCN, WWF and Ramsar, or local or national non-governmental organizations (NGOs)). It also illustrates that knowledge of the processes and the cause–effect relationships are required at all levels, as well as participation, communication and negotiation.

Challenges to Groundwater Management

The fact that groundwater use continues to grow on a global scale, only occasionally levelling off as a reactive rather than proactive response to perceived severe impacts in some areas, portends poorly for the overriding question, as raised in the introduction, of whether groundwater can be managed.

In developing countries, farmers make up the majority of the population. When their livelihoods, and sometimes their very lives, depend on groundwater, they understandably resist uncompensated measures to curtail use. In developed countries, though the number of groundwater-using farmers is lower, the continued operation of their farms, and the value of their often substantial investments, can sometimes only be maintained if the groundwater continues to flow. From another perspective, groundwater can fall within broader political agendas, making efforts to manage use secondary to other concerns, such as supporting a certain population or political group irrespective of obvious natural resource encroachments (see Allan, Chapter 4, this volume). To various degrees in each of these cases, the social and political will as well as the economic backing for effective groundwater management may not be in place.

As an alternative to direct management, water-saving irrigation techniques have been promoted to improve food production per unit of water input, but it is questionable whether such approaches significantly reduce stress on groundwater resources. This is because such methods are often associated with the shift to more intensive cultivation, using more water-intensive crops, higher levels of chemical input and better soil-conservation techniques. So crop yields increase per area under cultivation and per water input. But overall, groundwater use may have actually increased because of the intensification. The fundamental problem of how to feed an ever-increasing global population while at the same time maintaining or even decreasing the water requirements is one that puzzles planners as well as scientists (Comprehensive Assessment of Water Management in Agriculture Synthesis Report, forthcoming 2006).

In a sense, much wealth creation and poverty reduction has been derived on a loan that will never be directly paid back. There is a danger that the poverty-reducing potential of groundwater will be lost, making societies more vulnerable to climate changes and extreme events. There is a major challenge in securing basic water needs to people in developing countries who depended to a large extent on sustainably replenished shallow wells that are now out of reach. The primordial role of drinking water needs to ensured, for example, by having deep, protected wells for drinking and shallow wells for irrigation (and not the other way around as is often the case today in rural areas), or by zoning of areas with precedence for drinking water.

The link between groundwater use in agriculture and for urban areas is also becoming increasingly apparent and needs much more research and management focus. Realizing that irrigation generally poses less strict requirements on water quality compared to urban use (for domestic and industrial uses) obviously suggests prioritizing urban water use and making irrigated agriculture the second in line in a cycle of water reuse. The challenge in many cases becomes one of sending treated wastewater back upstream in the catchment as irrigation areas are often upstream while cities are located downstream along rivers or in coastal areas. Obviously, this is a complex and costly intervention, but one that can be further explored when economic and socio-economic conditions are right (see Turral and Fullagar, Chapter 15, this volume, for an example from Australia).

Conclusions

Many of the problems of groundwater management may seem insurmountable. However, paradigms for their solution are being articulated. These range from community management approaches (see Schlager, Chapter 7; and Sakthivadivel, Chapter10, this volume) hinging on local initiatives, social norms and informal agreements, to more formalized laws and associated formal rights and regulations within and outside the groundwater sector (see Kemper, Chapter 8; and Shah, Chapter 11, this volume), to a focus not on resource management itself but rather to people's adaptive ability to overcome stress caused by groundwater decline and degradation (see Moench, Chapter 9; and Mudrakartha, Chapter 12, this volume) and turn today's groundwater use into an opportunity for tomorrow's improved livelihood.

In considering our options for sustainably managing groundwater in the future, two key points should be remembered. First, it is as important to consider the socio-economic and sociopolitical characteristics of any groundwater-using society as the physical characteristics of groundwater resources in any proposed management solutions. A solution for places with large numbers of small farmers may be inappropriate for other locations with small numbers of large farmers and vice versa. Similarly, that which might work in a country with a strong central government and significant financial resources for enforcing regulations may work less well in a country where political power is more diffuse or financial resources are scarce. Second, the 'groundwater revolution' has had a short history. The development of institutions for resource management in general and for a complicated, often 'invisible', resource like groundwater can be expected to take time and experimentation. The initial growth in agricultural groundwater use has brought benefits to millions, perhaps billions, of farmers and consumers around the world. The goal now is to ensure that those benefits continue into the future as we shift the focus from groundwater development to long-term groundwater management. Although some opportunities have perhaps already been lost, there is still time to learn from experiences around the world on how to proceed, provided increased focus, awareness and political will is exercised.

Note

1 Local groundwater-irrigating farmers may ascribe the decrease in water availability to general drought phenomena. Managers may blame the farmers for the excessive pumping and feeling no responsibility towards the issue.

References

Burke, J.J. and Moench, M.H. (2000) *Groundwater and Society: Resources, Tensions and Opportunities*. United Nations Publications, Sales No. E.99.II.A.1, ISBN 92-1-104485-5.

Comprehensive Assessment of Water Management Synthesis Report (forthcoming 2006).

Custodio, E. (2002) Aquifer overexploitation: What does it mean? *Hydrogeology Journal* 10, 254–277.

Danielopol, D.L., Griebler, C., Gunatilaka, A. and Notenboom, J. (2003) Present state and future prospects for groundwater ecosystems. *Environmental Conservation* 30(2), 104–130.

FAO (2003) *Groundwater Management – The Search for Practical Approaches*. Water Reports 25. FAO (Food and Agricultural Organization), ISBN 92-5-104908-4.

Kaushal, S.S., Groffman, P.M., Likens, G.E., Belt, K.T., Stack, W.P., Kelly, V.R., Band, L.E. and Fisher, G.T. (2005) Increased salinization of fresh water in the northeastern United States. *National Academy of Sciences of the USA* 102(38), 13517–13520.

Llamas, R. and Custodio, E. (eds) (2003) *Intensive Use of Groundwater: Challenges and Opportunities*. A.A. Balkema, Rotterdam, The Netherlands, p. xii.

Margat, J. (1994). Groundwater operations and management. In: Gibert, J., Danielopol, D.L. and Stanford, J.A. (eds). *Groundwater Ecology*. Elsevier, Amsterdam, pp. 505–522.

Moench, M. and Dixit, A. (2004) *Adaptive Capacity and Livelihood Resilience: Adaptive Strategies for Responding to Floods and Droughts in South Asia*. ISET (Institute for Social and Environmental Transition), Nepal, India/Boulder, Colorado.

Moench, M., Burke, J. and Moench, Y. (2003) *Rethinking the Approaches to Groundwater and Food Security*. Water Reports 24. FAO (Food and Agricultural Organization), ISBN 92-5-104904-1.

Morris, B.L., Lawrence, A.R.L., Chilton, P.J.C., Adams, B., Calow, R.C. and Klinck, B.A. (2003) *Groundwater and Its Susceptibility to Degradation: A Global Assessment of the Problem and Options for Management*. Early Warning and Assessment Report Series, RS. 03-3. UNEP (United Nations Environment Programme), Nairobi, Kenya. ISBN 920807-2297-2.

Post, J.C. and Lundin, C.G. (1996) *Guidelines for Integrated Coastal Zone Management*. World Bank report, ISBN 0-8213-3735-1.

Shah, T., Villholth, K.G. and Burke, J.J. (2006) *Groundwater Use in Agriculture: A Global Assessment of Scale and Significance for Food, Livelihoods and Nature*. Comprehensive Assessment, IWMI, International Water Management Institute.

Stenbergen, F. van (2003) *Local Groundwater Regulation*. Water Praxis Document Nr. 14. Arcadis, Euroconsult.

UN/WWAP (United Nations/World Water Assessment Programme) (2003) *UN World Water Development Report: Water for People, Water for Life*. UNESCO (United Nations Educational, Scientific and Cultural Organization) and Berghahn Books, Paris.

Index